Naïve Set Theory

Naïve Set Theory: A Rigorous Approach aims to provide a complete and unitary presentation of naive set theory as the foundation of the whole mathematics. Suitable for undergraduate students, this book focuses on the main foundational issues, trying to clarify explicitly what is (necessarily) assumed as intuitively known and what is rigorously founded on more elementary concepts. It analyses in detail the nature of natural numbers, both as meta-theoretical objects and through their set-theoretical model. The author also pays particular attention to some topics that are not usually covered by the literature on naive set theory, like the universal properties of Cartesian product and disjoint union, that lead to a precise formulation of their basic features (associativity, commutativity, and distributivity). The exposition is organized coherently from the initial meta-theoretical notions to the construction of the universe of well-founded sets.

Features:

- Replete with exercises, partially spread within the text and partially listed at the end of each chapter with a solutions manual available on www.Routledge.com/9781032933047
- Covers foundational topics that are not usually discussed in the literature on naive set theory, such as universal properties of Cartesian product and disjoint union, through which one can properly state associativity, commutativity, and distributivity of these operations

Fabio Ferrari Ruffino received his bachelor's degree in mathematics from the University of Milan, his master's degree from the University of Pisa and at Scuola Normale Superiore, and his Ph.D. from SISSA (International School for Advanced Studies, Trieste, Italy). He is currently an Assistant Professor at São Carlos University, Brazil. In his research, he has been working mainly on geometry and topology, with a particular focus on the applications in theoretical physics (especially in string theory). He has always been particularly interested in foundational problems from a philosophical point of view in both his research and teaching activities.

Naïve Set Theory

A Rigorous Approach

Fabio Ferrari Ruffino

CRC Press
Taylor & Francis Group
Boca Raton London New York

CRC Press is an imprint of the
Taylor & Francis Group, an **informa** business
A CHAPMAN & HALL BOOK

Designed cover image: Fabio Ferrari Ruffino

First edition published 2025
by CRC Press
2385 NW Executive Center Drive, Suite 320, Boca Raton FL 33431

and by CRC Press
4 Park Square, Milton Park, Abingdon, Oxon, OX14 4RN

CRC Press is an imprint of Taylor & Francis Group, LLC

ISBN: 978-1-032-93304-7 (hbk)
ISBN: 978-1-032-93305-4 (pbk)
ISBN: 978-1-003-56529-1 (ebk)

DOI: 10.1201/9781003565291

Publisher's note: This book has been prepared from camera-ready copy provided by the authors.

Typeset in Latin Modern font
by KnowledgeWorks Global Ltd.

Access the Instructor and Student Resources: www.Routledge.com/9781032933047

To my wife Haydée and my sons Damián and Isabella

Contents

Preface xiii

Aims and Target Audience xxi

Author Bio xxiii

CHAPTER 1 ▪ Preliminaries 1

1.1 PROPOSITIONS, CONNECTIVES, AND QUANTIFIERS 1
1.1.1 Logical Connectives 1
1.1.2 Quantifiers 2
1.2 EQUALITY 2
1.3 SETS, ELEMENTS, AND SUBSETS 2
1.4 NATURAL NUMBERS AND INDUCTION 3
1.4.1 Fundamental Operations and Order 4
1.4.2 Other Versions of the Induction Principle 6
1.4.3 Sequences and Number-Theoretic Functions 7
1.4.4 Decimal Notation 10
1.5 EXERCISES 10

CHAPTER 2 ▪ Hereditary Sets and Basic Operations 13

2.1 HEREDITARY SETS 13
2.2 PROPERTIES AND WELL-FOUNDED FORMULAS 14
2.2.1 Well-Founded Formulas 14
2.2.2 Bounded and Free Variables 15
2.2.3 Set-Theoretical Properties 15
2.3 PATHOLOGIES AND RUSSELL'S PARADOX 16
2.3.1 Possible Pathologies 16
2.3.2 Russell's Paradox 17
2.4 CONSTRUCTING FINITE SETS 18
2.5 ORDERED FINITE SETS 19

2.6 OPERATIONS ON FINITE FAMILIES OF SETS 20
2.6.1 Unary and Binary Operations 20
2.6.2 n-Ary Operations 22
2.6.3 Basic Operations and the Universe 23
2.7 COMPREHENSION HYPOTHESIS 23
2.8 EXERCISES 24

CHAPTER 3 ■ Relations and Functions 27

3.1 RELATIONS 27
3.2 FUNCTIONS 28
3.2.1 Composition 29
3.2.2 Diagrams 29
3.2.3 Identity and Inversion 30
3.2.4 Restriction 31
3.2.5 Induced Functions between Power Sets 31
3.3 EQUIVALENCE RELATIONS 32
3.4 ORDER RELATIONS 33
3.5 EXERCISES 36

CHAPTER 4 ■ Sets and Indexed Sets 40

4.1 SETS 40
4.2 INDEXED SETS 43
4.3 COMPARISON 47
4.3.1 Union and Intersection 47
4.3.2 Cartesian Product 47
4.3.3 Disjoint Union 47
4.3.4 Separation 48
4.3.5 Summary 49
4.3.6 Operations and the Universe 50
4.4 COMPLEMENTS 50
4.4.1 Iterated Union 50
4.4.2 Generated Equivalence Relation 51
4.4.3 Increasing Families of Functions 52
4.5 EXERCISES 54
4.5.1 Part I: Sets 54
4.5.2 Part II: Indexed Sets 55

CHAPTER 5 ■ Universal Properties 58

5.1 UNIVERSAL PROPERTY OF CARTESIAN PRODUCT 59
5.1.1 Remarks on Cartesian Product 61
5.1.2 Induced Function between Cartesian Products 61
5.2 UNIVERSAL PROPERTY OF DISJOINT UNION 62
5.2.1 Remarks on Disjoint Union 63
5.2.2 Induced Function between Disjoint Unions 63
5.2.3 Independence of the Separating Sets 64
5.2.4 Disjoint Union of One Set 64
5.3 TWO RELEVANT PARTICULAR CASES 65
5.4 PROPERTIES OF THE OPERATIONS 66
5.4.1 Associativity 66
5.4.2 Commutativity 67
5.4.3 Distributivity 67
5.5 SETS AND INDEXED SETS 68
5.5.1 Cartesian Product 68
5.5.2 Disjoint Union 69
5.5.3 Properties of the Operations 71
5.6 EXERCISES 71

CHAPTER 6 ■ Sets and Classes 72

6.1 DEFINITIONS AND OPERATIONS AMONG CLASSES 72
6.1.1 Inclusion and Equality 72
6.1.2 Unary Operations 72
6.1.3 n-Ary Operations 73
6.1.4 Relations and Functions 73
6.1.5 Indexed Classes 74
6.2 STATING COMPREHENSION WITH CLASSES 74
6.3 REPLACEMENT HYPOTHESIS 75
6.4 WHAT IS A CLASS? 75
6.5 EXERCISES 76

CHAPTER 7 ■ Embedding Natural Numbers in Set Theory 78

7.1 FIRST ATTEMPT 78
7.2 THE CONSTRUCTION 79
7.2.1 Main Properties of Natural Numbers 80

7.2.2 Recursive Definitions 81
7.2.3 Operations and Order 83
7.2.4 Sequences 83
7.2.5 Second Form of Induction and Recursion 84
7.2.6 Finite and Infinite Sets 85
7.2.7 Alternative Approach 87
7.3 META-THEORETICAL CONSIDERATIONS 87
7.3.1 N *vs* $\mathbb{N}$ 87
7.3.2 What is a Natural Number? 89
7.4 EXERCISES 90

CHAPTER 8 ■ Well-Orders 93

8.1 DEFINITION AND FIRST PROPERTIES 93
8.2 WELL-ORDER ISOMORPHISMS 95
8.3 INITIAL SEGMENTS 97
8.4 COMPARING WELL-ORDERS 99

CHAPTER 9 ■ Axioms of Choice 102

9.1 THE AXIOM OF CHOICE 102
9.1.1 Formulation of the Axiom 103
9.1.2 Partition Principle 104
9.1.3 The Axiom of Choice in Mathematics 104
9.2 ZORN'S LEMMA AND THE HMP 105
9.2.1 Zorn's Lemma 105
9.2.2 Hausdorff Maximal Principle 107
9.3 WELL-ORDERING PRINCIPLE 108
9.4 EQUIVALENCE 109
9.5 EXTENSION TO CLASSES 113

CHAPTER 10 ■ Cardinality 115

10.1 COMPARISON OF CARDINALITIES 115
10.1.1 Comparison through Surjectivity 117
10.1.2 Some Relevant Examples 118
10.2 COUNTABLE SETS 120
10.3 FUNDAMENTAL RESULTS 124
10.3.1 Cantor-Bernstein Theorem 124

10.3.2 Cantor's Theorem on Power Set 126
10.3.3 Comparing Sets 127
10.4 EXERCISES 127

CHAPTER 11 ■ Ordinals 129

11.1 TRANSITIVE SETS 129
11.1.1 Equivalent Characterisations 130
11.1.2 Proper and Hereditary Transitivity 131
11.2 DEFINITION AND FIRST PROPERTIES 132
11.2.1 Structure of the Class of Ordinals 133
11.2.2 Ordinals and Well-Orders 136
11.3 SUCCESSOR AND NATURAL NUMBERS 138
11.3.1 Ordinals and Natural Numbers 139
11.4 TRANSFINITE INDUCTION AND RECURSION 141
11.4.1 Transfinite Induction 142
11.4.2 Transfinite Recursion 143
11.4.3 Proof of AC $\Rightarrow$ ZL by Transfinite Recursion 146
11.5 OPERATIONS ON ORDINALS 147
11.5.1 Ordinal Operations on Natural Numbers 150
11.6 EXERCISES 151

CHAPTER 12 ■ Cardinals 153

12.1 DEFINITION AND FIRST PROPERTIES 153
12.2 SUCCESSORS, LIMITS AND THE FUNCTION $\aleph$ (OR ω) 155
12.2.1 Successor and Limit Cardinals 156
12.2.2 The Function $\aleph$ (or ω) 159
12.3 TRANSFINITE INDUCTION AND RECURSION 160
12.3.1 Transfinite Induction 161
12.3.2 Transfinite Recursion 162
12.4 OPERATIONS ON CARDINALS 163
12.5 OPERATIONS ON NATURAL NUMBERS 167
12.6 COFINALITY 169
12.6.1 Basic Properties 170
12.6.2 Regular Ordinals 173
12.6.3 König's Lemma 174
12.7 UNDECIDABLE STATEMENTS 175
12.8 EXERCISES 175

CHAPTER 13 ■ Well-Founded Sets 177

13.1 DEFINITION AND FIRST PROPERTIES 177
13.1.1 Rank 178
13.1.2 The Structure of the Class wf 179
13.1.3 Equivalent Characterisation under AC 181
13.2 WELL-FOUNDED SETS AS THE UNIVERSE 181
13.3 WELL-FOUNDED RELATIONS 182
13.3.1 Well-Founded Relations and Strictly Decreasing Sequences 183
13.3.2 Well-Founded Relations *vs* Well-Founded Sets 184
13.3.3 An Alternative Proof of Proposition 13.26 184
13.3.4 Transitive Closure 185
13.4 THE FOUNDATION ASSUMPTION 187
13.5 WELL-FOUNDED SETS AND ORDINALS 188
13.6 WELL-FOUNDED SETS AND CARDINALS 189
13.7 EXERCISES 191

Conclusions 192

APPENDIX A ■ Hereditary Sets 196

A.1 Definition of Hereditary Set 196
A.2 Reducing the universe 197

APPENDIX B ■ Fundamental Properties 198

B.1 Basic n-Ary Operations 198
B.1.1 Unary and Binary Operations 198
B.1.2 n-Ary Operations 200
B.2 Operations on Sets and Indexed Sets 201
B.3 Relations and Functions 203

APPENDIX C ■ Formulations of Recursion 205

C.1 First Form 205
C.2 Second Form 208

Bibliography 209

Index 211

Preface

This is a course about naive set theory. What does it mean? Why is it *naive*? Let us try to reply to these questions in the clearest way we can. First of all, it is important to keep in mind that, following the most common approach in our age (which is not necessarily the best one, nor is it necessarily eternal), set theory is thought of as the foundation of the whole mathematics. This means that every definition, every statement, and every proof in any area of mathematics should be rigorously formulated within the framework of set theory. Nevertheless, before analysing this issue more in detail, we have to focus on the word "foundation". What is its meaning? When can we say that a definition or a reasoning is "rigorously formulated" in a certain foundational setting? The answer to these questions is not easy at all, hence we need a preliminary analysis on this problem before dealing with set theory.

THE PROBLEM OF FOUNDATIONS

Following common sense, mathematics is the most rigorous and objective area of human thought. Almost everybody recognises that we can have different opinions about a work of art or a landscape, often about a moral judgement too, but we cannot disagree about the fact that $2+2=4$. Two distinct points of view about a picture or a film can be both respectable, while any other result of $2+2$ is classified as a mistake, not as an opinion. Why do we feel that mathematics is so objective? We can reply at two levels. First of all, when a statement is as simple as "$2+2=4$", our certainty about it, independently of our scientific background, is due to an empirical observation: we take two apples, we add two apples more, and we count four apples. People who agree with this computation form such a huge majority that we have no doubts in considering wrong any alternative result. Of course, this observation does not explain the intrinsic meaning of the terms involved, like '2' or '+', but we have no doubts anyway. Unfortunately, this kind of certainty is not applicable to deep mathematical results. In this case, a second level of reliability is usually present in people's mind: almost everybody, even without a deep scientific knowledge, shares the opinion that in mathematics "everything is completely rigorous". This means that every concept is founded with precision on more elementary concepts, and that any reasoning is a sequence of logical steps that are impossible to deny. Actually, this common belief is usually considered the reason why mathematics is so difficult to understand for the majority of students and non-specialists. Is this belief true? Isn't it only a bias of common sense? It is difficult to provide a definitive answer, but we think it is reasonably true in our age, provided that we are aware of some delicate foundational issues, that are particularly relevant for our purposes here.

Intuitive Starting Points

We said that in mathematics every concept is supposed to be founded with precision on more elementary concepts. Nevertheless, the reader will have probably noticed by herself or himself that this affirmation cannot be true. In fact, let us suppose that a reasoning relies on a concept A. Since the latter must be properly defined, we need to found it on more elementary concepts, say B and C. Again, B and C must be properly defined on the basis of more elementary concepts, say D, E, and F. Keeping on in this way, we get an infinite regress, hence we never begin to state anything. This argument shows the necessity of fixing some starting points—that is, some notions that are considered intuitively known—without even trying to define them. Of course we try to *minimise* the quantity of these intuitive non-defined notions, but we cannot eliminate them completely.

We anticipated above that all of contemporary mathematics is thought of as founded on set theory, but now we are aware that any attempt of foundation cannot be free from some intuitive starting points. This does not mean that the whole set theory has to be assumed as intuitively known (otherwise, this course would be only a list of preliminaries). On the contrary, set theory itself can be constructed as a mathematical theory, but we have to clarify which are its starting points. We will briefly list them below, and we will dedicate chapter 1 to settle this issue.

Linguistic and Logical Frame

If we want to start a reasoning, we do not only need some notions to start with: we also need a language and some logical tools through which the reasoning is performed. If we aim to found them, we encounter the same problem discussed above—that is, an infinite regress. Therefore, we need a linguistic and logical frame that is assumed to be known. The language we use in definitions, statements, and proofs is ordinary English, not in its full extent, but in the (almost) minimal way in order to be understandable and rigorous (this does not apply to comments and remarks for clarification, that can be a bit richer linguistically). Actually, this is a quite vague assumption, but we do not carry on this discussion here, since it is more suitable for a course on philosophy of mathematics. About the logical tools, again we will briefly list them below, and more in detail in chapter 1.

Meta-Theory

We usually call *meta-theory* the framework formed by all of the notions and tools that we assume as intuitively known. Hence, our meta-theory is formed by the basic English language, the logical preliminary tools, and the intuitive concepts we start from. The theory we are constructing (that is, set theory and consequently the whole mathematics) is constructed on the basis of the meta-theory. Hence, we will frequently distinguish between a *meta-theoretical* object or tool, that is part of the background, and a *set-theoretical* one, that has been rigorously constructed as a part of the theory, starting from the background.

PRELIMINARIES OF NAIVE SET THEORY

Now we sketch how the previous general discussion will be applied specifically to set theory. The logical tools we are going to use are the basic propositional connectives (not, and, or, implies, if and only if), the basic quantifiers (for every, exists, exits unique), and their elementary properties. These tools can be formalised by constructing the theories of *propositional calculus* and *first-order logic*, but these theories require some basic tools of naive set theory, hence, in order to avoid circularity, we have to choose where we start from. Here we prefer to organise the foundations of mathematics in such a way that propositional calculus and first-order logics are developed within naive set theory, and not vice-versa.

The intuitive notions we start from are the following ones:

- the notion of *proposition*, depending if necessary on one or more *variables*
- the notion of *equality* and its elementary properties
- the notions of *set* and *element* of a set
- in a certain sense, to be clarified with precision, the elementary theory of *natural numbers*

The notion of equality cannot be really defined. We will see that it is an equivalence relation, but, when we will define equivalence relations, we will have used plenty of times the concept of equality, therefore it is indeed an intuitive preliminary.

The concepts of set and element cannot be really defined either. Sometimes it is possible to find attempts of definition through synonyms—for example, by affirming that a set is a collection of elements or a family of elements. Nevertheless, synonyms can be useful only to strengthen our intuition, not to formalise it. The same considerations apply to the concept of element.

With respect to natural numbers, some comments are necessary, since this is a quite delicate issue. We will see in chapters 7 and 11 that it is possible to construct natural numbers within set theory. Nevertheless, when we will reach those points of our path, we will have used meta-theoretical natural numbers plenty of times (as for equality). It is not easy to establish whether set-theoretical natural numbers can replace the meta-theoretical ones (that is, if it is possible to reorganise all of the exposition without assuming any meta-theoretical knowledge of natural numbers), but we will see in chapter 7 that the meta-theoretical knowledge is necessary at least for the sake of clarity of the exposition. Actually, we will not assume as intuitively known all of the theory of natural numbers, but we will construct it from the intuitive notions of *zero* and *successor*, by stating the meta-theoretical Peano's axioms. We will also need the intuitive notion of *sequence* of objects, that will be formalised within set theory, but again after having used it in the background.

NAIVE SET THEORY AND FOUNDATIONS OF MATHEMATICS

After this brief analysis of meta-theory, let us try to understand more deeply the role of set theory as the foundation of mathematics. First of all, we have to clarify the meaning of the word "naive".

Naive *vs* Axiomatic Set Theory

Naive set theory is so-called to counterpose it to *axiomatic* set theory. The main difference between these two approaches is essentially the following one: in naive set theory we do not have a precise and rigorous way to establish when a statement or a proof is precise and rigorous. This means that the rigour and the correctness of an argument will be based on a "shared feeling", that cannot be confirmed objectively. The main aim of the axiomatic setting consists in overcoming this problem. In fact, at least in principle, it allows to write down a computer program that verifies objectively if a proof is correct or a statement is well-formulated.

This brief comparison seems to indicate that the axiomatic setting is an improvement of the intuitive one. Nevertheless, we must be careful in drawing such a conclusion. In fact, axiomatic set theory (we are mainly thinking of the Zermelo-Fraenkel formulation) is a first-order theory, and the development of first-order logic requires certain notions of naive set theory (mainly the semantic approach, but partially also the syntactic one). Hence, we can informally say that axiomatic set theory, instead of overcoming the naive one, assumes it as a part of its background. Maybe the reader is thinking that this argument is quite cumbersome, therefore we make an example that hopefully helps to clarify it. Every reader of this book studied the basic grammar of her or his own native language(s). Here we suppose it is English, but it could be any other one. Thus, at a certain point of her or his education, the reader listened to a professor explaining what is a noun, what is an adjective, and so on. In these lectures, the professor was formalizing the English language, but (s)he was speaking in English. Therefore, when (s)he was explaining what is a noun or an adjective, (s)he was using English sentences containing in particular nouns and adjectives. Similarly, when (s)he was explaining what is the subject or the predicate of a phrase, (s)he was using phrases with a subject and a predicate. Thus, the professor was assuming as intuitively known the same language (s)he was formalizing. Of course, we could avoid this partial circularity by teaching English grammar in French, but then the students are supposed to have learnt French grammar before. As above, if we want to avoid an infinite regress, we have to start from a language that we use for its own formalisation. This is exactly what happens with set theory: we can axiomatise it by using first-order logic, but such an axiomatisation requires an intuitive knowledge of set theory itself.

This shows that the fact of dealing with naive set theory is not an initial simplification, intended to be overcome in future, but it is an intrinsic necessity that, following our contemporary approach, will remain the base of any mathematical theory, including axiomatic set theory.

The Foundational Structure

Coherently with the previous discussion, we choose the following path:

1. We assume the intuitive knowledge of propositions, variables, logical connectives and quantifiers, sets and elements, equality, and metatheoretical natural numbers.

2. We develop naive set theory.

3. We use the two previous items to develop formal propositional calculus and first-order logic.

4. Within first-order logic, we construct axiomatic set theory.

5. Within axiomatic set theory, we construct (in principle) all of mathematics.

It would be interesting to establish the position of *category theory* in this framework, but we prefer to avoid this issue here. The present setting is resumed figure 1.

This book concerns the second item of the previous list (hence, necessarily, the first one too). We remark that the steps of formal first-order theories and axiomatic set theory can be skipped if we accept not to have an objective criterion to establish if a proof is correct. This would be a relevant foundational lack, but it is what concretely happens in the majority of cases. In fact, mathematicians, in their everyday work, usually rely on naive set theory, without concerning with the formal presentation of their results within axiomatic set theory (that would be definitely too burdensome, even for elementary results). Nevertheless, in principle, this presentation could be performed, and, within the framework described above, it provides a test of validity of any statement and proof.

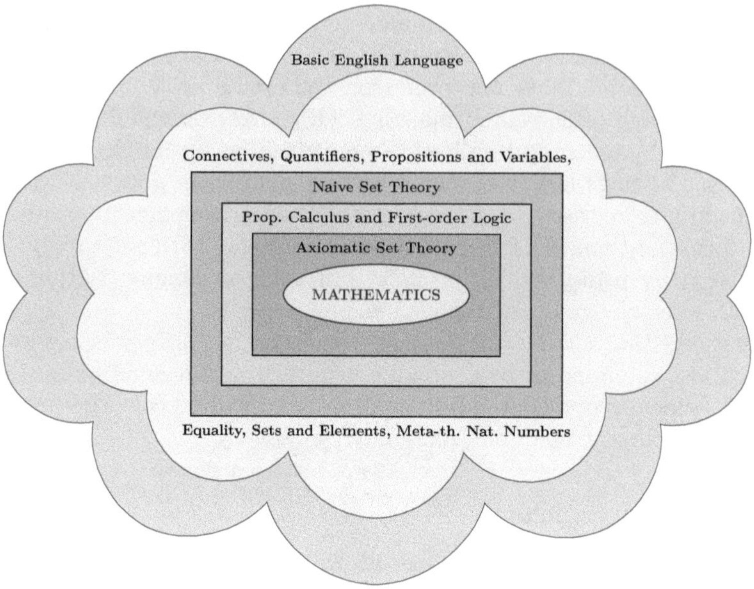

Figure 1 Foundational setting we have in mind.

THE UNIVERSE OF SETS

We conclude this introduction by describing how this course is organised. Basically, in order to provide a unitary presentation of naive set theory, we will follow a path towards the construction of a suitable set-theoretical universe. In fact, after having discussed the meta-theory in chapter 1, we will have to face the following problem: which sets and which elements are admissible? Intuitively, in order to form a set, we have to be able to establish without any doubt which elements belong to it. Nevertheless, this is not sufficient: since we are elaborating a mathematical theory, and we have no mathematical definition of the objects of real world, we need to rule out these objects from our universe. Hence, the first step will consist of assuming that *everything is a set*. This means that the elements of every set are sets too, so that no object of other kind is present.

This choice seems to be sufficient to rule out any undesired element, but this is not the case. In fact, we will see in chapter 2 that, even in a universe in which everything is a set, we cannot exclude some pathologies, that are annoying in set theory and useless in ordinary mathematics, and (even worse) some contradictions, like the famous Russell's paradox. Hence, we will encounter the necessity of suitably reducing the universe we work in, reaching a smaller universe that on one side is small enough to rule out pathologies and contradictions, and on the other side is big enough to host the whole mathematics.

We construct this reduced universe as follows. We allow for the presence of the empty set, since it is so simple that it can be considered a very safe starting point. Afterwards, we require that all of the operations and the constructions that are necessary in ordinary mathematics can be performed within the universe. The reader can imagine to start from the empty set with a torch in the hand, trying to illuminate the surrounding universe as far as (s)he needs, but being aware that (s)he cannot go too far, otherwise (s)he will meet dangers (s)he is not able to deal with. In this exploration, we will make a series of assumptions about our universe (nine in total), that will guarantee that it satisfies the properties we require. The construction of the reduced universe will be completed only in the last chapter, defining *well-founded sets.*

Now a comment is in order. If the reader could talk with us, probably (s)he would argue as follows: "You told that, even in a universe in which everything is a set, it is not possible to rule out certain pathologies and contradictions. Nevertheless, you anticipated that we will be able to construct a suitably reduced universe, in which ordinary mathematics can be founded avoiding those problems. Therefore, at the end of this path, you will prove in a clear way that the reduced universe is coherent and does not host any annoying pathology." Unfortunately, this legitimate expectation will be frustrated. In fact, we are not able at all to prove the coherence of the universe of well-founded sets. We can show that it rules out the pathologies and the contradictions we encountered before, but we cannot exclude that there are some new ones, that we have not detected. Hence, we can only say that the reduced universe is coherent as far as we know. This is the best we can do up to now.

Aims and Target Audience

The book is intended for undergraduate students who aim to deeply understand naive set theory. It is also suitable for graduate students, in order to strengthen their knowledge on this topic or to recall in detail some definitions and statements they are not familiar with. In principle, it might be interesting for a philosophy student too, provided that (s)he acquired (or (s)he desires to acquire) a good level of familiarity with rigorous mathematical language.

As we said in the introduction, the exposition is organised in a unitary way, as a path from the initial meta-theoretical concepts to the construction of the universe of well-founded sets. There are no substantial prerequisites, since the whole mathematics is supposed to be founded on naive set theory, but a certain familiarity with the basic notions of arithmetic (in particular, with the induction principle) is recommended, as well as a basic knowledge of logical connectives and quantifiers.

There are plenty of exercises, partially spread along the text and partially listed at the end of each chapter, most of which are fundamental to understand in detail the topics discussed here. All of the resolutions are available in a separate volume, hence we strongly encourage the reader to solve the exercises by herself or himself, and to compare her or his resolutions with the ones of the author.

The reader is invited to compare the present book with the following ones (among many others): [3], that introduces the same subject, but in a quite informal way (almost conversational, as the author himself says); [2, 4, 8, 9, 11], which insist less on foundational problems, but discuss various interesting topics and applications that are not covered here; [10], that contains a philosophically-oriented presentation of set theory; and [12], which is a basic guide to the elaboration of mathematical proofs, partially including naive set theory. About ordinals, cardinals and well-founded sets, the reader can see [5, chapters 2 and 3], [6, chapters I and III], and [7, chapter I].

Author Bio

Fabio Ferrari Ruffino received his Bachelor's Degree in Mathematics at the University of Milan, his Master's Degree at the University of Pisa and at Scuola Normale Superiore, and his Ph.D. at SISSA (International School for Advanced Studies, Trieste, Italy). He is currently an Assistant Professor at São Carlos University (Brazil). In his research, he has been working mainly on geometry and topology, with particular attention to the applications in theoretical physics (especially in string theory). Both in research and in teaching activities he has always been particularly interested in foundational problems, also from a philosophical point of view.

CHAPTER 1

Preliminaries

In this chapter we summarise the logical and mathematical notions that we assume to be intuitively known, stating some basic definitions and results related to them. Such notions constitute a logical frame and a necessary starting point in order to avoid an infinite regress in the formulation of rigorous definitions and statements.

1.1 PROPOSITIONS, CONNECTIVES, AND QUANTIFIERS

We deal with *propositions* and *variables* intuitively. A proposition with free variables $x_1, \dots, x_n$, that we denote by $\alpha(x_1, \dots, x_n)$ or simply by α, is a statement that is true or false, depending on the value of each variable x_i. Such value can range within a fixed set S_i or it can be undetermined a priori. If a proposition does not depend on any free variable, then it is absolutely true or false.

1.1.1 Logical Connectives

We use the symbol '$\sim$' to denote *not*. This means that $\sim\alpha$ is true if and only if α is false. We use the symbols '$\vee$' and '$\wedge$' to denote respectively *or* and *and*, the "or" being inclusive. This means that the proposition $\alpha \vee \beta$ is false if and only if both α and β are false, and the proposition $\alpha \wedge \beta$ is true if and only if both α and β are true. We use the symbols '$\Rightarrow$' and '$\Leftrightarrow$' to denote respectively *implies* and *if and only if*. This means that the proposition $\alpha \Rightarrow \beta$ is false if and only if α is true and β is false, and the proposition $\alpha \Leftrightarrow \beta$ is true if and only if α and β are both true or both false.

Two propositions β and γ, constructed from $\alpha_1, \dots, \alpha_n$ by applying iteratively the logical connectives, are called *equivalent* if their truth values, depending on the ones of $\alpha_1, \dots, \alpha_n$, coincide. For example, $\sim\sim\alpha$ is equivalent to α, and $\alpha \Leftrightarrow \beta$ is equivalent to $(\alpha \Rightarrow \beta) \wedge (\beta \Rightarrow \alpha)$. We remark in particular that the proposition $\alpha \Rightarrow \beta$ is equivalent to its *contrapositive* $(\sim\beta) \Rightarrow (\sim\alpha)$.

DOI: 10.1201/9781003565291-1

The connective '$\wedge$' is *associative*—that is, $(\alpha \wedge \beta) \wedge \gamma$ is equivalent to $\alpha \wedge (\beta \wedge \gamma)$. It follows by a standard inductive argument that $\alpha_1 \wedge \ldots \wedge \alpha_n$ is well-defined up to equivalence for any n, independently of the position of the parenthesis (see exercise 1.21 in section 1.4). Moreover, the connective '$\wedge$' is *commutative*—that is, $\alpha \wedge \beta$ is equivalent to $\beta \wedge \alpha$. It follows that $\alpha_1 \wedge \ldots \wedge \alpha_n$ does not depend on the order of $\alpha_1, \ldots, \alpha_n$ up to equivalence (see exercise 1.23 in section 1.4). The same considerations hold replacing '$\wedge$' by '$\vee$'.

1.1.2 Quantifiers

We use the symbols '$\forall$', '$\exists$', and '$\exists!$' to denote respectively *for every*, *exist(s)*, and *exist(s) unique*. This means that the proposition $\forall x_1, \ldots, x_n\, \alpha$ is true if and only if for every value of the variables $x_1, \ldots, x_n$ the proposition α is true, and the proposition $\exists x_1, \ldots, x_n\, \alpha$ (respectively, $\exists! x_1, \ldots, x_n\, \alpha$) is true if and only if there exits a value (respectively, there exists a unique value) for each variable $x_1, \ldots, x_n$ that makes the proposition α true.

1.2 EQUALITY

We consider the concept of *equality* as intuitively known. We denote by "$a = b$" the fact that the objects a and b are equal (equivalently, they coincide) and by "$a \neq b$" the opposite statement. It follows that $a \neq b$ has the same meaning of $\sim(a = b)$. The following properties hold for any objects a, b, and c:

- $a = a$
- $a = b$ if and only if $b = a$
- if $a = b$ and $b = c$, then $a = c$

We assume the following basic principle: if $a = b$ and we replace a by b in any statement depending on a, then the truth value of the statement remains unchanged.

1.3 SETS, ELEMENTS, AND SUBSETS

We consider the notions of *set* and *element* as intuitively known. We denote by $y \in x$ the fact that y is an element of x and by $y \notin x$ the opposite statement. It follows that $y \notin x$ has the same meaning as $\sim (y \in x)$. We assume the following fundamental principle:

(A1) *Two sets coincide if and only if they contain the same elements, that is:*

$$(x = y) \Leftrightarrow \forall z\big((z \in x) \Leftrightarrow (z \in y)\big).$$

This means that a set is completely determined by its elements. It follows that there exists a unique set with no elements called *empty set* and denoted by '$\emptyset$'.

Definition 1.1. We say that y is a *subset* of x if every element of y is also an element of x. We use the notation $y \subset x$, therefore

$$(y \subset x) \Leftrightarrow \forall z\big((z \in y) \Rightarrow (z \in x)\big).$$

If $y \subset x$ and $y \neq x$, then we say that y is a *proper* subset of x and we use the notation $y \subsetneq x$. It follows that $y \subsetneq x$ is equivalent to $(y \subset x) \wedge (y \neq x)$. ♢

Exercise 1.2. Prove that $\emptyset \subset x$ and $x \subset x$ for every set x.

Exercise 1.3. Prove that $x = y$ if and only if $x \subset y$ and $y \subset x$.

We denote by $x := \{y_1, \dots, y_n\}$ the (finite) set whose elements are $y_1, \dots, y_n$, forgetting the order and any repetition, and we denote by $x := \{y : \varphi(y)\}$ the set such that $y \in x$ if and only if $\varphi(y)$ is true, where φ is a proposition with one free variable. A proposition with one free variable is also called *property*. We remark that, with this notation, $\emptyset = \{\,\}$ or $\emptyset = \{x : x \neq x\}$.

1.4 NATURAL NUMBERS AND INDUCTION

For the moment we consider the notion of *natural number* as intuitively known, but we will analyse this assumption later on (without really removing it). More precisely, we assume that the following two notions are intuitively clear:

- 0 is a natural number.
- Given any natural number n, there exists its *successor* $s(n)$.

Every natural number is of the form 0, $s(0)$, $s(s(0))$, $s(s(s(0)))$, etc. We call 'N' the set of natural numbers and 'N*' the subset obtained by excluding 0 (the reason why at this stage we use the symbol 'N', instead of the more common 'ℕ', will be clarified in Chapter 7). The following principles (*meta-theoretical Peano's axioms*) hold and are considered intuitively clear too:

(P1) $s(n) \neq 0$ for every n.

(P2) $s(n) = s(m)$ if and only if $n = m$.

(P3) (Ind1) *Induction principle:* Let φ be a *number-theoretic property*—that is, a proposition with one free variable n, whose range is N. If $\varphi(0)$ is true and $\varphi(n) \Rightarrow \varphi(s(n))$ is true for every n, then $\varphi(n)$ is true for every n.

(Rec1) *Recursion principle:* It is possible to *define* an object $f(x, n)$, depending on $n \in$ N and on any object(s) x, recursively (or inductively) as follows. We define $f(x, 0)$ for every x and, assuming to know the definition of $f(x, n)$ for every x, we define $f(x, s(n))$ for every x. In this way, $f(x, n)$ turns out to be defined for every x and n.

We consider `(Ind1)` and `(Rec1)` are two parts of the same principle. We observe that, in `(Rec1)`, the object x can be undetermined or vary within a fixed set. If we let it vary within the empty set, then we get $f(n)$, that only depends on $n \in \mathrm{N}$. Moreover, it can happen that, for any fixed x, the object $f(x, s(n))$ only depends on $f(x, n)$, not on $f(y, n)$ for any y. This will be the case in defining the basic operations in the next section.

1.4.1 Fundamental Operations and Order

We set $1 := s(0)$. We define the operations of *sum*, *product*, and *power* in N recursively as follows, the recursion being applied on m for any fixed n:

- $n + 0 := n$ and $n + s(m) := s(n + m)$;
- $n \cdot 0 := 0$ and $n \cdot s(m) := n \cdot m + n$;
- $n^0 := 1$ and $n^{s(m)} := n^m \cdot n$.

Exercise 1.4. Prove that:

- For every $n \in \mathrm{N}$, we have that $n = 0$ or n is the successor of another unique natural number.
- $s(n) \neq n$ for every $n \in \mathrm{N}$.

Exercise 1.5. Prove that:

- The sum in N is associative and commutative—that is, $n+(m+k) = (n+m)+k$ and $n + m = m + n$.
- The product in N is distributive with respect to the sum—that is, $n(m+k) = nm + nk$ and $(n + m)k = nk + mk$.
- $n + 1 = s(n)$; $n \cdot 1 = n$; $1 \cdot n = n$.
- The product in N is associative and commutative—that is, $n(mk) = (nm)k$ and $nm = mn$.
- $n^1 = n$; $n^{m+k} = n^m n^k$; $n^{mk} = (n^m)^k$.

Exercise 1.6. Prove that:

- $n + m = 0$ if and only if $n = m = 0$.
- $nm = 0$ if and only if $n = 0$ or $m = 0$.

Exercise 1.7. Prove that cancellation law holds with respect to the sum in N—that is, $n + k = m + k$ implies $n = m$. In particular, $n + k = k$ implies $n = 0$.

We define $n \leq m$ if and only if there exists k such that $n + k = m$. In this case, k is unique (see Exercise 1.7), hence we set

$$m - n := k. \tag{1.1}$$

It immediately follows from commutativity that (equivalently) $m - k = n$. We remark that, given $n \in \mathrm{N}^*$, the unique number m such that $s(m) = n$ (see ex. 1.4) is $m = n - 1$, since $m + 1 = n$ (see ex. 1.5, third item). From the definition of '$\leq$' we can deduce that:

1. $n \leq n$.
2. If $n \leq m$ and $m \leq n$, then $n = m$.
3. If $n \leq m$ and $m \leq k$, then $n \leq k$.
4. For every $n, m \in \mathrm{N}$, we have $n \leq m$ or $m \leq n$.

We define $n < m$ if and only if $n \leq m$ and $n \neq m$. We have:

1. $n \not< n$.
2. If $n < m$ and $m < k$, then $n < k$.
3. For every $n, m \in \mathrm{N}$, one of the following alternatives holds: (i) $n < m$; (ii) $n = m$; (iii) $m < n$.

The notation $n \geq m$ (respectively, $n > m$) is equivalent to $m \leq n$ (respectively, $m < n$); moreover, given n and m such that $n \leq m$, we set $\{n, \ldots, m\} := \{k \in \mathrm{N} : n \leq k \leq n\}$. Given a subset $A \subset \mathrm{N}$, we say that $a \in A$ is the *minimum* of A if $a \leq b$ for every $b \in A$, and we say that it is the *maximum* of A if $b \leq a$ for every $b \in A$.

Exercise 1.8. Prove that $n < m$ if and only if there exists $k \in \mathrm{N}^*$ such that $m = n + k$.

Exercise 1.9. Prove the properties (1)–(4) of the relation '$\leq$' and the properties (1)–(3) of the relation '$<$' stated above.

Exercise 1.10. Prove that, if $n < m$, then $m \not< n$.

Exercise 1.11. Prove that, if $n \leq m$ and $m < k$, or $n < m$ and $m \leq k$, then $n < k$.

Exercise 1.12. Prove that cancellation law holds with respect to the product in N—that is, $nk = mk$ and $k \neq 0$ implies $n = m$. In particular, $nk = k$ and $k \neq 0$ implies $n = 1$.

Exercise 1.13. Prove that, for every $n, m, k \in \mathrm{N}$:

- $n \leq m$ (respectively, $n < m$) if and only if $n + k \leq m + k$ (respectively, $n + k < m + k$).
- If $k \neq 0$, then $n \leq m$ (respectively, $n < m$) if and only if $nk \leq mk$ (respectively, $nk < mk$).
- If $n \leq m$, then $nk \leq mk$.

Exercise 1.14. Prove that, for every $n \in \mathrm{N}^*$ and $m \in \mathrm{N}$:

- $m \leq n$ if and only if $m \leq n - 1$ or $m = n$.
- $m < n$ if and only if $m \leq n - 1$.
- There exists no $k \in \mathrm{N}$ such that $n - 1 < k < n$.

Equivalently, for every $n, m \in \mathrm{N}$:

- $m \leq s(n)$ if and only if $m \leq n$ or $m = s(n)$.
- $m < s(n)$ if and only if $m \leq n$.
- There exists no $k \in \mathrm{N}$ such that $n < k < s(n)$.

1.4.2 Other Versions of the Induction Principle

We can state the induction principle in the following equivalent way:

- (**Ind2**) If $\varphi(0)$ holds and $\big(\forall m < n(\varphi(m))\big) \Rightarrow \varphi(n)$ for every n, then $\varphi(n)$ holds for every n.

  (**Rec2**) We define $f(x, 0)$ for every x and, assuming to know the definition of $f(x, m)$ for every x and $m < n$, we define $f(x, n)$ for every x; in this way, $f(x, n)$ turns out to be defined for every x and n.

Exercise 1.15. Prove that the two formulations (**Ind1**) and (**Ind2**) of the induction principle are equivalent—that is, one of the two implies the other one. Prove the same statement about recursive definitions.

Remark: In the present setting, the statement of (**Ind2**) relies on (**Ind1**). In fact, the relation $m < n$ was defined by using the operation of sum, that was defined through (**Rec1**) and whose main properties were proven by applying (**Ind1**). Hence, only the implication (**Ind1**) $\Rightarrow$ (**Ind2**) really holds without circularity. In order to assume (**Ind2**) and deduce (**Ind1**), we have to construct the order relation in N and prove its properties independently of the induction principle (this will happen by considering the set-theoretical approach to natural numbers as finite ordinals or cardinals, shown in Chapters 11 and 12, but we will have used meta-theoretical natural numbers plenty of times at that stage). However, the logical equivalence between (**Ind1**) and (**Ind2**) is conceptually interesting in itself.

Exercise 1.16. *Well-ordering principle for natural numbers:* Prove that every nonempty subset of N admits a minimum.

Exercise 1.17. Prove that (**Ind2**) and the well-ordering principle for natural numbers, that we denote by (**WON**), are equivalent—that is, one of the two implies the other one.

Remark: It follows from Exercise 1.15 that (**Ind1**), (**Ind2**), and (**WON**) are equivalent. Nevertheless, because of the remark in Exercise 1.15, only the implications (**Ind1**) $\Rightarrow$ (**Ind2**) $\Rightarrow$ (**WON**) hold in the present setting without circularity. In principle, we could also prove that (**Ind1**) $\Rightarrow$ (**WON**) $\Rightarrow$ (**Ind2**), but the deduction (**Ind1**) $\Rightarrow$ (**WON**), realised by adapting the resolution of Exercise 1.16, would only "hide" the implication (**Ind1**) $\Rightarrow$ (**Ind2**) within the proof.

Exercise 1.18. Prove that, in both formulations (**Ind1**) and (**Ind2**) of the induction principle, both for proofs and definitions, we can start from any n_0 instead of 0—for example, if $\varphi(n_0)$ holds and $\varphi(n) \Rightarrow \varphi(s(n))$ for any $n \geq n_0$, then $\varphi(n)$ holds for any $n \geq n_0$.

Exercise 1.19. Let us fix $n \in \mathrm{N}^*$ and consider the subset $\{0, \ldots, n\} \subset \mathrm{N}$. We state the two following finite versions of the induction principle:

- (**FInd1**) if $\varphi(0)$ holds and, for every $m \in \{0, \ldots, n-1\}$, we have $\varphi(m) \Rightarrow \varphi(s(m))$, then $\varphi(m)$ holds for every $m \in \{0, \ldots, n\}$;
- (**FInd2**) if $\varphi(0)$ holds and, for every $m \in \{1, \ldots, n\}$, we have $\big(\forall k < m(\varphi(k))\big) \Rightarrow \varphi(m)$, then $\varphi(m)$ holds for every $m \in \{0, \ldots, n\}$.

Prove that (**FInd1**) and (**FInd2**), for any fixed $n \in \mathrm{N}^*$, follow respectively from (**Ind1**) and (**Ind2**). Moreover, prove that it is possible to start from any $n_0 < n$ instead of 0, as in ex. 1.18. The same results hold about recursive definitions.

1.4.3 Sequences and Number-Theoretic Functions

In stating the concept of recursive definition, we assumed as intuitively known the notion of an object $f(n)$ depending on $n \in \mathrm{N}$ (here we leave the dependence on any other fixed object(s) x implicit). The assignment of $f(n)$ to each $n \in \mathrm{N}$ is called *sequence*, and it is also denoted by $x_0, x_1, x_2, \ldots$, where $x_n = f(n)$. If we impose that $n \in \{0, \ldots, k\}$, then we get the *finite sequence* $x_0, \ldots, x_k$. Sometimes we start from any fixed n_0 (frequently 1) instead of 0. If the objects of the sequence are natural numbers, then the sequence is also called *number-theoretic function* and it is denoted by $f\colon \mathrm{N} \to \mathrm{N}$. This means that we associate $f(n) \in \mathrm{N}$ to each $n \in \mathrm{N}$. More generally, if we fix two subsets $A, B \subset \mathrm{N}$ and we associate $f(a) \in B$ to each $a \in A$, then we get a number-theoretic function of the form $f\colon A \to B$. In particular, a finite sequence of natural numbers is a function of the form $f\colon \{0, \ldots, k\} \to \mathrm{N}$. We can further generalise the notion of number-theoretic function. For any fixed $k \in \mathrm{N}$, we denote by N^k the meta-theoretical set of finite sequences of natural numbers of the form $n_1, \ldots, n_k$. If we fix $A \subset \mathrm{N}^k$ and $B \subset \mathrm{N}^h$, and we assign to each sequence $(n_1, \ldots, n_k) \in A$ a sequence $(m_1, \ldots, m_h) \in B$, then we get a number-theoretic function of the form $f\colon A \to B$.[1] In particular, a *number-theoretic k-ary operation* is a number-theoretic function of the form $*\colon \mathrm{N}^k \to \mathrm{N}$. If $k = 2$, it is called *binary* operation. Sum, product, and power are the main examples of number-theoretic binary operations. We usually denote the natural number $*(n_1, \ldots, n_k)$ as $n_1 * \cdots * n_k$.

Given a number-theoretic function $f\colon A \to B$, if $b = f(a)$, then we also use the notation $f\colon a \mapsto b$. The function f is called *injective* if $f(a) = f(a')$ implies $a = a'$. The *image* of f is the subset $\mathrm{Im}(f) = \{b \in B : \exists a \in A : b = f(a)\}$. The function f is called *surjective* if $\mathrm{Im}(f) = B$ and *bijective* if it is both injective and surjective. We remark that the notion of successor can be thought of as a function $s\colon \mathrm{N} \to \mathrm{N}$, injective but not surjective, or as a bijective function $s\colon \mathrm{N} \to \mathrm{N}^*$. Given $f\colon A \to B$ and $g\colon B \to C$, the *composition* $g \circ f\colon A \to C$ is defined by $(g \circ f)(a) := g(f(a))$. It is straightforward to verify that:

- Composition of functions is associative—that is, given $f\colon A \to B$, $g\colon B \to C$, and $h\colon C \to D$, we have $h \circ (g \circ f) = (h \circ g) \circ f$.
- The composition of injective (respectively, surjective or bijective) functions is injective (respectively, surjective or bijective).

Moreover, the *identity function* of A is the bijective function $\mathrm{id}_A\colon A \to A$, $a \mapsto a$. It follows that $f \circ \mathrm{id}_A = \mathrm{id}_B \circ f = f$ for any $f\colon A \to B$. Moreover, there exists a function $f^{-1}\colon B \to A$, such that $f^{-1} \circ f = \mathrm{id}_A$ and $f \circ f^{-1} = \mathrm{id}_B$, if and only if f is bijective. In this case, f^{-1} is unique and it is called *inverse*

[1] Equivalently, we can define a number-theoretic function $f\colon A \to B$, where $A \subset \mathrm{N}^k$ and $B \subset \mathrm{N}^h$, as a subset $f \subset \mathrm{N}^{k+h}$ such that, for every $(n_1, \ldots, n_k) \in A$, there exists a unique $(n_{k+1}, \ldots, n_{k+h}) \in B$ such that $(n_1, \ldots, n_{k+h}) \in f$. A function $f\colon \mathrm{N}^1 \to \mathrm{N}^1$, or $f\colon \{0, \ldots, k\} \to \mathrm{N}^1$, defined in this way, can be naturally identified with a (finite) sequence of natural numbers.

function of f. All of these concepts will be analysed more deeply in chapter 3, where the notion of function will be studied formally within set theory.

With these notions we can state the basic results of combinatorics. In particular, we recall that a *permutation of* n *elements* is a bijective function $\sigma\colon \{1,\ldots,n\} \to \{1,\ldots,n\}$; if σ exchanges h and k and fixes all the other elements, then it is called *transposition* and it is denoted by $\tau_{h,k}$ (leaving n implicit).

Moreover, we say that a subset $A \subset \mathrm{N}$ is *finite* if it is empty or it is the image of a finite sequence $f\colon \{0,\ldots,k\} \to \mathrm{N}$ for a fixed $k \in \mathrm{N}$. In the latter case, we use the notation $A = \{n_0,\ldots,n_k\}$, where $n_i = f(i)$. We say that A is *infinite* if it is not finite.

Exercise 1.20. Prove that every non-empty finite subset $\{n_0,\ldots,n_k\} \subset \mathrm{N}$ admits a maximum.

Exercise 1.21. Let '$*$' be an *associative* number-theoretic binary operation— that is, such that $(n_1 * n_2) * n_3 = n_1 * (n_2 * n_3)$ for every $n_1, n_2, n_3 \in \mathrm{N}$. Prove by induction that, for any finite sequence $n_1,\ldots,n_k$ of natural numbers, the natural number $n_1 * \cdots * n_k$ is well-defined independently of the position of parentheses.

More formally, this means the following. Given a sequence $n_1,\ldots,n_k$, such that $k > 1$, and $h \in \{1,\ldots,k-1\}$, we call $\varphi_h(n_1,\ldots,n_k)$ the sequence $n_1,\ldots,n_h * n_{h+1},\ldots,n_k$. Hence, given two sequences $n_1,\ldots,n_k$ and $h_1,\ldots,h_{k-1}$, such that $1 \leq h_1 < k$ and $1 \leq h_{i+1} < h_i$ for every i (thus, $h_{k-1} = 1$), we obtain the natural number $\varphi_{h_{k-1}}(\cdots(\varphi_{h_1}(n_1,\ldots,n_k)))$. If '$*$' is associative, then this number does not depend on the sequence $h_1,\ldots,h_{k-1}$. (See exercise 10 in section 5.6 for a formalisation of this statement within set theory.)

Remark: It follows that $n_1 + \cdots + n_k$ and $n_1 \cdots n_k$ are well-defined for any $n_1,\ldots,n_k \in \mathrm{N}$ and $k \in \mathrm{N}^*$. Hence, sum and product can be treated as k-ary number-theoretical operations $+, \cdot : \mathrm{N}^k \to \mathrm{N}$ for every $k \geq 2$. The same argument, adapted in the straightforward way, proves that the propositions $\alpha_1 \wedge \ldots \wedge \alpha_k$ and $\alpha_1 \vee \ldots \vee \alpha_k$ are well-defined up to equivalence. Similarly, given a sequence of number-theoretic functions $f_1 : A_0 \to A_1, \ldots, f_k : A_{k-1} \to A_k$, the composition $f_k \circ \cdots \circ f_1 : A_0 \to A_k$ is well-defined.

Exercise 1.22. Prove that any permutation of n elements, with $n \geq 2$, is the composition of transpositions of consecutive elements (that is, of the form $\tau_{k,k+1}$ with $1 \leq k < n$).

Exercise 1.23. Let '$*$' be an associative number-theoretic binary operation that is also *commutative*—that is, $n_1 * n_2 = n_2 * n_1$ for every $n_1, n_2 \in \mathrm{N}$. Prove that $n_1 * \cdots * n_k$ does not depend on the order of $n_1,\ldots,n_k$—that is, $n_{\sigma(1)} * \cdots * n_{\sigma(k)} = n_1 * \cdots * n_k$ for any permutation σ of k elements.

Remark: It follows that $n_1 + \cdots + n_k$ and $n_1 \cdots n_k$ do not depend on the order of $n_1,\ldots,n_k$. The same argument, adapted in the straightforward way, implies that $\alpha_1 \wedge \ldots \wedge \alpha_k$ and $\alpha_1 \vee \ldots \vee \alpha_k$ do not depend on the order of $\alpha_1,\ldots,\alpha_k$ up to equivalence.

Exercise 1.24. Prove that, for every $k \in \mathrm{N}^*$ and $n_1,\ldots,n_k,m_1,\ldots,m_k \in \mathrm{N}$:

1. $n_1 + \cdots + n_k = 0$ if and only if $n_1 = \cdots = n_k = 0$.

2. $n_1 \cdots n_k = 0$ if and only if there exists $h \in \{1, \ldots, k\}$ such that $n_h = 0$.
3. If $n_1 \leq m_1, \ldots, n_k \leq m_k$, then $n_1 + \cdots + n_k \leq m_1 + \cdots + m_k$ and $n_1 \cdots n_k \leq m_1 \cdots m_k$.
4. In the previous item, if there exists $h \in \{1, \ldots, k\}$ such that $n_h < m_h$, then $n_1 + \cdots + n_k < m_1 + \cdots + m_k$; if $m_1, \ldots, m_k \neq 0$ too, then $n_1 \cdots n_k < m_1 \cdots m_k$.
5. If $n_1 \leq \cdots \leq n_k$, then $n_1 \leq n_k$; in this case, if $k \geq 2$ and there exists $h \in \{1, \ldots, k-1\}$ such that $n_h < n_{h+1}$, then $n_1 < n_k$.

Exercise 1.25. Prove that there exists a bijective function $f\colon \{0, \ldots, n\} \xrightarrow{\approx} \{0, \ldots, m\}$ if and only if $n = m$. The same result holds about $f\colon \{n_0, \ldots, n\} \xrightarrow{\approx} \{n_0, \ldots, m\}$ for any $n_0 \leq \min\{n, m\}$. *Remark:* Cfr. exercise 4-vii in section 1.5.

Exercise 1.26. (i) Prove that there exists no bijective function $f\colon \{0, \ldots, n\} \xrightarrow{\approx} \mathrm{N}$ for any fixed $n \in \mathrm{N}$. The same result holds about $f\colon \{n_0, \ldots, n\} \xrightarrow{\approx} \mathrm{N}$ for any $n_0 \leq n$. (ii) Deduce that N is infinite. *Remark:* Cfr. exercise 4-viii in section 1.5.

We defined above the concepts of *finite* and *infinite* subset of N. Let us generalise them to any set. We say that a set x is *finite* if it is empty or it can be represented in the form $x = \{y_1, \ldots, y_n\}$ (see section 1.3), where $n \in \mathrm{N}^*$ and $y_1, \ldots, y_n$ is a finite sequence of objects. We say that x is *infinite* if it is not finite.

It follows from exercise 1.25 that the *number of elements* of a finite set x is well-defined. In fact, we establish that the empty set contains 0 elements. Otherwise, by definition we can represent x in the form $x = \{y_1, \ldots, y_n\}$. If the elements of the fixed sequence are all distinct (that is, $y_i \neq y_j$ for every $i \neq j$), then we define the number of elements of x as n. If we consider another representation $x = \{z_1, \ldots, z_m\}$, formed by distinct elements, then we get the bijection $f\colon \{1, \ldots, n\} \to \{1, \ldots m\}$ defined by $y_i = z_{f(i)}$, therefore $n = m$ by exercise 1.25. Thus, the number of elements of x is well-defined. In general, it is not restrictive to suppose that the elements of the sequence $y_1, \ldots, y_n$ are all distinct. In fact, let us call $A \subset \mathrm{N}$ the set of numbers n such that there exists a representation of x of the form $x = \{y_1, \ldots, y_n\}$. By hypothesis, $A \neq \emptyset$, hence, by exercise 1.16, there exists $k := \min(A)$. We have $x = \{y_1, \ldots, y_k\}$. If by contradiction there exist i and j distinct such that $y_i = y_j$, then, by removing y_j, we represent x in the form $x = \{z_1, \ldots, z_{k-1}\}$, contradicting the minimality of k. Therefore, by definition, the number of elements of x is k.

We remark that, up to now, we have not considered sets represented by an infinite sequence of objects (that is, of the form $x = \{y_0, y_1, y_2, \ldots\}$), since we have no necessity to deal with them meta-theoretically. We will introduce them formally in chapter 7.

Exercise 1.27. Prove that, if x is infinite, then, for every $n \in \mathrm{N}$, there exists a finite subset $y \subset x$ that contains n elements. *Remark:* See exercise 4-x at the end of the chapter for the converse.

1.4.4 Decimal Notation

We have seen that every natural number is of the form 0, $s(0)$, $s(s(0))$, $s(s(s(0)))$, etc. We can also represent it in *decimal notation*, defined recursively as follows. We set:

$$1 := s(0) \qquad 2 := s(1) \qquad \cdots \qquad 9 := s(8). \tag{1.2}$$

Let us consider a finite sequence of digits between 0 and 9, that we denote by $l_n\#\cdots\#l_0$ (for example, $3\#5\#0$ coincides with 350). When $n = 0$, we have already defined its value as a natural number in (1.2). For a generic n, we inductively set

$$l_{n+1}\#l_n\#\cdots\#l_0 := l_{n+1}\cdot(s(9))^{n+1} + l_n\#\cdots\#l_0. \tag{1.3}$$

Exercise 1.28. Prove that

$$l_n\#\cdots\#l_0 = l_n\cdot(s(9))^n + \cdots + l_1\cdot s(9) + l_0.$$

Exercise 1.29. Prove that every $n \in \mathrm{N}$ can be uniquely represented in decimal notation.

Of course there are other possible representations beyond the decimal one, that can be obtained by considering another basis $b \geq 2$ instead of $s(9)$. For example, the *binary representation*, with $b = 2$, is commonly used in computer science. If $b > s(9)$, then it is necessary to introduce new symbols to represent with one digit the numbers between $s(9)$ and $b - 1$.

1.5 EXERCISES

1. Prove that $2 + 2 = 2\cdot 2 = 2^2 = 4$.

2. Prove that:

1. $n \neq 0$ if and only if $n \geq 1$.
2. $n \neq 0, 1$ if and only if $n \geq 2$ if and only if $n > 1$.
3. $n_1\cdots n_k = 1$, where $k \in \mathrm{N}^*$, if and only if $n_1 = \cdots = n_k = 1$.
4. $n_1 + \cdots + n_k = 1$, where $k \in \mathrm{N}^*$, if and only if there exists $h \in \{1, \ldots, k\}$ such that $n_h = 1$ and $n_i = 0$ for every $i \neq h$.
5. $n \leq m$ if and only if $n < m + 1$.

3. Prove that:

1. If $m \geq k$, then $n + (m - k) = (n + m) - k$.
2. If $n \geq m$, then $n - (n - m) = m$.
3. If $m \geq k$, then $n(m - k) = nm - nk$.
4. If $n + m = k + h$ and $n \leq k$ (or $n < k$), then $m \geq h$ (respectively, $m > h$) and $m - h = k - n$.
5. If $n \leq m \leq k$ (or $n \leq m < k$), then $m - n \leq k$ (respectively, $m - n < k$).

4. Prove that:

(i) There exists an injective function $f\colon \{0,\dots,n\} \hookrightarrow \{0,\dots,m\}$ if and only if $n \le m$.

(ii) There exists a surjective function $f\colon \{0,\dots,n\} \twoheadrightarrow \{0,\dots,m\}$ if and only if $n \ge m$.

(iii) An injective function $f\colon \{0,\dots,n\} \hookrightarrow \{0,\dots,m\}$, with $n \le m$, is bijective if and only if $n = m$.

(iv) A surjective function $f\colon \{0,\dots,n\} \twoheadrightarrow \{0,\dots,m\}$, with $n \ge m$, is bijective if and only if $n = m$.

(v) There exists no injective function $f\colon \mathrm{N} \hookrightarrow \{0,\dots,n\}$.

(vi) There exists no surjective function $f\colon \{0,\dots,n\} \twoheadrightarrow \mathrm{N}$.

Deduce that:

(vii) There exists a bijection $f\colon \{0,\dots,n\} \xrightarrow{\approx} \{0,\dots,m\}$ if and only if $n = m$.

(viii) There exists no bijection $f\colon \{0,\dots,n\} \xrightarrow{\approx} \mathrm{N}$.

(ix) If x contains n elements and m elements at the same time, then $n = m$.

(x) A set x is infinite if and only if, for every $n \in \mathrm{N}$, there exists a finite subset $y \subset x$ that contains n elements.

(xi) If x is finite and $y \subset x$, then y is finite; in this case, if x contains n elements and y contains k elements, then $k \le n$; moreover, $k = n$ if and only if $y = x$.

(xii) If y is infinite and $y \subset x$, then x is infinite.

All of these results keep on holding replacing $\{0,\dots,n\}$ and $\{0,\dots,m\}$ respectively by $\{n_0,\dots,n\}$ and $\{n_0,\dots,m\}$ for any $n_0 \le \min\{n,m\}$.

Remark: Items (vii), (viii), and (ix) had already been stated respectively in exercise 1.25, in exercise 1.26 and in the comments after the latter. Here we easily deduced them from items (i)–(vi).

5. We say that m *divides* n (or m is a *divisor* of n, or n is a *multiple* of m), and we write $m \mid n$, if and only if there exists $k \in \mathrm{N}$ such that $n = mk$. Prove that:

1. $n \mid n$.
2. $n \mid m$ and $m \mid n$ if and only if $n = m$.
3. If $n \mid m$ and $m \mid h$, then $n \mid h$.
4. $1 \mid n$ and $n \mid 0$ for every $n \in \mathrm{N}$.
5. If $k \mid n$ and $k \mid m$, then $k \mid (n+m)$ and, if $n \ge m$, then $k \mid (n-m)$.
6. If $m \mid n$ and $n \ne 0$, then $1 \le m \le n$.

We say that n is *even* if $2 \mid n$ and *odd* otherwise.

6. (*Euclid's division lemma*) Prove that, for every $n \in \mathrm{N}$ and $m \in \mathrm{N}^*$, there exist unique $q, r \in \mathrm{N}$ such that $n = mq + r$ and $r < m$. *Remark:* Usually the numbers q and r are called respectively *quotient* and *reminder* of the division of n by m. We observe that $r = 0$ if and only if $m \mid n$.

7. We say that $n, m \in \mathrm{N}^*$ are *coprime* if they have no common divisors different from 1. Prove that, for every $n, a, b \in \mathrm{N}^*$, if $n \mid ab$ and n and a are coprime, then $n \mid b$. (*Hint:* Induction on ab. The result is trivial if $ab = 1$. Suppose it holds for every $n, \alpha, \beta \in \mathrm{N}^*$ such that $1 \leq \alpha\beta < ab$ and prove it for every n and the fixed a and b. By hypothesis there exists $m \in \mathrm{N}^*$ such that $ab = nm$. Distinguish three cases: I. $n = a$: Trivial. II. $n < a$: Compute $(a-n)b$. III. $n > a$: Compute $(n-a)m$.)

8. We set $\mathrm{N}^{**} := \{n \in \mathrm{N} : n \neq 0, 1\}$. We say that $n \in \mathrm{N}^{**}$ is *prime* if $(m \mid n) \Rightarrow \big((m = 1) \vee (m = n)\big)$ for every $m \in \mathrm{N}^*$. We denote by P the set of prime numbers. Prove that:

(i) (*Euclid's lemma*) If $p \in \mathrm{P}$, $a, b \in \mathrm{N}^*$ and $p \mid ab$, then $p \mid a$ or $p \mid b$.

(ii) If $p, q \in \mathrm{P}$, $b \in \mathrm{N}^*$ and $p \mid q^b$, then $p = q$.

(iii) If $k \in \mathrm{N}^*$, $p, q_1, \ldots, q_k \in \mathrm{P}$, $b_1, \ldots, b_k \in \mathrm{N}^*$ and $p \mid q_1^{b_1} \cdots q_k^{b_k}$, then there exists $i \in \{1, \ldots, k\}$ such that $p = q_i$.

(iv) If $k \in \mathrm{N}^*$, $p \in \mathrm{P}$, $q_1, \ldots, q_k \in \mathrm{P}$ are all distinct, $a, b_1, \ldots, b_k \in \mathrm{N}^*$ and $p^a \mid q_1^{b_1} \cdots q_k^{b_k}$, then there exists $i \in \{1, \ldots, k\}$ such that $p = q_i$ and $a \leq b_i$.

9. Prove that:

(i) 2 is prime.

(ii) For every $n \in \mathrm{N}^{**}$ there exist $k \in \mathrm{N}^*$ and $p_1, \ldots, p_k \in \mathrm{P}$ such that $n = p_1 \cdots p_k$.

(iii) (*Fundamental theorem of arithmetic* or *Unique factorisation theorem*) Every $n \in \mathrm{N}^{**}$ can be written as

$$n = p_1^{a_1} \cdots p_k^{a_k}, \tag{\#}$$

where $k \in \mathrm{N}^*$, $p_1, \ldots, p_k \in \mathrm{P}$ are all distinct and $a_1, \ldots, a_k \in \mathrm{N}^*$. Moreover, the expression (#) is unique for a fixed n, up to the order of $p_1, \ldots, p_k$.

We call (#) the *decomposition in prime factors* of n.

10. Given $n, m \in \mathrm{N}^*$, we call:

- *Greatest common divisor* of n and m the maximum of the finite set $\{k \in \mathrm{N}^* : (k \mid n) \wedge (k \mid m)\}$. We denote it as $\mathrm{GCD}(n, m)$.
- *Lowest common multiple* of n and m the minimum of the set $\{k \in \mathrm{N}^* : (n \mid k) \wedge (m \mid k)\}$. We denote it as $\mathrm{LCM}(n, m)$.

(i) Let $n = p_1^{\alpha_1} \cdots p_k^{\alpha_k}$ and $m = q_1^{\beta_1} \cdots p_h^{\beta_h}$ be the decompositions (#) of $n, m \in \mathrm{N}^{**}$. We set $\{r_1, \ldots, r_l\} := \{p_1, \ldots, p_k, q_1, \ldots, q_h\}$, assuming that $r_1, \ldots, r_l$ are all distinct. It follows that $n = r_1^{\gamma_1} \cdots r_l^{\gamma_l}$ and $m = r_1^{\delta_1} \cdots r_l^{\delta_l}$, where $\gamma_1, \ldots, \gamma_l, \delta_1, \ldots, \delta_l \in \mathrm{N}$ (that is, they can be zero as well). In particular, $\gamma_i = \alpha_j$ if $r_i = p_j$ and $\gamma_i = 0$ if $r_i \notin \{p_1, \ldots, p_k\}$. Similarly, $\delta_i = \beta_j$ if $r_i = q_j$ and $\delta_i = 0$ if $r_i \notin \{q_1, \ldots, q_h\}$. Prove that:

$$\mathrm{GCD}(n, m) = r_1^{\min\{\gamma_1, \delta_1\}} \cdots r_l^{\min\{\gamma_l, \delta_l\}}$$
$$\mathrm{LCM}(n, m) = r_1^{\max\{\gamma_1, \delta_1\}} \cdots r_l^{\max\{\gamma_l, \delta_l\}}.$$

(ii) Prove that $\mathrm{GCD}(n, 1) = 1$ and $\mathrm{LCM}(n, 1) = n$ for every $n \in \mathrm{N}^*$.
(iii) Given $n, m \in \mathrm{N}^*$, prove that $d = \mathrm{GCD}(n, m)$ if and only if the following condition holds: $d \mid n$, $d \mid m$, and, for every $k \in \mathrm{N}^*$, if $k \mid n$ and $k \mid m$, then $k \mid d$. Similarly, prove that $M = \mathrm{LCM}(n, m)$ if and only if the following condition holds: $n \mid M$, $m \mid M$, and, for every $k \in \mathrm{N}^*$, if $n \mid k$ and $m \mid k$, then $M \mid k$.

CHAPTER 2

Hereditary Sets and Basic Operations

In this chapter we begin to describe the set-theoretical universe by establishing that every object is a set. In this way, even if the notion of set remains intuitive, we do not need to include any other kind of non-defined object. The background introduced in chapter 1 remains intuitive too, but it is considered as the meta-theory, not as a part of the universe itself.

Nevertheless, we will show that this universe is too big to avoid contradictions and pathologies, therefore it is necessary to strongly reduce it. The reduced universe has to be sufficiently rich to "host" the whole mathematics, and sufficiently narrow to be (hopefully) coherent and not pathological. Only in the last chapter will we be able to complete this reduction process, but here we take the first steps. In particular, we begin from the empty set, since it is so simple that we are allowed to consider it a very safe starting point. Then, we state some suitable assumptions, which allow to perform within the universe some basic constructions and operations that are unavoidable in every area of ordinary mathematics.

2.1 HEREDITARY SETS

Intuitively, in order to define a set, we must be able to establish without doubts if an element belongs to it or not. Hence, the set of "interesting books" is not well-defined, since the property of being interesting is not objectively definable. On the contrary, the set of football teams that participated in the world-cup of 2022 is well-defined, since it has 32 clearly established elements. Nevertheless, we aim to construct set theory as the foundation of mathematics, and we have no mathematical definition of "football team". Therefore, the latter set, in spite of being well-defined, must be ruled out anyway. Thus, in order to minimise the intuitive notions we start from, we establish the following fundamental principle:

DOI: 10.1201/9781003565291-2

In our universe every object is a set.

Consequently, we consider sets whose elements are other sets, whose elements are other sets, and so on. Such a set is called *hereditary*. We will analyse more in detail this notion in appendix A, but we do not need more details here.

From now on, when we say "x is a set", we are always assuming that it is hereditary (except for meta-theoretical sets like N, but, in these cases, we will specify that the set is meta-theoretical). In particular, the notation $x := \{y : \varphi(y)\}$ indicates that x is the set formed by the hereditary sets y that satisfy the property φ. It easily follows that x is hereditary too.

2.2 PROPERTIES AND WELL-FOUNDED FORMULAS

In this section we focus on the notion of *property*. In the meta-theory, we defined a property as a proposition with one free variable. Now we need to be more formal about this concept, since it will be used to state a fundamental assumption called *comprehension hypothesis*. In particular, we have to establish which properties φ are admissible when considering a set of the form $x := \{y : \varphi(y)\}$ in the universe. We settle this problem by providing the definition of *well-founded formula* (wf) in the framework of naïve set theory.

2.2.1 Well-Founded Formulas

We define wfs in three steps:

(i) An *atomic formula* is a formula of the form $(y \in x)$ or $(y = x)$, where x and y are sets (previously defined) or variables taking values among sets (it is possible that one of the two is a variable and the other one is a set). Every atomic formula is well-founded by hypothesis.

(ii) If φ and ψ are well-founded formulas, then the following formulas are well-founded too by hypothesis: $(\sim \varphi)$, $(\varphi \vee \psi)$, $(\varphi \wedge \psi)$, $(\varphi \Rightarrow \psi)$, $(\varphi \Leftrightarrow \psi)$, $(\forall x_1, \dots, x_n\, \varphi)$, $(\exists x_1, \dots, x_n\, \varphi)$, $(\exists! x_1, \dots, x_n\, \varphi)$.

(iii) A formula is *well-founded* if and only if it can be constructed starting from atomic formulas, as defined in step (i), by applying iteratively step (ii). More precisely, φ is a wf if and only if there exists a finite sequence $\varphi_0, \dots, \varphi_n$, where $n \in$ N, such that:

- φ_0 is an atomic formula.
- For every i such that $1 \leq i \leq n$, φ_i is an atomic formula or it has been constructed by applying item (ii) to one or two formulas among $\varphi_0, \dots, \varphi_{i-1}$.
- $\varphi_n = \varphi$.

Some remarks are in order.

Remark 2.1. Item (ii) is not minimal, since some of the formulas considered in the statement are equivalent to suitable combinations of the other ones. However, we are not interested in minimality here. ◇

Remark 2.2. Following the definition above, parentheses are put in such a way that there is no need to establish a priority order among the operators. Nevertheless, we frequently omit parenthesis when there is no ambiguity, with the following priority order: $\sim$, $\wedge$ and $\vee$, $\forall$ and $\exists$ and $\exists!$, $\Rightarrow$ and $\Leftrightarrow$. Hence, for example, $\forall x \sim \varphi(x) \wedge \psi(x) \Rightarrow \eta(x)$ is equivalent to $\big((\forall x((\sim \varphi(x)) \wedge \psi(x))) \Rightarrow \eta(x)\big)$. We could be more formal about parenthesis elimination, but it is not necessary, since we will always put them in such a way that it will be immediately clear which is the order to follow. ◇

Remark 2.3. It may happen that we use some definitions in order to shorten a wf. For example, $\exists y(y \subset x)$ is well-founded, since $y \subset x$ has been defined as $\forall z((z \in y) \Rightarrow (z \in x))$. We can also use meta-theoretical natural numbers to express in one formula a family of wfs—for example, $(y \in x_1) \vee \ldots \vee (y \in x_n)$ is a wf for every fixed $n \in \mathrm{N}^*$; similarly, the formula $\exists i \in \{1, \ldots, n\}(y = x_i)$ is equivalent to $(y = x_1) \vee \ldots \vee (y = x_n)$, that is a wf for every fixed $n \in \mathrm{N}$ (while it is not a wf in itself, since the existential quantifier is applied to a meta-theoretical natural number). ◇

2.2.2 Bounded and Free Variables

Let us consider the wf $\varphi(x) := (x = \emptyset) \vee \forall y(y \in x)$. We say that $(y \in x)$ is the *scope* of the quantifier $\forall y$. The variable y is said to be *bounded* in φ, since a quantifier has been applied to it, and the variable x is said to be *free* in φ, since it appears outside the scope of any quantifier applied to it. When we write $\varphi(x_1, \ldots, x_n)$, we assume that the free variables of φ are $x_1, \ldots, x_n$. Hence, when we define a set as $x := \{y : \varphi(y)\}$, we assume that y is the unique free variable of φ. We remark that a variable can be both free and bounded—for example, in the formula $\varphi(x, y) := (x \neq \emptyset) \vee \forall x(x \in y)$, the variable x is both free and bounded. Nevertheless, we will always avoid such situations by replacing the variable in one of the two cases with another one—for example, we can write the previous formula as $\varphi(x, y) := (x \neq \emptyset) \vee \forall z(z \in y)$.

2.2.3 Set-Theoretical Properties

We define a *set-theoretical property* as a wf with one free variable. Intuitively, this means that we allow for properties expressed by a formula constructed only through the relations '$\in$' and '$=$', the basic propositional operators and the quantifiers.

At this point, it seems reasonable to assume that, if φ is a set-theoretical property, then the set $x := \{y : \varphi(y)\}$ is well-defined. We will see in the next section that, surprisingly for a beginner, this assumption is not reasonable at all, nor is it coherent.

2.3 PATHOLOGIES AND RUSSELL'S PARADOX

By establishing that every object in the universe is a set, we excluded sets that contain non-mathematical objects, even if they are intuitively well-defined. Nevertheless, this is not enough to exclude some pathologies and contradictions, that we are going to analyse.

2.3.1 Possible Pathologies

We "define" $x := \{x\}$—that is, x is the set whose unique element is x itself, as figure 2.1 shows.

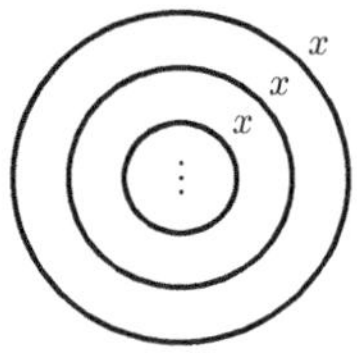

Figure 2.1 The set $x = \{x\}$.

Actually, the formula $x := \{x\}$ is not a definition, since we are defining x through x itself, hence it is circular. Therefore, we should better ask ourselves if there exists a set x such that $x = \{x\}$. In this case, it would belong to the universe of hereditary sets, but it would be clearly pathological. We neither try to reply to this question, since, even admitting that such a set exists, it is completely useless in ordinary mathematics. Therefore, we desire to rule it out anyway.

There are other analogous situations. For example, there exist two sets x and y such that $x = \{y\}$ and $y = \{x\}$?

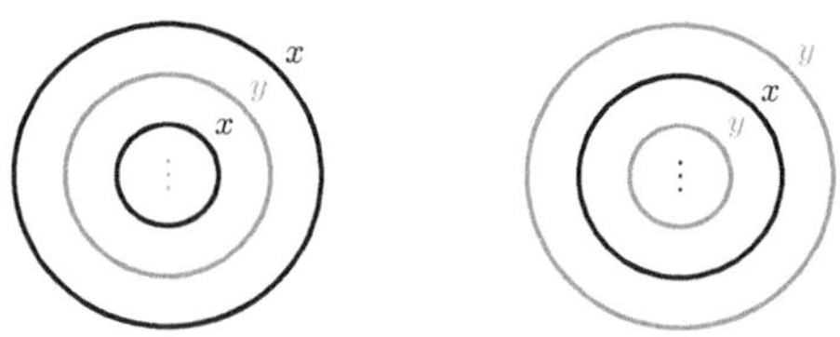

Figure 2.2 Two sets such that $x = \{y\}$ and $y = \{x\}$.

Again, we are not interested at all in sets of this kind.

2.3.2 Russell's Paradox

The (possible) pathologies shown above are unpleasant, but they are not contradictory. Therefore, ruling them out from the universe is preferable, but not strictly necessary. Now we are going to show that, if we assume that the set $x := \{y : \varphi(y)\}$ is well-defined for any set-theoretical property φ, as it seems reasonable at a first glance, then we get a contradiction. Hence, the reduction of the universe will become a necessity for the sake of coherence.

In general, it may happen that a set belongs to itself or not. For example, if u is the set of all sets—that is, $u := \{x : x = x\}$—then $u \in u$, since u itself is a set. On the contrary, $\emptyset \notin \emptyset$.[1] Let us now consider the set formed by those sets that do not belong to themselves:

$$x := \{y : y \notin y\}.$$

It seems that x is well-defined, since $\varphi(y) := (y \notin y)$ is a wf with one free variable.[2]

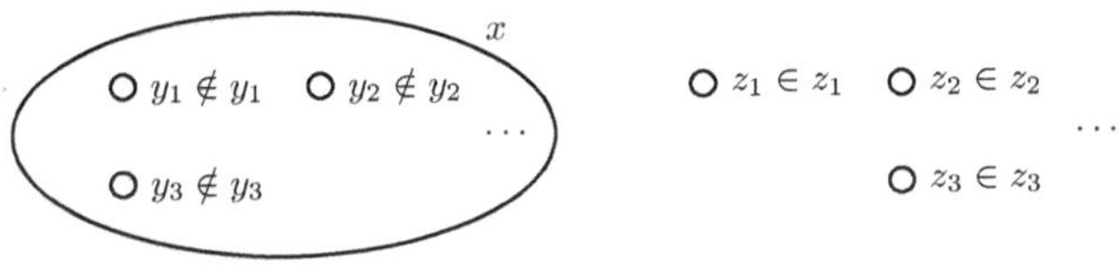

Figure 2.3 The set $x := \{y : y \notin y\}$, leading to Russell's paradox.

We ask ourselves if x belongs to itself. Let us suppose that the answer is positive—that is, $x \in x$. Since $y \in x$ if and only if $y \notin y$, by setting $y = x$ we deduce that $x \notin x$, a contradiction.

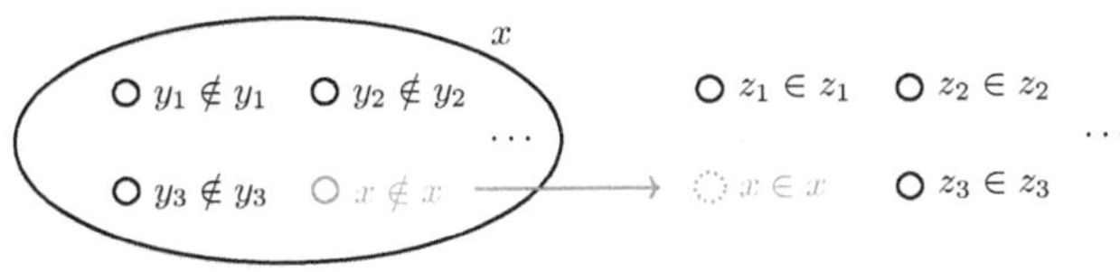

Figure 2.4 Contradiction assuming that $x \in x$.

Similarly, let us suppose that the answer is negative—that is, $x \notin x$. Since $y \in x$ if and only if $y \notin y$, by setting $y = x$ we deduce that $x \in x$, a contradiction.

---

[1] We observe that these two examples are opposite: $u := \{x : x = x\}$ and $\emptyset := \{x : x \neq x\}$. The former contains every set, hence in particular it contains itself, while the latter contains no sets, hence in particular it does not contain itself.

[2] In order to prove the latter statement, we construct the finite sequence φ_0, φ_1, where $\varphi_0 := (y \in y)$ is an atomic formula and $\varphi_1 := {\sim}\varphi_0$. Since $\varphi_1 = (y \notin y)$, we get the result.

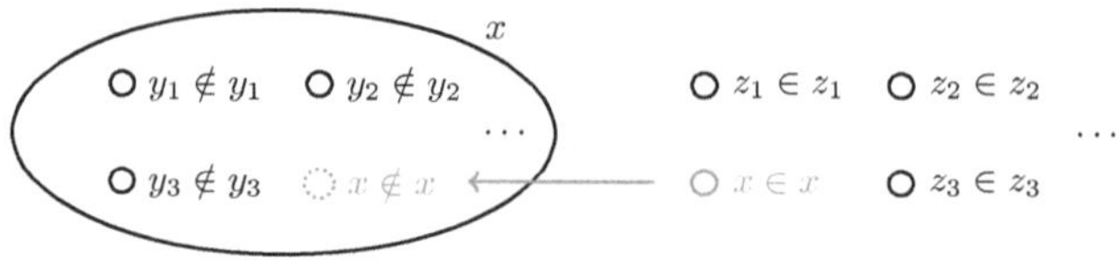

Figure 2.5 Contradiction assuming that $x \notin x$.

We proved that $x \in x \Leftrightarrow x \notin x$, hence the existence of x as a set leads to a contradiction. This argument, called *Russell's paradox*, shows that constructing a set through a set-theoretical property is not possible in general.

As we anticipated above, in order to eliminate these pathologies and contradictions, we will simply avoid to deal with "problematic sets", by suitably restricting our universe. In particular, we will rule out every set belonging to itself. Actually, the word "restriction" assumes the we are really neglecting some sets outside the small universe. Nevertheless, since we are not sure about their existence, it is also possible that we do not have to reduce anything, but only to establish which hereditary sets really exist. Since this is not a relevant issue for us, we only affirm that the fact of being a hereditary set will not be sufficient a priori to be an object of our universe, as we have been assuming up to now. Anyway, we will use the expression "reduced universe" to simplify the exposition.

2.4 CONSTRUCTING FINITE SETS

In order to begin the construction of the reduced universe, we make the following fundamental assumption:

(A2) *The empty set belongs to our universe and, given a finite sequence of sets $y_1, \ldots, y_n$ in the universe, the (finite) set $x := \{y_1, \ldots, y_n\}$ belongs to the universe too.*

We stress the two following particular cases:

- If $n = 2$, then we get the *unordered pair* (or simply *pair*) $x = \{y_1, y_2\}$.
- If $n = 1$, then we get the *singleton* $x = \{y_1\}$.

We remark that:

- The singleton $\{y_1\}$ is different from the set y_1 in general (it will always be different in our reduced universe).
- In defining the notion of unordered pair, we did not exclude the possibility $y_1 = y_2$. In this case, $\{y_1, y_1\} = \{y_1\}$, hence a singleton is a particular case of pair (contrary to the usual meaning of the word "pair").

- We pointed out in section 1.3 that, when defining $x := \{y_1, \ldots, y_n\}$, we forget the order and any repetition. Moreover, we have seen in section 1.4.3 that, if $y_i \neq y_j$ for every $i \neq j$, then n is the *number of elements* of x and it is well-defined. Otherwise, the number of elements of x can be any $k \in \{1, \ldots, n\}$ and it can be computed by eliminating repetitions.

We do not need to assume the existence of sets of the form $x := \{y_1, y_2, \ldots\}$, where $y_1, y_2, \ldots$ is an *infinite* sequence. We will deal with such sets in section 7.3.1.

Now we can construct plenty of sets through (A2), starting from the empty one and forming iteratively new finite sets, with elements chosen among the ones previously constructed. We get the following hierarchy:

- Level 0: $\emptyset$
- Level 1: $\{\emptyset\}$
- Level 2: $\{\{\emptyset\}\}, \{\emptyset, \{\emptyset\}\}$
- Level 3: $\Big\{\{\{\emptyset\}\}\Big\}, \Big\{\{\emptyset, \{\emptyset\}\}\Big\}, \Big\{\{\{\emptyset\}\}, \{\emptyset, \{\emptyset\}\}\Big\},$
  $\Big\{\{\emptyset\}, \{\{\emptyset\}\}, \{\emptyset, \{\emptyset\}\}\Big\}, \ldots$
- ...

At each level $n > 0$ we have all of the distinct finite sets formed by any number of elements chosen from level 0 to level $n - 1$, in such a way that the level of at least one element is $n - 1$. It follows that these sets are all distinct. In fact, if the level of a set x is lower than n, then x contains no elements of level bigger or equal to $n - 1$. On the contrary, if the level of x is bigger or equal to n, then x contains at lest one element of level bigger or equal to $n - 1$.

By solving exercise 12 at the end of the chapter, the reader will see the formal construction of each level and (s)he will compute the corresponding number of elements.

2.5 ORDERED FINITE SETS

Given a pair $x := \{y, z\}$, if we want to fix an order between its elements, we have to specify which is the first one. This is enough, since the other one is obviously the second. Hence, a pair with a fixed order is given by an unordered pair and a fixed element. For this reason, given two sets y and z, we define the *ordered pair* $(y, z) := \{\{y\}, \{y, z\}\}$. It follows from assumption (A2), applied twice, that (y, z) belongs to our universe if y and z belong to it.

Exercise 2.4. Prove that $(y, z) = (y', z')$ if and only if $y = y'$ and $z = z'$.

In this way, the additional piece of information provided by the order is encoded in the set-theoretical language. We define the *ordered triple*

$(x, y, z) := ((x, y), z)$. In general, we define recursively the *ordered n-uple* $(x_1, \ldots, x_n)$ as follows. For $n = 1$, we set $(x_1) := x_1$, and, for any $n \geq 2$, we set $(x_1, \ldots, x_n) := ((x_1, \ldots, x_{n-1}), x_n)$.

Exercise 2.5. Prove that $(x_1, \ldots, x_n) = (y_1, \ldots, y_n)$ if and only if $x_1 = y_1$, ..., $x_n = y_n$.

Exercise 2.6. Prove that, if $x_1, \ldots, x_n$ belong to our universe, then $(x_1, \ldots, x_n)$ belongs to it too.

We remark that, when $n = 1$, the ordered singleton $(x) = x$ does not coincide with the unordered one $\{x\}$. This distinction is necessary to make the recursive definition $(x_1, x_2) = ((x_1), x_2)$ hold for $n = 2$ as well.

Exercise 2.7. Suppose to define an ordered triple as follows: $(x, y, z) := \{\{x\}, \{x, y\}, \{x, y, z\}\}$. Can we affirm that $(x, y, z) = (x', y', z')$ if and only if $x = x'$, $y = y'$ e $z = z'$? *Hint:* Consider the case in which x, y e z are *not* all distinct.

2.6 OPERATIONS ON FINITE FAMILIES OF SETS

In every area of ordinary mathematics we need to apply basic set-theoretical operations like union, intersection, Cartesian product, etc. In this section we define them and we state by hypothesis that they can be applied within our universe.

2.6.1 Unary and Binary Operations

The first definition concerns operations that are *unary* (that is, they take one set and give another one) or *binary* (that is, they take two sets and give another one).

Definition 2.8. Given a set x, we define the following unary operations:

- the *power set* $\mathcal{P}(x) := \{y : y \subset x\}$
- fixing a bigger set X, such that $x \subset X$, the *complement* of x in X as $x^{\complement} := \{y \in X : y \notin x\}$

Moreover, given two sets x and y, we define the following binary operations:

- the *union* $x \cup y := \{z : (z \in x) \vee (z \in y)\}$
- the *intersection* $x \cap y := \{z : (z \in x) \wedge (z \in y)\}$
- the *Cartesian product* $x \times y := \{(z, w) : z \in x, w \in y\}$
- the *disjoint union* $x \sqcup y := (x \times \{\emptyset\}) \cup (y \times \{\{\emptyset\}\})$
- the *set difference* $x \setminus y := \{z : (z \in x) \wedge (z \notin y)\}$
- the *symmetric difference* $x \,\Delta\, y := (x \cup y) \setminus (x \cap y) = (x \setminus y) \cup (y \setminus x)$ ◇

We set $\mathcal{P}^*(x) := \{y \in \mathcal{P}(x) : y \neq \emptyset\}$. It follows that $\mathcal{P}(\emptyset) = \{\emptyset\}$, while $\mathcal{P}^*(\emptyset) = \emptyset$. We remark that, using set difference, we have $x^{\complement} = X \setminus x$ and $\mathcal{P}^*(x) = \mathcal{P}(x) \setminus \{\emptyset\}$. Moreover, the operation of complement is considered unary because we think of the set X as fixed and we let x vary. If we let both X and x vary, then it becomes binary and it coincides with set difference, under the hypothesis $x \subset X$.

We represent the binary operations introduced above graphically. Let us consider the following sets x and y.

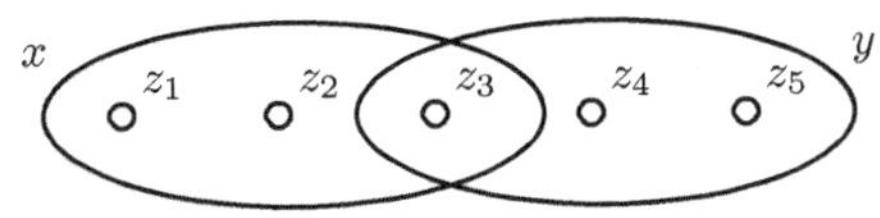

Figure 2.6 Two sets.

The union $x \cup y$ contains the elements of x and the ones of y, while the intersection $x \cap y$ contains the common elements.

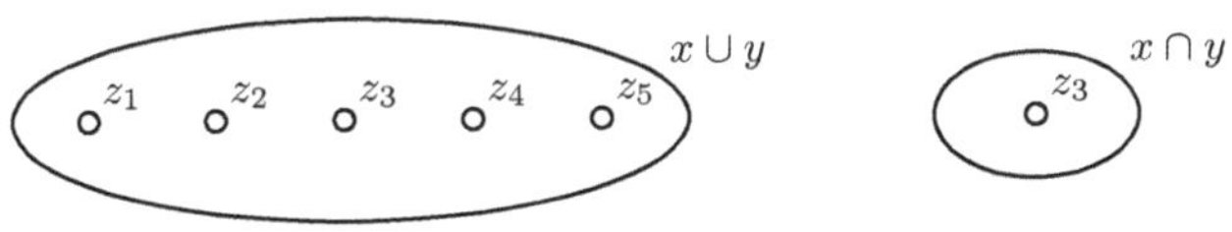

Figure 2.7 Union and intersection.

The Cartesian product $x \times y$ is the set of the ordered pairs, formed by one element of x and one element of y. We can represent it trough a discrete Cartesian diagram as in figure 2.8.

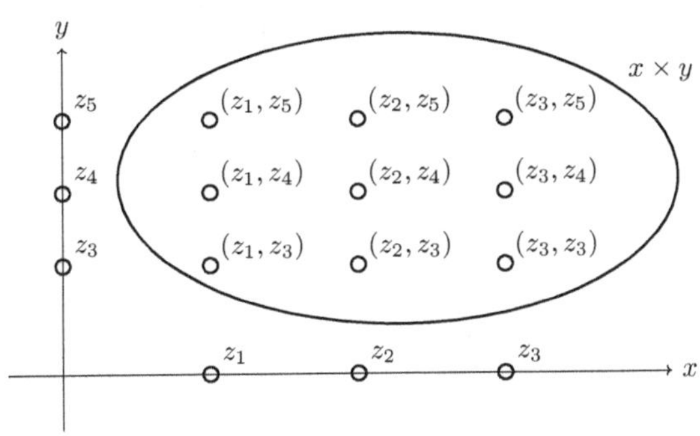

Figure 2.8 Cartesian product.

The disjoint union $x \sqcup y$ contains the elements of x and the elements of y, but separating them, so that the elements of the intersection are counted twice: the first time as elements of x, and the second time as elements of y. In order to realise this separation, we use $\emptyset$ as a "tag" for the elements of x, and $\{\emptyset\}$ as a "tag" for the elements of y.

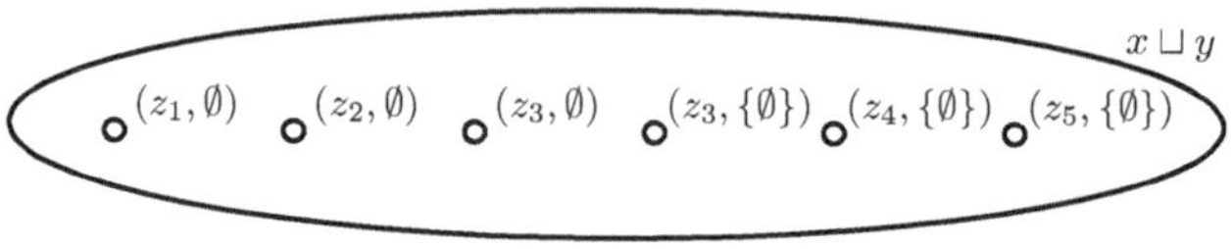

Figure 2.9 Disjoint union.

The set difference $x \setminus y$ contains the elements of x that do no belong to y, while the symmetric difference $x \,\Delta\, y$ contains the elements that belong to x or to y, but not to both.

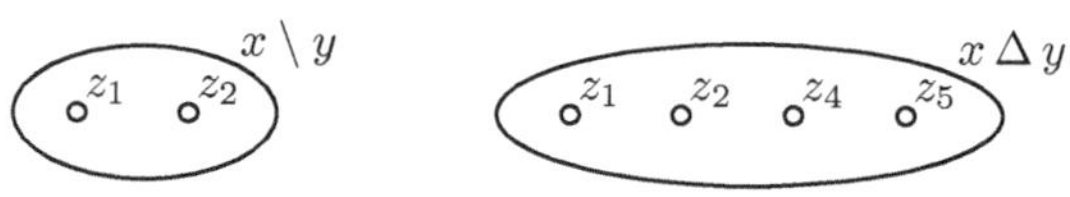

Figure 2.10 Difference and Symmetric difference.

2.6.2 n-Ary Operations

We extend the first four binary operations of definition 2.8 to *n-ary* operations (that is, operations that take n sets, for any $n \in \mathrm{N}^*$, and give another one).

Definition 2.9. Given a finite sequence of sets $x_1, \ldots, x_n$, we define:

- the *union* $x_1 \cup \ldots \cup x_n := \{y : (y \in x_1) \vee \ldots \vee (y \in x_n)\}$;
- the *intersection* $x_1 \cap \ldots \cap x_n := \{y : (y \in x_1) \wedge \ldots \wedge (y \in x_n)\}$;
- the *Cartesian product* $x_1 \times \cdots \times x_n := \{(y_1, \ldots, y_n) : y_1 \in x_1, \ldots, y_n \in x_n\}$;
- the *disjoint union* $x_1 \sqcup \ldots \sqcup x_n := (x_1 \times \{\emptyset\}) \cup \ldots \cup (x_n \times \{\cdots\{\emptyset\}\cdots\})$. More precisely, we recursively set $u_0 := \emptyset$ and $u_{n+1} := \{u_n\}$ and we define $x_1 \sqcup \ldots \sqcup x_n := (x_1 \times u_1) \cup \ldots \cup (x_n \times u_n)$. ♢

When $n = 2$, we recover the corresponding binary operations as particular cases. When $n = 1$, the union, the intersection, and the Cartesian product coincide with the trivial operation that takes x_1 and gives x_1 itself. We will discuss the disjoint union for $n = 1$ in section 5.2.4.

Remark 2.10. The reader will find various important properties of these basic operations in the exercises at the end of the chapter. ♢

2.6.3 Basic Operations and the Universe

As we anticipated at the beginning of this section, the operations defined above are essential in every area of ordinary mathematics, hence we make the following fundamental assumption:

(A3) *All of the operations defined in 2.8 and 2.9, when applied to sets in our universe, produce sets that belong to our universe as well.*

Assumptions (A2) and (A3) together are definitely not minimal. In fact, stating (A3) as we did, we could simply state (A2) with respect to singletons (that is, if y is an element of the universe, then $\{y\}$ is too), since $\{y_1, \ldots, y_n\} = \{y_1\} \cup \ldots \cup \{y_n\}$. Moreover, we could only state (A3) for binary operations and extend it to n-ary operations by induction (cfr. exercise 2 at the end of the chapter about union and intersection, and chapter 5 about Cartesian product and disjoint union). Nevertheless, since we are not dealing with the axiomatic approach, we are not particularly interested in the minimality of each statement.[3]

Remark 2.11. In chapter 4 we will state an assumption similar to (A3), but about arbitrary unions, intersections, etc. Thanks to arbitrary unions in particular, it would be sufficient to state (A3) only with respect to power set and (A2) only with respect to (unordered) pairs. This is the common approach in the axiomatic framework, mainly because it does not explicitly rely on the meta-theoretical knowledge of natural numbers. In fact, n-ary operations are excluded from the statement of the assumptions and, in the case of a pair $x := \{y, z\}$, we do not need to stress that y and z are *two* sets. Nevertheless, as we anticipated in the introduction, the necessity of the meta-theoretical knowledge of natural numbers is a delicate matter, that we will analyse in section 7, without really providing a solution. In any case, at this stage, we consider much more natural and elegant to deal with generic finite sets, with n-ary operations, with ordered n-uples, and so on. We think it would be quite unnatural and cumbersome not to have this possibility now. ◇

2.7 COMPREHENSION HYPOTHESIS

We have seen in section 2.3 that, if we define a set in the form $x := \{y : \varphi(y)\}$, then we can get a contradiction, even if φ is a set-theoretical property. For this reason, we impose the following limitation: instead of considering *any* set y verifying φ, we fix a set a, belonging to our universe by hypothesis, and we impose that $y \in a$. This means that we select the elements of a (not of the entire universe) that satisfy the property φ, defining the set $x := \{y \in a : \varphi(y)\}$.

[3] About power set, we remark that $\mathcal{P}(x)$ contains the subsets $y \subset x$ *that belong to the universe*. In principle, it is possible that some subsets of x exist in the bigger universe of hereditary sets, but not in the small one we are constructing. In this case, they do not belong to $\mathcal{P}(x)$. Hence, assumption (A3) does not specify which subsets of x belong to the universe; it only states that they form a new set in the universe itself. Assumption (A4) below will characterise some of them.

With this restriction, we assume that x is well-defined and it belongs to our universe. Actually, nobody can really prove that this assumption (together with all the others that we made and we will make) avoids any contradiction, but it does as far as we know and it is necessary to perform plenty of constructions in ordinary mathematics. We call this assumption *comprehension hypothesis*, and we state it as follows:

(A4) *If a is a set in our universe and φ is a set-theoretical property, then the set $x := \{y \in a : \varphi(y)\}$ belongs to our universe.*

From (A4) we get the following meaningful result.

Proposition 2.12. *In our universe there not exists the set of all sets in the universe.*

Proof. Let us suppose by contradiction that $u := \{x : x = x\}$ belongs to the universe. Then, by applying (A4), we can construct the set $x := \{y \in u : y \notin y\}$, that is equivalent to $x = \{y : y \notin y\}$, leading to Russell's paradox. □

Assumption (A4) makes the statement of (A3) not minimal. In fact, by applying (A4), the intersection can be defined as $x_1 \cap \ldots \cap x_n := \{y \in x_1 : (y \in x_2) \wedge \ldots \wedge (y \in x_n)\}$; the set difference can be defined as $x \setminus y := \{z \in x : z \notin y\}$, and, through these two operations, we define complement, disjoint union and symmetric difference. Nevertheless, we repeat that we are more interested in the clarity and naturality of the exposition than in the minimality of each statement.

2.8 EXERCISES

1. Prove that binary union and intersection are associative and commutative, that is:

(i) $(x \cup y) \cup z = x \cup (y \cup z)$ and $(x \cap y) \cap z = x \cap (y \cap z)$

(ii) $x \cup y = y \cup x$ and $x \cap y = y \cap x$

2. It follows from exercise 1.21 (adapted in the straightforward way) and exercise 1-i that the sets $x_1 \cup \ldots \cup x_n$ and $x_1 \cap \ldots \cap x_n$, defined by applying iteratively the corresponding binary operations, are well-defined independently of the position of parentheses. Moreover, because of exercises 1.23 and 1-ii, they do not depend on the order of $x_1, \ldots, x_n$. Prove that they coincide with the sets $x_1 \cup \ldots \cup x_n$ and $x_1 \cap \ldots \cap x_n$ defined in 2.9.

3. Prove that binary union and intersection are distributive with respect to each other, that is:

- $x \cap (y \cup z) = (x \cap y) \cup (x \cap z)$
- $x \cup (y \cap z) = (x \cup y) \cap (x \cup z)$

Conclude that:

- $x \cap (y_1 \cup \ldots \cup y_n) = (x \cap y_1) \cup \ldots \cup (x \cap y_n)$
- $x \cup (y_1 \cap \ldots \cap y_n) = (x \cup y_1) \cap \ldots \cap (x \cup y_n)$

Remark: Exercises 1–3 show important properties of union and intersection. The analogous properties of Cartesian product and disjoint union are more elaborated to state. We will discuss them in detail in chapter 5.

4. Prove that:

1. $x \cup \emptyset = x \cup x = x \cap x = x$ and $x \cap \emptyset = \emptyset$
2. $x \cap y \subset x \subset x \cup y$ (it follows that $x \cap y \subset x \cup y$)
3. $x \cap (x \cup y) = x \cup (x \cap y) = x$
4. $(y \subset x) \Leftrightarrow (x \cup y = x) \Leftrightarrow (x \cap y = y)$
5. $x \setminus y = x \setminus (x \cap y)$
6. $x \setminus y = x$ if and only if $x \cap y = \emptyset$
7. $(x \cup y) \setminus x = y \setminus (x \cap y)$

It follows from (6) and (7) that $(x \cup y) \setminus x = y$ if and only if $x \cap y = \emptyset$.

5. Prove the following items, assuming that x and y are subsets of a fixed set X:

1. $(x^{\complement})^{\complement} = x$
2. $\emptyset^{\complement} = X$ and $X^{\complement} = \emptyset$
3. $x \cap x^{\complement} = \emptyset$ and $x \cup x^{\complement} = X$
4. $(x \subset y) \Leftrightarrow (y^{\complement} \subset x^{\complement})$
5. $(x \cup y)^{\complement} = x^{\complement} \cap y^{\complement}$ and $(x \cap y)^{\complement} = x^{\complement} \cup y^{\complement}$ (*De Morgan's laws*)
6. $x \setminus y = x \cap y^{\complement}$

Moreover, extend De Morgan's laws to n-ary operations—that is, assuming that $x_1, \ldots x_n \subset X$:

7. $(x_1 \cup \ldots \cup x_n)^{\complement} = x_1^{\complement} \cap \ldots \cap x_n^{\complement}$ and $(x_1 \cap \ldots \cap x_n)^{\complement} = x_1^{\complement} \cup \ldots \cup x_n^{\complement}$

6. Prove that:

1. If $a \subset x$ and $b \subset y$, then $a \cup b \subset x \cup y$ and $a \cap b \subset x \cap y$, but in general the converse does not hold.
2. If $a \subset x$ and $b \subset y$, then $a \times b \subset x \times y$. If $a, b \neq \emptyset$, then the converse holds too; in this case, $a = x$ e $b = y$ if and only if $a \times b = x \times y$.
3. $a \subset x$ and $b \subset y$ if and only if $a \sqcup b \subset x \sqcup y$. This implies that $a = x$ and $b = y$ if and only if $a \sqcup b = x \sqcup y$.
4. If $a \subset x$, then $a \setminus y \subset x \setminus y$. If $b \subset y$, then $x \setminus y \subset x \setminus b$.
5. If $y \subset x$, then $\mathcal{P}(y) \subset \mathcal{P}(x)$.

7. Prove that:

1. $\mathcal{P}(x_1 \cap \ldots \cap x_n) = \mathcal{P}(x_1) \cap \ldots \cap \mathcal{P}(x_n)$.
2. $\mathcal{P}(x_1 \cup \ldots \cup x_n) \supset \mathcal{P}(x_1) \cup \ldots \cup \mathcal{P}(x_n)$, but in general the converse does not hold.
3. $\mathcal{P}(x \setminus y) \subset \mathcal{P}(x) \setminus \mathcal{P}^*(y)$, or equivalently $\mathcal{P}^*(x \setminus y) \subset \mathcal{P}(x) \setminus \mathcal{P}(y)$, but in general the converse does not hold.
4. In general none of the two sets $\mathcal{P}^*(x \,\Delta\, y)$ and $\mathcal{P}(x) \,\Delta\, \mathcal{P}(y)$ is included in the other one.

8. Prove that, if x contains n elements, then $\mathcal{P}(x)$ contains 2^n elements.

9. Prove that, if x contains n elements and $y \subset x$ contains k element, then we can express x in the form $x = \{z_1, \ldots, z_n\}$, in such a way that $y = \{z_1, \ldots, z_k\}$.

10. Prove that, if x contains n elements, y contains m elements and $x \cap y$ contains k elements, then

- $x \cup y$ contains $n + m - k$ elements;
- $x \times y$ contains nm elements;
- $x \sqcup y$ contains $n + m$ elements;
- $x \setminus y$ contains $n - k$ elements; and
- $x \Delta y$ contains $n + m - 2k$ elements

11. Prove that $x \sqcup \overset{n \text{ times}}{\cdots} \sqcup x = x \times \{u_0, \ldots, u_{n-1}\}$ for every $n \geq 2$, where we recall that $u_0 := \emptyset$ and $u_{n+1} := \{u_n\}$.

12. Let L_n be the set of finite sets of level n (see section 2.4). Prove that L_n is a finite set in our universe and compute the corresponding number of elements. Which is the level of L_n?

13. (i) Prove that symmetric difference is associative and commutative. (ii) Because of the previous item, we can extend symmetric difference to n sets, by defining $x_1 \Delta \ldots \Delta x_n$ as the iterated application of the binary operation. Moreover, since $x \Delta y = (x \setminus y) \cup (y \setminus x)$, we can also consider the following n-ary extension:

$$x_1 \dot{\Delta} \ldots \dot{\Delta} x_n := \{y : \exists! i \in \{1, \ldots, n\} : y \in x_i\}.$$

There is one more possibility. Since $x \Delta y = (x \cup y) \setminus (y \cap x)$, we can define the following n-ary extension:

$$x_1 \ddot{\Delta} \ldots \ddot{\Delta} x_n := (x_1 \cup \ldots \cup x_n) \setminus (x_1 \cap \ldots \cap x_n).$$

Prove that:

1. Δ, $\dot{\Delta}$ e $\ddot{\Delta}$ are three distinct operations for every $n \geq 3$.
2. $x_1 \dot{\Delta} \ldots \dot{\Delta} x_n \subset x_1 \Delta \ldots \Delta x_n$ and $x_1 \dot{\Delta} \ldots \dot{\Delta} x_n \subset x_1 \ddot{\Delta} \ldots \ddot{\Delta} x_n$ for every n.
3. If n is even, then $x_1 \Delta \ldots \Delta x_n \subset x_1 \ddot{\Delta} \ldots \ddot{\Delta} x_n$.

14. Prove that, by applying assumptions `(A1)`–`(A4)`, we can construct all of the finite sets described in section 2.4, at any level n, and only them.

Remark: Exercise 14 shows that *up to now we have no way to construct an infinite set.* In fact, existence of infinite sets will be a direct consequence of another assumption, that we will introduce in chapter 7. Hence, the sets we can construct through `(A1)`–`(A4)` are the same we can construct through `(A1)` and `(A2)`. This does not mean that `(A3)` and `(A4)` are useless, since they will keep on holding when infinite sets will be included in the universe, and they will become essential.

15. Prove that, coherently with proposition 2.12, no finite set can be the set of all sets in our universe.

CHAPTER 3

Relations and Functions

In section 1.4 we introduced the meta-theoretical notion of *number-theoretic function*. Here we introduce the formal notion of *function* within set theory, as a particular case of the more general concept of *relation*. We also analyse two other particular cases of relations—that is, *equivalence relations* and *order relations*—that are commonly applied in ordinary mathematics. In the case of functions, the formalisation we are going to show will not replace the corresponding meta-theoretical notion, even after embedding natural numbers in set theory (see chapter 7), but it provides a rigorous foundations of this concept in the universe of sets.

3.1 RELATIONS

Definition 3.1. Given two sets x and y, a *graph of a binary relation* or *binary pre-relation* on the ordered pair (x, y) is a subset $\rho \subset x \times y$. A *binary relation* is an ordered triple (x, y, ρ) such that x and y are two sets and ρ is a binary pre-relation on (x, y). If $x = y$, we say that the relation is *internal* to x and we denote it by (x, ρ); in this case, ρ is called pre-relation *in* x. ◇

We usually write $a\,\rho\,b$ to denote $(a, b) \in \rho$.

Definition 3.2. The *domain* of a binary relation is defined as $\mathrm{dom}(x, y, \rho) := \{a \in x : \exists b \in y : a\,\rho\,b\}$ and the *image* is defined as $\mathrm{Im}(x, y, \rho) := \{b \in y : \exists a \in x : a\,\rho\,b\}$. ◇

It is common to identify the concepts of pre-relation and relation, indicating the relation (x, y, ρ) simply by ρ, but in general x and y cannot be deduced from ρ (when x and y are clear from the context, we too will use this abuse of notation). On the contrary, the domain and the image only depend on ρ itself, hence we can simply write $\mathrm{dom}(\rho)$ and $\mathrm{Im}(\rho)$ without ambiguity.

Definition 3.3. Given a binary relation (x, y, ρ) and two subsets $z \subset x$ and $w \subset y$, we denote by $(z, w, \rho|_{(z,w)})$ the *restriction* of (x, y, ρ) to z and w, defined by $\rho|_{(z,w)} := \rho \cap (z \times w)$. ◇

DOI: 10.1201/9781003565291-3

We generalise the concept of binary relation as follows.

Definition 3.4. Given n sets $x_1, \ldots, x_n$, a *graph of an n-ary relation* or *n-ary pre-relation* on the ordered n-uple $(x_1, \ldots, x_n)$ is a subset $\rho \subset x_1 \times \cdots \times x_n$. An *n-ary relation* is an ordered $(n+1)$-uple $(x_1, \ldots, x_n, \rho)$ such that $x_1, \ldots, x_n$ are sets and ρ is an n-ary pre-relation on $(x_1, \ldots, x_n)$. Given n subsets $y_1 \subset x_1, \ldots, y_n \subset x_n$, we denote by $(y_1, \ldots, y_n, \rho|_{(y_1,\ldots,y_n)})$ the corresponding *restriction* of ρ—that is, $\rho|_{(y_1,\ldots,y_n)} := \rho \cap (y_1 \times \cdots \times y_n)$. ◊

As a particular case, a *unary pre-relation* or *pre-property* in a set x is a subset $\rho \subset x$. A *unary relation* or *property* is an ordered pair (x, ρ), with $\rho \subset x$. This is a rigorous definition of property, but it does not replace the corresponding meta-theoretical concept, that has already been used several times (in chapter 1, as a proposition with one free variable, and more formally in chapter 2, trough the definition of wf).[1]

Exercise 3.5. Let (x, y, ρ) be a relation and let $z \subset x$ and $w \subset y$. Prove that $\mathrm{dom}(\rho|_{(z,w)}) \subset \mathrm{dom}(\rho) \cap z$ and $\mathrm{Im}(\rho|_{(z,w)}) \subset \mathrm{Im}(\rho) \cap w$, but in general the inverse inclusions do not hold.

3.2 FUNCTIONS

Definition 3.6. Given two sets x and y, a *graph of a function* or *pre-function* from x to y is a binary pre-relation $f \subset x \times y$ such that, for every $a \in x$, there exists a unique $b \in y$ such that $(a, b) \in f$. A *function* is a binary relation (x, y, f) such that f is a pre-function from x to y. ◊

We usually denote the function (x, y, f) by $f : x \to y$, and the statement $(a, b) \in f$ by $f(a) = b$ or $f : a \mapsto b$.

Definition 3.7. The *codomain* of the function (x, y, f) is the set y. The *domain* of f is defined as for any relation, hence it is the whole x by definition. The *image* of f is defined as for any relation as well—that is, $\mathrm{Im}(f) := \{b \in y : \exists a \in x : f(a) = b\}$. ◊

We denote by $\mathrm{Func}(x, y)$ the set of functions from x to y.

Definition 3.8. We say that a function $f : x \to y$ is *injective* if $f(a_1) = f(a_2)$ implies $a_1 = a_2$ for every $a_1, a_2 \in x$. An injective function is also called *embedding*. We say that $f : x \to y$ is *surjective* if $\mathrm{Im}(f) = y$. We say that $f : x \to y$ is *bijective* (or that it is a *bijection*) if it is both injective and surjective. ◊

---

[1]The statement of (A4) would be tautological if the property φ were defined as a subset of a, as we are doing here. About the induction principle, we could replace the notion of number-theoretic property with the one of meta-theoretical subset of N (as we will do in chapter 7 within set theory). Nevertheless, establishing when such a subset is well-defined is as intuitive as the meta-theoretical notion of property. Hence, the two concepts are interchangeable, but both are distinct from the definition of property given here.

We denote an injective function by $f\colon x \hookrightarrow y$, a surjective function by $f\colon x \twoheadrightarrow y$ and a bijection by $f\colon x \xrightarrow{\approx} y$.

As we said about relations, it is common to indicate the function $f\colon x \to y$ simply by f, but in general y cannot be deduced from f, unless the latter is surjective (again, we too will use this abuse of notation when y is clear from the context). On the contrary, we can write $\operatorname{dom}(f)$ and $\operatorname{Im}(f)$ without ambiguity.

Exercise 3.9. Show that, for every set x:

- $\operatorname{Func}(\emptyset, x) = \{(\emptyset, x, \emptyset)\}$.
- If $x \neq \emptyset$, then $\operatorname{Func}(x, \emptyset) = \emptyset$.

Exercise 3.10. Let u be a set with one element. Show that, for every set x:

- $\operatorname{Func}(x, u)$ contains one element.
- There exists a (natural) bijection between $\operatorname{Func}(u, x)$ and x.

3.2.1 Composition

Given two functions $f\colon x \to y$ and $g\colon y \to z$, we define the *composition* $g \circ f\colon x \to z$ by $u \mapsto g(f(u))$.

Exercise 3.11. Prove that composition is associative—that is, given $f\colon x \to y$, $g\colon y \to z$ and $h\colon z \to w$, we have $h \circ (g \circ f) = (h \circ g) \circ f$.

Remark: It follows from exercise 1.21 (adapted in the straightforward way) that, given $f_1\colon x_0 \to x_1, \ldots, f_n\colon x_{n-1} \to x_n$, the composition $f_n \circ \cdots \circ f_1\colon x_0 \to x_n$ is well-defined.

Exercise 3.12. (i) Prove that the composition of injective/surjective/bijective functions is respectively injective/surjective/bijective. (ii) Prove that, if $g \circ f$ is injective, then f is injective. Show that g is not necessarily injective. (iii) Prove that, if $g \circ f$ is surjective, then g is surjective. Show f is not necessarily surjective.

We set $\operatorname{Func}(x) := \operatorname{Func}(x, x)$. In this case, given $f, g\colon x \to x$, both $g \circ f\colon x \to x$ and $f \circ g\colon x \to x$ are defined, but in general they are different. This means that composition in $\operatorname{Func}(x)$ is not commutative in general.

Exercise 3.13. Show with an example that composition in $\operatorname{Func}(x)$ is not commutative in general. For which sets x is it commutative?

3.2.2 Diagrams

We call *diagram* of sets and functions a finite family of sets, usually represented by points or point-wise positions, and of functions between these sets, usually represented by arrows. If we fix a path of consecutive arrows, that starts from x and ends in y, then the corresponding functions can be composed, producing a function from x to y. We say that a diagram *commutes* or *is commutative* if, fixing any two sets x and y and choosing any two paths of arrows from x to

y, the corresponding compositions coincide. For example, let us consider the following diagram:

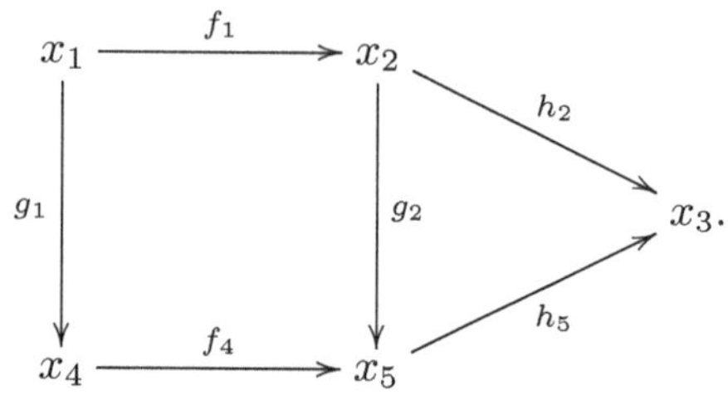

By definition, this diagram is commutative if $g_2 \circ f_1 = f_4 \circ g_1$ and $h_5 \circ g_2 = h_2$.

Exercise 3.14. Consider the following diagram:

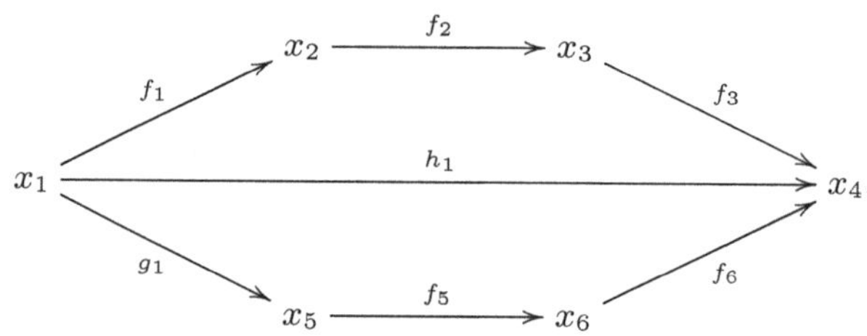

What does it mean that it is commutative?

Diagrams are useful to represent in a clear way a configurations of various sets and functions. Moreover, it is frequently natural to require (or to verify) that commutativity holds.

3.2.3 Identity and Inversion

The *identity function* of a set x is the function that associates each element of x to itself—that is, $\mathrm{id}_x : x \to x$, $a \mapsto a$. Given any function $f : x \to y$, it immediately follows that $f \circ \mathrm{id}_x = \mathrm{id}_y \circ f = f$.

Exercise 3.15. Prove that id_x is the *unique* function from x to x such that $f \circ \mathrm{id}_x = f$ for every $f : x \to y$ and $\mathrm{id}_x \circ g = g$ for every $g : y \to x$.

We say that $g : y \to x$ is *left inverse* to $f : x \to y$ if $g \circ f = \mathrm{id}_x$ and *right inverse* to f if $f \circ g = \mathrm{id}_y$. We say that g is *inverse* to f if it is both left and right inverse.

Exercise 3.16. Given $f : x \to y$, prove that:

(i) f admits a left inverse if and only if it is injective.

(ii) If f admits a right inverse, then it is surjective. (Assuming the axiom of choice, that we will discuss in chapter 9, the converse holds too; indeed, this is one of the possible formulations of the axiom of choice itself. Otherwise, the converse can be proven only assuming that y is finite.)

(iii) f admits an inverse if and only if it is bijective.

(iv) A left or right inverse, when it exists, is not necessarily unique. On the contrary, if an inverse exists, then it is unique. More precisely, if f admits a left inverse g and a right inverse h, then $g = h$ and g (or h) is the inverse of f.

(v) If g is a left inverse of f (hence f is injective) and x contains more than one element, then g is the unique left inverse if and only if f is surjective (hence bijective). In this case, $g = f^{-1}$.

(vi) If g is a right inverse of f (hence f is surjective), then g is the unique right inverse if and only if f is injective (hence bijective). In this case, $g = f^{-1}$.

It follows from exercise 3.16-iv that f is invertible (equivalently, bijective) if and only if it is both left and right invertible. In this case, the inverse function is unique, hence we denote it by f^{-1} without ambiguity. When there exists a bijection between x and y, we use the notation $x \approx y$.

3.2.4 Restriction

Given a function $f \colon x \to y$ and two subsets $z \subset x$ and $w \subset y$, the restriction $(z, w, f|_{(z,w)})$ is always well-defined as a relation, but it is a function if and only if $f(a) \in w$ for every $a \in z$. Of course, this is always the case if $w = y$.

We usually denote by $f|_z \colon z \to y$ the restriction of f to (z, y) and by $f\|_w \colon x \to w$ the restriction of f to (x, w), the latter being a function if and only if $\mathrm{Im}(f) \subset w$ (in particular, $f\|_{\mathrm{Im}(f)} \colon x \to \mathrm{Im}(f)$ is always well-defined and it is surjective by construction). It follows that $f|_{(z,w)} = f|_z\|_w = f\|_w|_z$, the latter being a function if and only if $\mathrm{Im}(f|_z) \subset w$.

Given a subset $y \subset x$, we call *inclusion* of y in x the injective function $\iota_{y,x} \colon y \hookrightarrow x$, $a \mapsto a$. It follows that $\iota_{x,x} = \mathrm{id}_x$. Moreover, given $f \colon x \to y$ and two subsets $z \subset x$ and $w \subset y$, the latter containing $\mathrm{Im}(f)$, we have $f|_z = f \circ \iota_{z,x}$ and $f = \iota_{w,y} \circ f\|_w$. More generally, if $\mathrm{Im}(f|_z) \subset w$, we have $f|_{(z,w)} = f\|_w \circ \iota_{z,x}$ and $f|_z = \iota_{w,y} \circ f|_{(z,w)}$.

3.2.5 Induced Functions between Power Sets

A function $f \colon x \to y$ induces the following two functions: $\dot{f} \colon \mathcal{P}(x) \to \mathcal{P}(y)$, $a \mapsto \{f(z) : z \in a\}$, and $\ddot{f} \colon \mathcal{P}(y) \to \mathcal{P}(x)$, $b \mapsto \{z \in x : f(z) \in b\}$ (see figures 3.1 and 3.2). We will state some important properties of these induced functions in the exercises at the end of the chapter.

Exercise 3.17. (i) Prove that $\dot{\mathrm{id}}_x = \ddot{\mathrm{id}}_x = \mathrm{id}_{\mathcal{P}(x)}$. (ii) Given $f \colon x \to y$ and $g \colon y \to z$, prove that $(g \circ f)^{\cdot} = \dot{g} \circ \dot{f}$ and $(g \circ f)^{\cdot\cdot} = \ddot{f} \circ \ddot{g}$.

Observing figures 3.1 and 3.2, we see that $\ddot{f}(\dot{f}(a)) = \{z_1, \ldots, z_4\} \supsetneq a$ and $\dot{f}(\ddot{f}(b)) = \{w_2\} \subsetneq b$. In fact, by resolving exercise 7 at the end of the chapter, the reader will verify that $a \subset \ddot{f}(\dot{f}(a))$ and $\dot{f}(\ddot{f}(b)) \subset b$ for every f, a and b, but these inclusions are not invertible in general. In particular, the equality $a = \ddot{f}(\dot{f}(a))$ holds for every a if and only if f is injective, and the equality $\dot{f}(\ddot{f}(b)) = b$ holds for every b if and only if f is surjective.

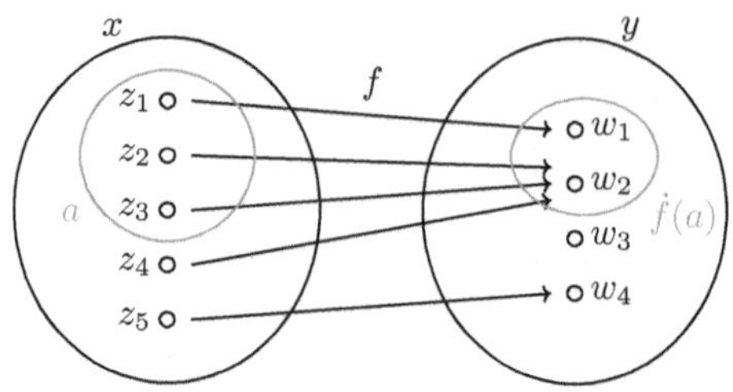

Figure 3.1 The induced function $\dot{f}\colon \mathcal{P}(x) \to \mathcal{P}(y)$.

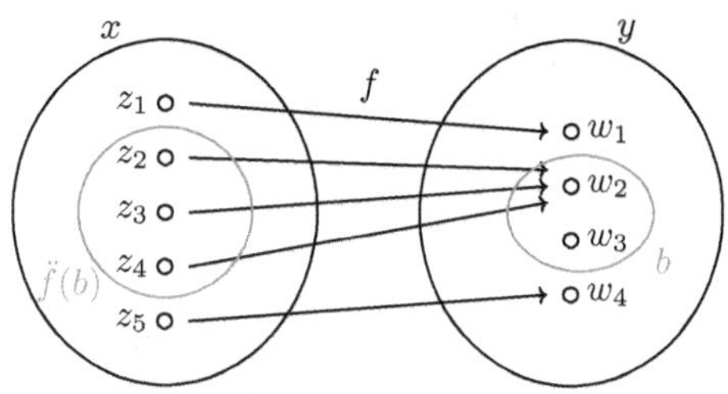

Figure 3.2 The induced function $\ddot{f}\colon \mathcal{P}(y) \to \mathcal{P}(x)$.

3.3 EQUIVALENCE RELATIONS

Definition 3.18. A binary relation (x, ρ), internal to a set x by hypothesis, is called

- *reflexive* if $a \,\rho\, a$ for every $a \in x$;
- *symmetric* if $(a \,\rho\, b) \Leftrightarrow (b \,\rho\, a)$ for every $a, b \in x$; and
- *transitive* if $\big((a \,\rho\, b) \wedge (b \,\rho\, c)\big) \Rightarrow (a \,\rho\, c)$ for every $a, b, c \in x$.

It is called *equivalence relation* if it is reflexive, symmetric, and transitive. Fixing a set x, the same terminology holds for pre-relations in x. ◇

When dealing with equivalence (pre-)relations, the symbol '$\simeq$' or '$\sim$' is more common than 'ρ'.

Exercise 3.19. Show that there exist binary relations satisfying any one of the three properties stated in definition 3.18, but not the other two. Similarly, show that there exist binary relations satisfying any two of these properties, but not the other one.

Definition 3.20. Given an equivalence relation $(x, \simeq)$, the *equivalence class* of $y \in x$ is the subset $[y] \subset x$ defined by $[y] := \{z \in x : z \simeq y\}$. ◇

When we want to emphasise that the equivalence class $[y]$ refers to the relation $\simeq$, we use the notation $[y]_{\simeq}$.

Definition 3.21. A *partition* of a set x is a subset $p \subset \mathcal{P}^*(x)$ such that,

- for every $y \in x$, there exists $a \in p$ such that $y \in a$; and,
- for every $a, b \in p$ such that $a \neq b$, we have $a \cap b = \emptyset$.

These two properties can be stated jointly as follows: for every $y \in x$, there exits a *unique* $a \in p$ such that $y \in a$. ♢

The concepts of partition and equivalence relation are equivalent, as the following proposition shows.

Proposition 3.22. *Fixing a set x, we denote by* $\mathrm{EqRel}(x)$ *the set of equivalence pre-relations in x and by* $\mathrm{Part}(x)$ *the set of partitions of x. We have the following (canonical) bijection:*

$$\begin{aligned} \Theta_x \colon \mathrm{EqRel}(x) &\xrightarrow{\approx} \mathrm{Part}(x) \\ \simeq &\mapsto \{[y]_{\simeq} : y \in x\} \end{aligned} \qquad \begin{aligned} \Theta_x^{-1} \colon \mathrm{Part}(x) &\xrightarrow{\approx} \mathrm{EqRel}(x) \\ p &\mapsto \simeq_p, \end{aligned}$$

where '$\simeq_p$' is defined by $y \simeq_p z$ if and only if there exists $a \in p$ such that $y \in a$ and $z \in a$.

Exercise 3.23. Prove proposition 3.22. In particular, prove that Θ_x and Θ_x^{-1} are well-defined functions and that they are inverse to each other.[2]

Definition 3.24. Given an equivalence relation $(x, \simeq)$, the partition $p := \Theta_x(\simeq)$ is also called *quotient set* and it is usually denoted by $x/\simeq$. We call *projection* the surjective function $\pi \colon x \twoheadrightarrow x/\simeq$, $y \mapsto [y]_{\simeq}$. ♢

Given any set x, the equality in x is an equivalence pre-relation. As a subset of $x \times x$, it coincides with the diagonal $\Delta = \{(y, y) : y \in x\}$. In this case, the equivalence class of $y \in x$ is $\{y\} \in \mathcal{P}(x)$, hence the projection $\pi \colon x \twoheadrightarrow x/=$, $y \mapsto \{y\}$, is a bijection. Moreover, the trivial pre-relation $\simeq \,=\, \mathcal{P}(x \times x)$, that identifies any two elements of x, is an equivalence pre-relation too. As a subset of $x \times x$, it coincides with $x \times x$ itself. In this case, the equivalence class of every $y \in x$ is the whole x, hence the projection $\pi \colon x \twoheadrightarrow x/\simeq$, $y \mapsto x$, is a constant function.

3.4 ORDER RELATIONS

Definition 3.25. A binary relation (x, ρ), internal to a set x by hypothesis, is called

[2] Of course, the notation 'Θ_x^{-1}' is justified since we already know that it is the inverse of Θ_x. In principle, we should denote it by another symbol, like Θ'_x, and, after proving that Θ'_x is well-defined and inverse to Θ_x, denote Θ'_x by Θ_x^{-1}.

- *antireflexive* if $a \not\rho a$ for every $a \in x$; and
- *antisymetric* if $\big((a \rho b) \wedge (b \rho a)\big) \Rightarrow (a = b)$ for every $a, b \in x$.

The relation (x, ρ) is called

- *order relation* (or *partial order relation* or *(partially) ordered set*) if it is reflexive, antisymmetric, and transitive; in this case, it is called *total order relation* if, for every $a, b \in x$, we have $a \rho b$ or $b \rho a$;
- *strict order relation* (or *partial strict order relation* or *(partially) strictly ordered set*) if it is antireflexive and transitive; in this case, it is called *total strict order relation* if, for every $a, b \in x$, one of the three following alternatives holds: $a \rho b$, $b \rho a$ or $a = b$.

Fixing a set x, the same terminology holds for pre-relations in x. ◇

When dealing with order relations (respectively, strict order relations), the symbol '$\preceq$' (respectively, '$\prec$') is more common than 'ρ'. It is straightforward to verify that, if $(x, \prec)$ is a strict order relation, then $a \prec b$ and $b \prec a$ cannot hold at the same time.

Exercise 3.26. Show that there exist binary relations satisfying any one of the three properties defining an order relation (reflexivity, antisymmetry, and transitivity) but not the other two. Similarly, show that there exist binary relations satisfying any two of these properties, but not the other one.

Exercise 3.27. Show that a binary relation can be antireflexive and not transitive and vice versa.

The notions of (total) order relation and strict (total) order relation are equivalent, as the following proposition shows.

Proposition 3.28. *Fixing a set x, we denote by* $\mathrm{OrdRel}(x)$ *the set of order pre-relations in x and by* $\mathrm{StrOrdRel}(x)$ *the set of strict order pre-relations in x. We have the following (canonical) bijection:*

$$\Xi_x \colon \mathrm{OrdRel}(x) \xrightarrow{\approx} \mathrm{StrOrdRel}(x) \qquad \preceq \;\mapsto\; \prec \qquad\qquad \Xi_x^{-1} \colon \mathrm{StrOrdRel}(x) \xrightarrow{\approx} \mathrm{OrdRel}(x) \qquad \prec \;\mapsto\; \preceq,$$

where '$\prec$' on the l.h.s. is defined by $y \prec z$ if and only if $y \preceq z$ and $y \neq z$, and '$\preceq$' on the r.h.s. is defined by $y \preceq z$ if and only if $y \prec z$ or $y = z$. In both directions, the image of a total relation is total too.

Exercise 3.29. Prove proposition 3.28. In particular, prove that Ξ_x and Ξ_x^{-1} are well-defined functions and that they are inverse to each other.[3]

Exercise 3.30. (i) Given an order relation $(x, \preceq)$, prove that, if $z \preceq w \prec u$ or $z \prec w \preceq u$, then $z \prec u$. (ii) Deduce that, if $z_1 \preceq \cdots \preceq z_k$ for any $k \geq 2$, and there exists $i \in \{1, \ldots, k-1\}$ such that $z_i \prec z_{i+1}$, then $z_1 \prec z_k$.

[3]See footnote 2 about the notation Ξ_x^{-1}.

Given any set x, the equality in x is an order pre-relation. The corresponding strict order pre-relation is the empty one. On the contrary, the trivial relation $\mathcal{P}(x \times x)$, that identifies any two elements of x, is not an order pre-relation, unless x contains at most one element, since it is not antisymmetric. Moreover, the set inclusion in $\mathcal{P}(x)$ is an order pre-relation (see exercise 14 at the end of the chapter). The corresponding strict order pre-relation is strict inclusion.

Definition 3.31. Given an ordered set $(x, \preceq)$, we say that $y \in x$ is the *minimum* (respectively, the *maximum*) of $(x, \preceq)$ if $y \preceq z$ (respectively, $z \preceq y$) for every $z \in x$. Furthermore, we say that $y \in x$ is a *minimal* (respectively, *maximal*) element if there exist no elements $z \in x$ such that $z \prec y$ (respectively, $y \prec z$). ♢

Exercise 3.32. Let $(x, \preceq)$ be an order relation.

1. Prove that, if there exists a minimum (respectively, a maximum), then it is unique and it is the unique minimal (respectively, maximal) element.
2. Show that in general a minimal (respectively, maximal) element is not unique and, even if it is unique, in general it is not the minimum (respectively, the maximum).

Exercise 3.33. Prove that, if $(x, \preceq)$ is a totally ordered *finite* set, then it admits a minimum and a maximum.

We now consider an order relation $(x, \preceq)$ and a subset $a \subset x$.

Definition 3.34. We say that $y \in x$ is an *upper bound* of a if $z \preceq y$ for every $z \in a$. We say that y is a *strict* upper bound if the strict inequality $z \prec y$ holds for every $z \in a$. Moreover, a is *(strictly) bounded above* if it admits a (strict) upper bound. ♢

Definition 3.35. We say that $y \in x$ is a *lower bound* of a if $y \preceq z$ for every $z \in a$. We say that y is a *strict* lower bound if the strict inequality $y \prec z$ holds for every $z \in a$. Moreover, a is *(strictly) bounded below* if it admits a (strict) lower bound. ♢

Notation 3.36. We denote by $\mathrm{ubs}_x(a)$ the set of upper bounds of a in x and by $\mathrm{lbs}_x(a)$ the set of lower bounds of a in x. If the set x is clear from the context, we simply write $\mathrm{ubs}(a)$ and $\mathrm{lbs}(a)$.

Definition 3.37. An element $y \in x$ is called:

- *supremum* of a in x, and it is denoted by $\sup_x(a)$, if it is the minimum of $\mathrm{ubs}_x(a)$;
- *infimum* of a in x, and it is denoted by $\inf_x(a)$, if it is the maximum of $\mathrm{lbs}_x(a)$.

If the set x is clear from the context, we simply write $\sup(a)$ and $\inf(a)$. ♢

Exercise 3.38. Prove that a admits a maximum if and only if it admits a supremum and $\sup(a) \in a$. In this case, $\sup(a) = \max(a)$. Similarly, a admits a minimum if and only if it admits an infimum and $\inf(a) \in a$. In this case, $\inf(a) = \min(a)$.

Exercise 3.39. (i) Show an example of a subset a that admits a supremum, but not a maximum (hence $\sup(a) \notin a$ by exercise 3.38). Similarly, show an example of a subset a that admits an infimum, but not a minimum (hence $\inf(a) \notin a$ by exercise 3.38). (ii) Show an example of a subset a that does not admit a supremum and/or an infimum, even if the sets $\text{ubs}(a)$ and $\text{lbs}(a)$ are not empty.

Exercise 3.40. Prove that, if $a \subset b \subset x$ and there exist $\sup(a)$ and $\sup(b)$, then $\sup(a) \preceq \sup(b)$. Similarly, if there exist $\inf(a)$ and $\inf(b)$, then $\inf(b) \preceq \inf(a)$.

3.5 EXERCISES

Remark: In the following exercises and in their resolutions, we will frequently denote a relation (x, y, ρ) simply by ρ and a function (x, y, f) simply by f.

1. Prove that, if x, y, $x_1, \dots, x_n$ are sets in our universe, then, because of the assumptions `(A1)`–`(A4)`, the following sets belong to the universe too:

- the set of (pre-)relations on $(x_1, \dots, x_n)$
- the domain and the image of a binary relation on (x, y)
- the set of (pre-)functions from x to y
- the set of injective/surjective/bijective (pre-)functions from x to y
- the sets of equivalence, (total) order, and strict (total) order (pre-)relations in x
- the quotient with respect to an equivalence relation in x

2. Let x be finite set with n elements, y be a finite set with m elements and z be an infinite set. We suppose $n, m > 0$. Prove that

(i) there exists an injective function $f : x \hookrightarrow y$ if and only if $n \leq m$; such a function is bijective if and only if $n = m$;

(ii) there exists a surjective function $f : x \twoheadrightarrow y$ if and only if $n \geq m$; such a function is bijective if and only if $n = m$;

(iii) there exists an injective function $f : x \hookrightarrow z$, but no injective function $f : z \hookrightarrow x$; and

(iv) there exists a surjective function $f : z \twoheadrightarrow x$, but no surjective function $f : x \twoheadrightarrow z$.

Deduce that

(v) if x and y are finite, then there exists a bijection between them if and only if they contain the same number of elements;

(vi) if x is finite and y is infinite, then there exists no bijection between them;

(vii) if x is a finite set with n elements, y is a finite set with m elements and $y \subset x$, then $m \leq n$; in this case, $x = y$ if and only if $n = m$ (cfr. exercise 4-xi of section 1.5);

(viii) a set x is infinite if and only if, for every $n \in \mathrm{N}$, there exists a finite subset $y \subset x$ that contains n elements (cfr. exercise 4-x of section 1.5);

(ix) if x is finite and $y \subset x$, then y is finite (cfr. exercise 4-xi of section 1.5); and

(x) if y is infinite and $y \subset x$, then x is infinite (cfr. exercise 4-xii of section 1.5).

Remark: The reader can easily verify that items (i), (iii), and (v)–(x) hold even without supposing $n, m > 0$. On the contrary, item (ii) does not hold if $n > 0$ and $m = 0$, since there is no (surjective) function from a non-empty set to the empty one. Similarly, the first part of item (iv) does not hold if $n = 0$.

3. Let x and y be finite sets containing respectively n and m elements.

1. Verify that the set of relations on (x, y) contains 2^{nm} elements.
2. Verify that the set of functions from x to y contains m^n elements.
3. Supposing that $n \leq m$, compute the number of elements of the set of injective functions from x to y.
4. Supposing $n = m$, compute the number of elements of the set of bijective functions from x to y.
5. Supposing $m \leq n$ and calling $S(n, m)$ the number of partitions of $\{1, \dots, n\}$ formed by m elements, compute the number of elements of the set of surjective functions from x to y.

4. Let $f\colon x \to y$ be a function.

- Show that f induces the equivalence pre-relation '$\simeq_f$' in x defined by $z \simeq_f w$ if and only if $f(z) = f(w)$.
- Show that f induces the *bijective* function $\bar{f}\colon x/\simeq_f \xrightarrow{\approx} \mathrm{Im}(f)$, $[z] \mapsto f(z)$, in such a way that, calling $\pi_f\colon x \twoheadrightarrow x/\simeq_f$ the projection and $\iota_f := \iota_{\mathrm{Im}(f),y}\colon \mathrm{Im}(f) \hookrightarrow y$ the inclusion, we have $f = \iota_f \circ \bar{f} \circ \pi_f$—that is, the following diagram commutes:

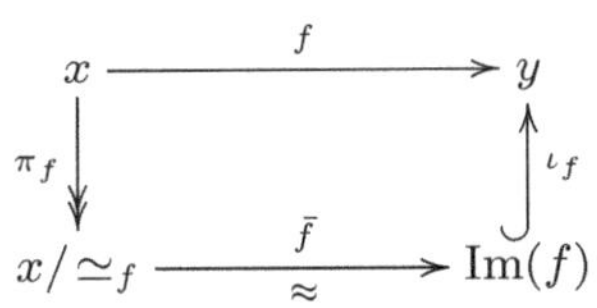

5. Let $f\colon x \to y$ and $g\colon y \to z$ be two functions.

1. Prove that $g \circ f$ is injective if and only if f and $g|_{\mathrm{Im}(f)}$ are injective.
2. Prove that $g \circ f$ is surjective if and only if $g|_{\mathrm{Im}(f)}$ is surjective if and only if g and $\pi_g \circ f$ are surjective, where $\pi_g\colon y \twoheadrightarrow y/\simeq_g$ is the projection defined in exercise 4.
3. Prove that $g \circ f$ is bijective if and only if f is injective and $g|_{\mathrm{Im}(f)}$ is bijective.
4. Extend the previous items to a composition of n functions for any $n \geq 2$.

6. Let us fix a set with two distinct elements $a = \{b_0, b_1\}$. Show that, for every set x, there exists a (natural) bijection between $\mathcal{P}(x)$ and $\mathrm{Func}(x, a)$. Deduce from exercise 3-ii, that, if x contains n elements, then $\mathcal{P}(x)$ contains 2^n elements (cfr. exercise 8 of section 2.8).

7. Given a function $f\colon x \to y$, prove that

1. $a \subset \ddot{f}(\dot{f}(a))$ for every $a \in \mathcal{P}(x)$;
2. $\dot{f}(\ddot{f}(b)) \subset b$ for every $b \in \mathcal{P}(y)$;
3. $a = \ddot{f}(\dot{f}(a))$ for every $a \in \mathcal{P}(x)$ if and only if f is injective; and
4. $\dot{f}(\ddot{f}(b)) = b$ for every $b \in \mathcal{P}(y)$ if and only if f is surjective.

8. Given a function $f\colon x \to y$, prove that

1. $\dot{f}(a_1 \cup a_2) = \dot{f}(a_1) \cup \dot{f}(a_2)$ and $\dot{f}(a_1 \cap a_2) \subset \dot{f}(a_1) \cap \dot{f}(a_2)$ for every $a_1, a_2 \subset x$;
2. $\dot{f}(a_1 \cap a_2) = \dot{f}(a_1) \cap \dot{f}(a_2)$ for every $a_1, a_2 \subset x$ if and only if f is injective; and
3. $\ddot{f}(b_1 \cup b_2) = \ddot{f}(b_1) \cup \ddot{f}(b_2)$ and $\ddot{f}(b_1 \cap b_2) = \ddot{f}(b_1) \cap \ddot{f}(b_2)$ for every $b_1, b_2 \subset y$.

9. Given a function $f\colon x \to y$, prove that

1. $\ddot{f}(y \setminus b) = x \setminus \ddot{f}(b)$ for every $b \subset y$;
2. in general, given $a \subset x$, there not exists any inclusion relation between $\dot{f}(x \setminus a)$ and $y \setminus \dot{f}(a)$;
3. $\dot{f}(x \setminus a) \subset y \setminus \dot{f}(a)$ for every $a \subset x$ if and only if f is injective; and
4. $y \setminus \dot{f}(a) \subset \dot{f}(x \setminus a)$ for every $a \subset x$ if and only if f is surjective.

10. Given $f\colon x \to y$, consider the functions $\dot{f}\colon \mathcal{P}(x) \to \mathcal{P}(y)$ and $\ddot{f}\colon \mathcal{P}(y) \to \mathcal{P}(x)$. Prove that

1. $\dot{f}$ is injective if and only if f is injective;
2. $\dot{f}$ is surjective if and only if f is surjective;
3. $\ddot{f}$ is injective if and only if f is surjective;
4. $\ddot{f}$ is surjective if and only if f is injective; and
5. if f is bijective, then $\ddot{f} = (\dot{f})^{-1} = (f^{-1})^{\cdot}$ and $\dot{f} = (\ddot{f})^{-1} = (f^{-1})^{\cdot\cdot}$.

11. Denoting by $\mathrm{Func}(x, y)$ the set of functions from x to y, consider the functions $\dot{\mathcal{F}}\colon \mathrm{Func}(x, y) \to \mathrm{Func}\big(\mathcal{P}(x), \mathcal{P}(y)\big)$, $f \mapsto \dot{f}$, and $\ddot{\mathcal{F}}\colon \mathrm{Func}(x, y) \to \mathrm{Func}\big(\mathcal{P}(y), \mathcal{P}(x)\big)$, $f \mapsto \ddot{f}$. Prove that $\dot{\mathcal{F}}$ and $\ddot{\mathcal{F}}$ are injective but in general not surjective.

12. (i) Given a function $f\colon x \to y$ in the universe, prove that $\dot{f}$ and $\ddot{f}$ belong to the universe too. (ii) Given two sets x and y in the universe, prove that the functions $\dot{\mathcal{F}}$ and $\ddot{\mathcal{F}}$, defined in exercise 11, belong to the universe too.

13. Let x be a finite set containing n elements. Compute:

1. the number of reflexive relations in x
2. the number of symmetric relations in x
3. the number of reflexive and symmetric relations in x

14. Fixing a set x, let us consider the relation $(\mathcal{P}(x), \subset)$.

1. Verify that it is an order relation, but in general not total.
2. Is there a minimum? Is there a maximum?

15. Supposing to have defined N as a set in our universe (see chapter 7) and denoting it by $\mathbb{N}$, let us consider the relation in $\mathbb{N}$ defined by $n \preceq m$ if and only if $n \mid m$.

1. Verify that it is an order relation, but not total.
2. Is there a minimum? Is there a maximum?

16. Let x be a set. Given two equivalence pre-relations $\sim_1$ and $\sim_2$ in x, we say that $\sim_1$ is *thinner* than $\sim_2$ if and only if $(y \sim_1 z) \Rightarrow (y \sim_2 z)$ for every $y, z \in x$ (equivalently, $\sim_1 \subset \sim_2$). Given two partitions p and q in x, we say that p is *thinner* than q if and only if for every $a \in p$ there exists $b \in q$ such that $a \subset b$.

1. Verify that $\sim_1$ is thinner than $\sim_2$ if and only if $\Theta_x(\sim_1)$ is thinner than $\Theta_x(\sim_2)$.
2. Consider the pre-relation $\preceq$ in the set EqRel(x), defined by $\sim_1 \preceq \sim_2$ if and only if $\sim_2$ is thinner than $\sim_1$. Prove that it is an order relation, which is not total in general. Obviously the same holds for Part(x).
3. Is there a minimum? Is there a maximum?

17. Let x be a set. Given two order pre-relations $\preceq_1$ and $\preceq_2$ em x, we say that $\preceq_1$ is *thinner* than $\preceq_2$ if and only if $(y \preceq_1 z) \Rightarrow (y \preceq_2 z)$ for every $y, z \in x$ (equivalently, $\preceq_1 \subset \preceq_2$).

1. Consider the pre-relation $\preceq$ in the set OrdRel(x), defined by $\preceq_1 \preceq \preceq_2$ if and only if $\preceq_2$ is thinner than $\preceq_1$. Prove that it is an order relation, which is not total in general. We may consider the same construction with respect to strict order relations.
2. Is there a minimum? Is there a maximum?

18. Let $\simeq$ be an equivalence pre-relation in a set x and let $\pi\colon x \twoheadrightarrow x/\simeq$ be the corresponding projection to the quotient. Prove that, given a set y and a function $f\colon x \to y$, there exists a function $\tilde{f}\colon (x/\simeq) \to y$, such that $f = \tilde{f} \circ \pi$, if and only if $(z \simeq w) \Rightarrow \big(f(z) = f(w)\big)$ for every $z, w \in x$. In this case, $\tilde{f}$ is unique.

19. Let $a \subset x$ and let $\iota\colon a \hookrightarrow x$ be the inclusion. Prove that, given a set y and a function $f\colon y \to x$, there exists a function $\tilde{f}\colon y \to a$, such that $f = \iota \circ \tilde{f}$, if and only if $\mathrm{Im}(f) \subset a$. In this case, $\tilde{f}$ is unique.

20. A relation $(x, \preceq)$ is called *pre-order relation* if it is reflexive and transitive. In this case, we consider the pre-relation in x defined by:

$$y \simeq z \Leftrightarrow \big((y \preceq z) \wedge (z \preceq y)\big).$$

1. Prove that $(x, \simeq)$ is an equivalence relation.
2. Consider the relation $(x/\simeq, \preceq)$ defined by:

$$[y] \preceq [z] \Leftrightarrow y \preceq z.$$

   Prove that $(x/\simeq, \preceq)$ is well-defined (that is, it does not depend on the representatives y e z in the definition) and it is an order relation.

CHAPTER 4

Sets and Indexed Sets

In this chapter we compare *sets* and *indexed sets*. In the latter case, "indexed set" simply means "surjective function", but it is common to use this expression when we are thinking of a "set with repetitions" (that is, a collection of elements in which some of them might coincide). We show some basic unary operations among sets on one side and among indexed sets on the other side, developing the two cases in a parallel way.

4.1 SETS

Definition 4.1. Given a set x, we define:

- the *union* $\bigcup x := \{z : \exists y \in x(z \in y)\}$
- if $x \neq \emptyset$, the *intersection* $\bigcap x := \{z : \forall y \in x(z \in y)\}$
- the *Cartesian product* $\prod x := \{f : x \to \bigcup x : \forall y \in x(f(y) \in y)\}$
- the *disjoint union* $\bigsqcup x := \{(z, y) : z \in y \in x\}$
- the *separation* $\hat{x} := \{y \times \{y\} : y \in x\}$ ◇

We observe that these operations are unary, as power set and complement (see definition 2.8). In the case of the intersection we supposed $x \neq \emptyset$, because $\bigcap \emptyset$ would be the set of all sets, that has been ruled out form our universe (see proposition 2.12).

Let us represent graphically the operations introduced in definition 4.1. We consider the set x shown in figure 4.1. The union of x contains the elements of the elements of x. Hence, in order to get $\bigcup x$, we have to "remove the skin" from the element of x. Moreover, the intersection of x contains the elements that belong to every element of x. Both $\bigcup x$ and $\bigcap x$ are represented in figure 4.2. The Cartesian product of x contains the functions that assign an element of y to each $y \in x$. Figure 4.3 represents *one element* $f \in \prod x$. The disjoint union of x, shown in figure 4.4, contains the elements of the elements of x, but separating them, so that, if a set z belongs to n elements of x, then it is counted

DOI: 10.1201/9781003565291-4

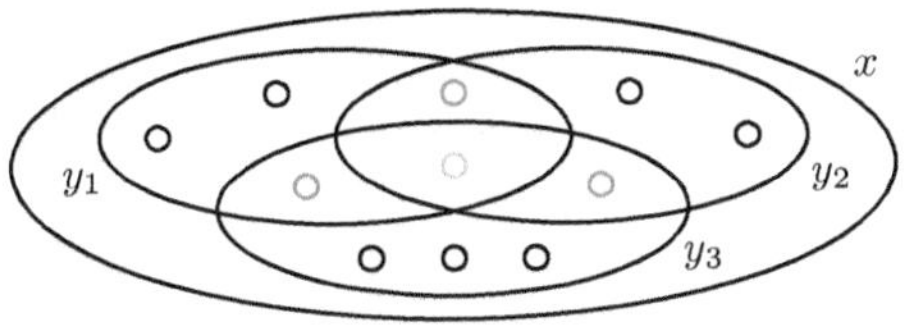

Figure 4.1 A set.

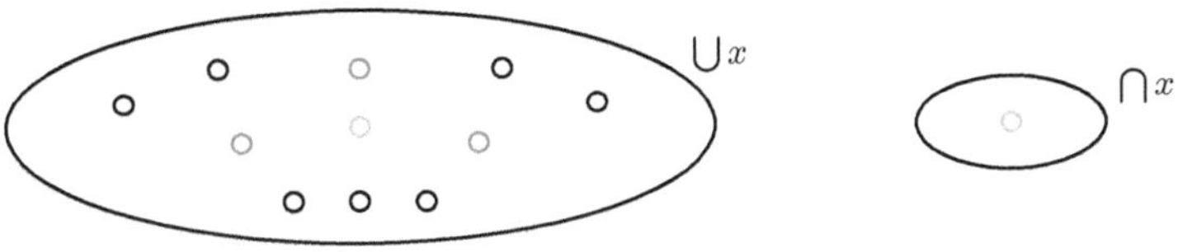

Figure 4.2 Union and intersection of a set.

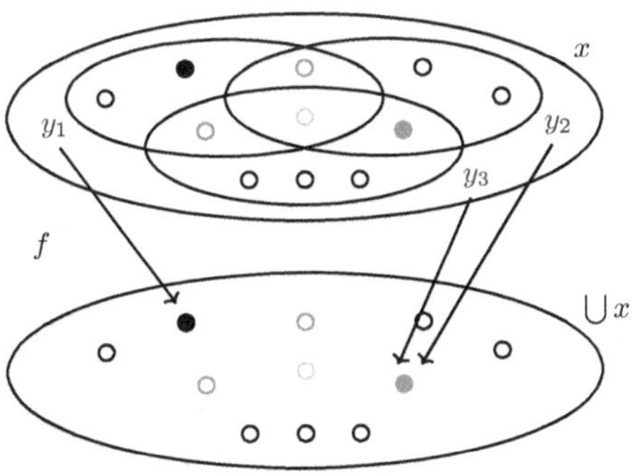

Figure 4.3 One element of the Cartesian product of a set.

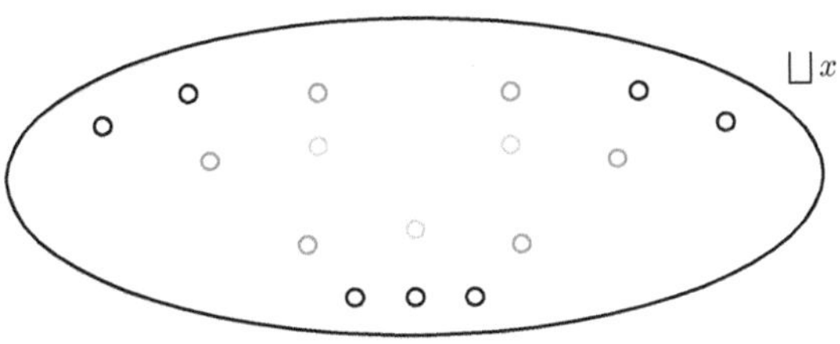

Figure 4.4 Disjoint union of a set.

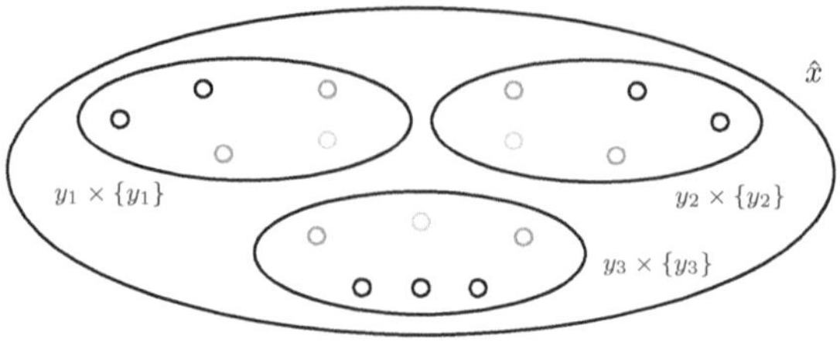

Figure 4.5 Separation of a set.

n times in $\bigsqcup x$. In order to realise this separation, if $z \in y \in x$, then we use y as a "tag" for z. The separation of x, as indicated by its own name, separates the element of x, but without taking the union. This is realised by endowing each $z \in y \in x$ with the "tag" y, as we did for disjoint union. In fact, it is straightforward to verify that $\bigsqcup x = \bigcup \hat{x}$. Figure 4.5 represents $\hat{x}$.

Remark 4.2. With respect to Cartesian product and disjoint union, we observe that

- if $\emptyset \in x$, then $\prod x = \emptyset$, since the condition $f(\emptyset) \in \emptyset$ cannot be satisfied;
- $\prod \emptyset = \{\emptyset\}$ and $\bigsqcup \emptyset = \emptyset$, as it trivially follows from the definitions;
- for any set x and any element $y \in x$, we have the projection $\pi_y \colon \prod x \to y,\ f \mapsto f(y)$; and
- for any set x and any element $y \in x$, we have the embedding $\iota_y \colon y \hookrightarrow \bigsqcup x,\ w \mapsto (w, y)$. ◇

One of the possible formulations of the axiom of choice, that we will discuss in chapter 9, states that, if $\emptyset \notin x$, then $\prod x \neq \emptyset$. Without assuming this axiom, we can only affirm that, if $\emptyset \notin x$ and x is finite, then $\prod x \neq \emptyset$.

Exercise 4.3. Prove that, if $\prod x \neq \emptyset$, then the projection π_y, defined in remark 4.2, is surjective for every y. (It is straightforward to verify that the embedding ι_y is injective for every y.)

Given a set x, we have the following surjective map:

$$\begin{aligned} \eta_x \colon \bigsqcup x &\twoheadrightarrow \bigcup x \\ (z, y) &\mapsto z. \end{aligned} \tag{4.1}$$

This function is injective (hence bijective) if and only if any two elements of x are disjoint. In general, it "forgets the tags", so that it identifies different copies of the same element. Considering the set x in figure 4.1, we represent η_x graphically in figure 4.6.

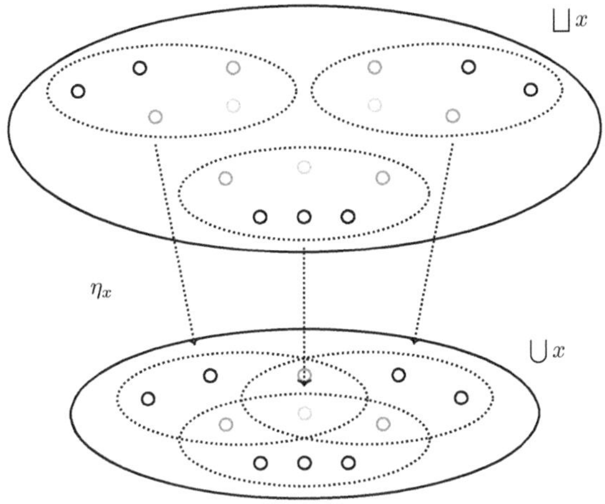

Figure 4.6 The projection from disjoint union to union of a set.

Dually, given $x \neq \emptyset$, we have the *diagonal embedding*:

$$\begin{aligned} \Delta_x \colon \bigcap x &\hookrightarrow \prod x \\ z &\mapsto (y \mapsto z). \end{aligned} \tag{4.2}$$

4.2 INDEXED SETS

Definition 4.4. An *indexed set* $x = \{x_i\}_{i \in I}$ is a surjective function $\varphi \colon I \to x$, $i \mapsto x_i$. Sometimes we say that x is indexed *over* I. ♢

The notation used in definition 4.4 might be confusing, since $x = \{x_i\}_{i \in I}$ seems to denote the set containing the elements x_i, which is the codomain (and the image) of the indexed set φ. The point is that we are actually emphasizing the codomain, since we think of the domain I only as a set of indices. This is a way to allow repetitions, since it can happen that $x_i = x_j$ but $i \neq j$. In this case, the pairs $(i, x_i), (j, x_j) \in \varphi$ are thought of as two distinct copies of x_i (or x_j).

Remark 4.5. Every set x can be considered an indexed set over itself by setting $I := x$ and $\varphi := \mathrm{id}_x$. Equivalently, $x = \{i\}_{i \in x}$. ♢

We defined the notion of function between sets in chapter 3. In the case of indexed sets, we give the following definition.

Definition 4.6. A *function* from the indexed set $x = \{x_i\}_{i \in I}$ to the indexed set $y = \{y_j\}_{j \in J}$ is a function $f \colon I \to J$. ♢

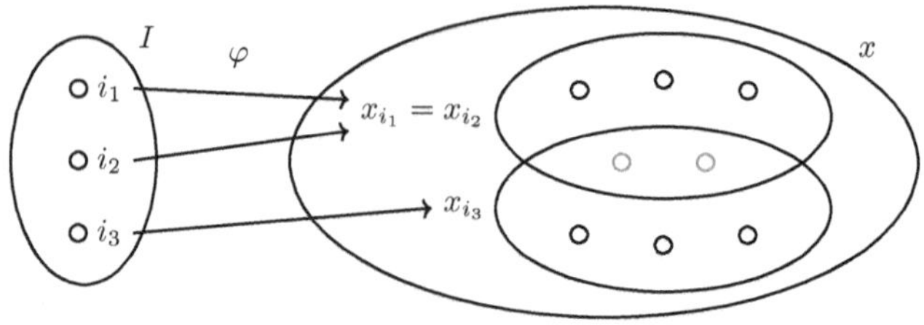

Figure 4.7 An indexed set.

The function f in definition 4.6 can be injective, surjective or bijective. We observe that, given two indexed sets $\varphi\colon I \to x$, $i \mapsto x_i$, and $\psi\colon J \to y$, $j \mapsto y_j$, a function $f\colon I \to J$ can be naturally identified with the function $\bar{f}\colon \varphi \to \psi$, $(i, x_i) \mapsto (f(i), y_{f(i)})$.

Definition 4.7. Given an indexed set $x = \{x_i\}_{i \in I}$, we define the following (non-indexed) sets:

- the *union* $\bigcup_i x_i := \{z : \exists i \in I(z \in x_i)\}$
- if $I \neq \emptyset$ (equivalently, $x \neq \emptyset$), the *intersection* $\bigcap_i x_i := \{z : \forall i \in I(z \in x_i)\}$
- the *Cartesian product* $\prod_i x_i := \{f\colon I \to \bigcup_i x_i : \forall i \in I(f(i) \in x_i)\}$
- the *disjoint union* $\bigsqcup_i x_i := \{(z, i) : i \in I, z \in x_i\}$
- the *separation* $\hat{\hat{x}} := \{x_i \times \{i\} : i \in I\}$ ◇

Let us represent graphically the operations introduced in definition 4.7. We consider the index set $x = \{x_i\}_{i \in I}$ shown in figure 4.7. The union and the intersection of $x = \{x_i\}_{i \in I}$ coincides with the union and the intersection of x as a (non-indexed) set. Both are represented in figure 4.8. The Cartesian product of x contains the functions that assign an element of x_i to each $i \in I$. Figure 4.9 represents *one element* $f \in \prod_i x_i$. The disjoint union of $x = \{x_i\}_{i \in I}$, shown in figure 4.10, contains the elements of the elements of x, but separating them, in such a way that each index induces a set disjoint from the others. In order to realise this separation, if $z \in x_i$, then we use i as a "tag" for z. The separation of $x = \{x_i\}_{i \in I}$, as indicated by its own name, separates the element of x, but without taking the union. This is realised by endowing each $z \in x_i$ with the "tag" i, as we did for disjoint union. In fact, it is straightforward to verify that $\bigsqcup_i x_i = \bigcup \hat{\hat{x}}$. Figure 4.11 represents $\hat{\hat{x}}$.

Remark 4.8. With respect to Cartesian product and disjoint union, we observe that

- if there exists $j \in I$ such that $x_j = \emptyset$ (equivalently, if $\emptyset \in x$), then $\prod_i x_i = \emptyset$, since the condition $f(j) \in \emptyset$ cannot be satisfied;

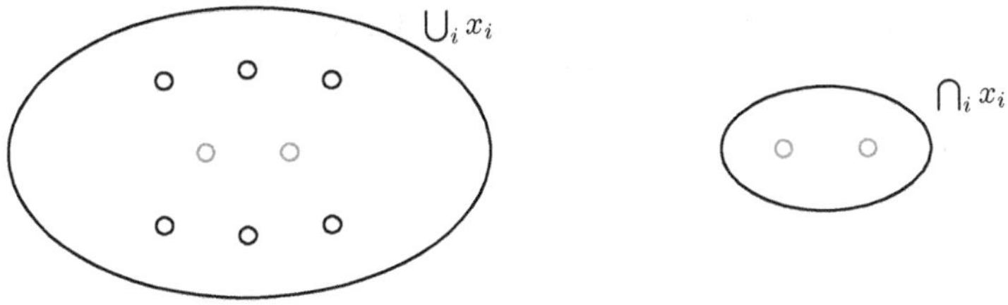

Figure 4.8 Union and intersection of an indexed set.

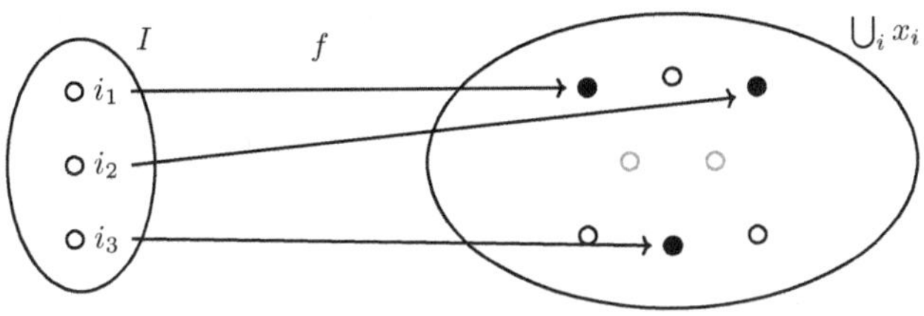

Figure 4.9 An element of the Cartesian product of an indexed set.

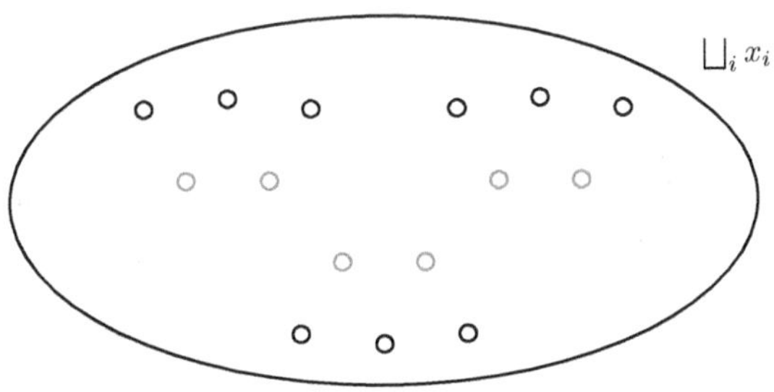

Figure 4.10 The disjoint union of an indexed set.

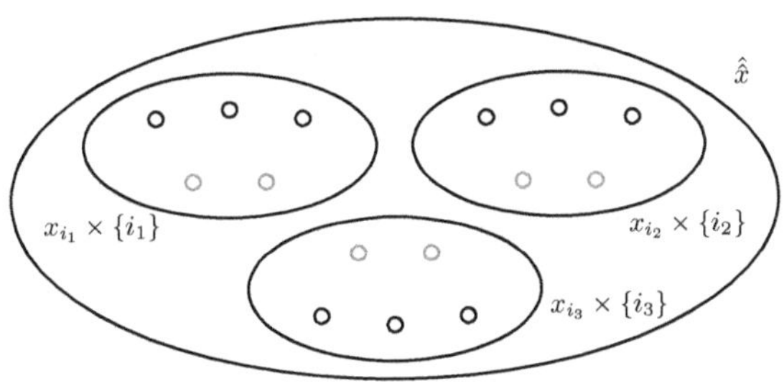

Figure 4.11 The separation of and indexed set.

- $\prod \emptyset = \{\emptyset\}$ and $\bigsqcup \emptyset = \emptyset$, where the indexed set $\emptyset$ is the only one over $I = \emptyset$ (that is, it is the surjective function $\emptyset\colon \emptyset \to \emptyset$);
- for any indexed set $x = \{x_i\}_{i \in I}$ and any index $j \in I$, we have the projection $\pi_j\colon \prod_i x_i \to x_j$, $f \mapsto f(j)$; and
- for any indexed set $x = \{x_i\}_{i \in I}$ and any index $j \in I$, we have the embedding $\iota_j\colon x_j \hookrightarrow \bigsqcup_i x_i$, $y_j \mapsto (y_j, j)$. ♢

One of the possible formulation of the axiom of choice, that we will discuss in chapter 9, states that, if $x_i \neq \emptyset$ for every i, then $\prod_i x_i \neq \emptyset$. Without assuming this axiom, we can only affirm that, if $x_i \neq \emptyset$ for every i and x is finite (not necessarily I), then $\prod_i x_i \neq \emptyset$.

Exercise 4.9. Prove that, if $\prod_i x_i \neq \emptyset$, then the projection π_i, defined in remark 4.8, is surjective for every i. (It is straightforward to verify that the embedding ι_j is injective for every j.)

Given an indexed set $x = \{x_i\}_{i \in I}$, we have the following surjective map:

$$\begin{aligned} \bar{\eta}_x\colon \bigsqcup_i x_i &\twoheadrightarrow \bigcup_i x_i \\ (z, i) &\mapsto z. \end{aligned} \tag{4.3}$$

This function is injective (hence bijective) if and only if for any pair $(i, j) \in I \times I$ such that $i \neq j$ the sets x_i and x_j are disjoint. In general, it "forgets the tags", so that it identifies different copies of the same element. Considering the indexed set $x = \{x_i\}_{i \in I}$ in figure 4.7, we represent $\bar{\eta}_x$ graphically as in figure 4.12.

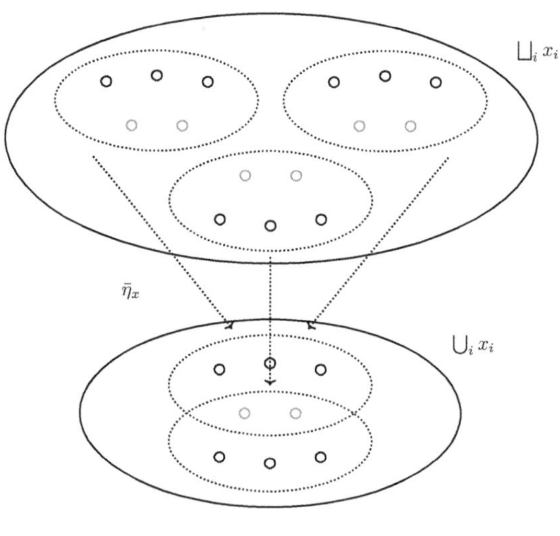

Figure 4.12 The projection from disjoint union to union of an indexed set.

Dually, if $I \neq \emptyset$ (equivalently, $x \neq \emptyset$), we have the *diagonal embedding*:

$$\begin{aligned} \bar{\Delta}_x \colon \bigcap_i x_i &\hookrightarrow \prod_i x_i \\ z &\mapsto (i \mapsto z). \end{aligned} \tag{4.4}$$

4.3 COMPARISON

Let us compare explicitly the framework of sets with the one of indexed sets.

4.3.1 Union and Intersection

The union and the intersection of an indexed set $x = \{x_i\}_{i \in I}$ coincide with the ones of x as a set—that is, $\bigcup_{i \in I} x_i = \bigcup x$ and $\bigcap_{i \in I} x_i = \bigcap x$.[1]

4.3.2 Cartesian Product

On the contrary, in the case of Cartesian product, we have to distinguish between sets and indexed sets. In fact, an element of $\prod_i x_i$ is a function from I to $\bigcup x$, while an element of $\prod x$ is a function from x to $\bigcup x$. The fundamental difference is that, if $x_i = x_j$ but $i \neq j$, a function from I picks up an element of x_i and an element of x_j (hence two elements of x_i or x_j), while a function from x picks up only one element of x_i or x_j. We have the following (canonical) injective function:

$$\begin{aligned} \upsilon_x \colon \prod x &\hookrightarrow \prod_i x_i \\ f &\mapsto f \circ \varphi. \end{aligned} \tag{4.5}$$

Such a function is surjective (hence bijective) if and only if φ is injective (hence bijective). The following diagram commutes:

$$\begin{array}{ccc} \bigcap x & \xhookrightarrow{\;\Delta_x\;} & \prod x \\ \Big\| & & \Big\downarrow \upsilon_x \\ \bigcap_i x_i & \xhookrightarrow{\;\bar{\Delta}_x\;} & \prod_{i \in I} x_i . \end{array} \tag{4.6}$$

4.3.3 Disjoint Union

Similarly, the two notions of disjoint union are distinct. In fact, let us suppose that $x_i = x_j$ but $i \neq j$. Then, in $\bigsqcup_i x_i$ we get the elements of the two copies

---

[1] We remark that $\bigcup x = \bigcup \{x_i\}$ and $\bigcap x = \bigcap \{x_i\}$, with curly brackets on the r.h.s., since we are taking the union and the intersection of the (non-indexed) set containing the elements x_i.

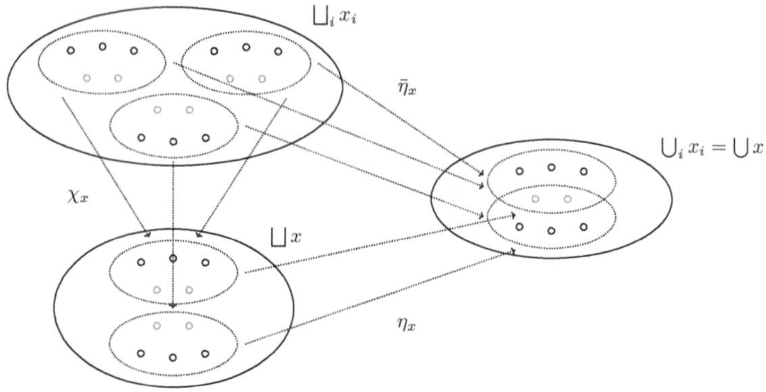

Figure 4.13 Disjoint unions and projections.

$x_i \times \{i\}$ and $x_j \times \{j\}$, while in $\bigsqcup x$ we just get one copy. We have the following (canonical) surjective function:

$$\begin{aligned} \chi_x \colon \bigsqcup_i x_i &\twoheadrightarrow \bigsqcup x \\ (z, i) &\mapsto (z, x_i). \end{aligned} \tag{4.7}$$

Such a function is injective (hence bijective) if and only if φ is. In this case, it identifies the two disjoint unions. The following diagram commutes:

$$\begin{array}{ccc} \bigsqcup_{i \in I} x_i & \xrightarrow{\bar{\eta}_x} & \bigcup_i x_i \\ {\scriptstyle \chi_x} \downarrow & & \| \\ \bigsqcup x & \xrightarrow{\eta_x} & \bigcup x. \end{array} \tag{4.8}$$

Considering the indexed set $x = \{x_i\}_{i \in I}$ in the pictures above, we can represent diagram (4.8) as in figure 4.13.

4.3.4 Separation

In the case of separation, the distinction between sets and indexed sets is meaningful as well. In fact, $\hat{\hat{x}} = \{x_i \times \{i\} : i \in I\}$, while $\hat{x} = \{x_i \times \{x_i\} : x_i \in x\}$. In the latter case, if $x_i = x_j$ but $i \neq j$, we get one element $x_i \times \{x_i\}$, while in the former case we get two elements $x_i \times \{i\}$ and $x_j \times \{j\}$. We have the following (canonical) surjective function:

$$\begin{aligned} \xi_x \colon \hat{\hat{x}} &\twoheadrightarrow \hat{x} \\ x_i \times \{i\} &\mapsto x_i \times \{x_i\}. \end{aligned} \tag{4.9}$$

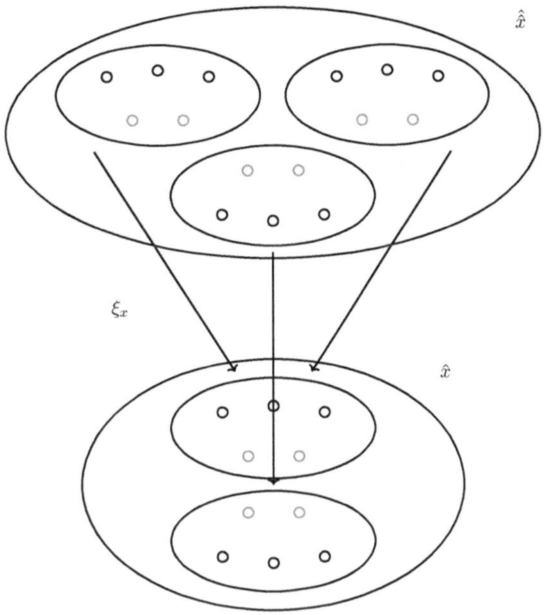

Figure 4.14 Separations.

The function (4.9) is injective (hence bijective) if and only if φ is. Considering the indexed set $x = \{x_i\}_{i \in I}$ in the pictures above, figure 4.14 represents such function graphically.

4.3.5 Summary

We can summarise the comparison between sets and indexed sets through the following functions between classes (the concept of *class* will be introduced in chapter 6, but here the intuitive meaning of the word is enough):

$$\{\text{Sets}\} \underset{\Psi}{\overset{\Phi}{\rightleftarrows}} \{\text{Indexed sets}\},$$

where $\Phi\colon x \mapsto x = \{i\}_{i \in x}$ and $\Psi\colon x = \{x_i\}_{i \in I} \mapsto x$. We have that $\Psi \circ \Phi = \mathrm{id}$, while $\Phi \circ \Psi$ forgets repetitions. The reader can easily verify that Φ respects all of the operations, while Ψ respects only union and intersection, since we have the natural functions (4.5), (4.7), and (4.9), that are not bijective in general.

4.3.6 Operations and the Universe

Since the operations introduced up to now are essential in every area of ordinary mathematics, we make the following fundamental assumption, analogous to (A3):

(A5) *The operations defined in 4.1 and 4.7, when applied to sets or indexed sets in our universe, produce sets that belong to our universe as well.*

Again the statement of (A5) if far from being minimal. Actually, it could be stated only with respect to union, generalizing the analogous assumption in (A3) (see exercise 2 at the end of the chapter). Anyway, as we already pointed out, we are not interested in minimality of the statements, but in their naturality as far as possible.

4.4 COMPLEMENTS

We state some meaningful definitions and results that require the operations of union and intersection of a set.

4.4.1 Iterated Union

The operations defined in 4.1 can be iterated. We are interested in iterated union, that we define recursively as follows:

- $\bigcup^0 x := x$;
- $\bigcup^{n+1} x := \bigcup(\bigcup^n x)$.

Obviously, $\bigcup^1 x = \bigcup x$. We clarify with a picture the definition of iterated union. Let us consider the set represented in figure 4.15. In order to compute $\bigcup^n x$, we have to "remove the skin" n times starting from the elements of x, as figure 4.16 shows.

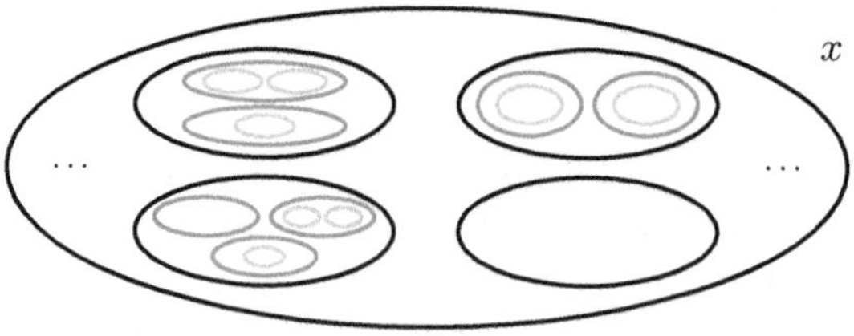

Figure 4.15 A set.

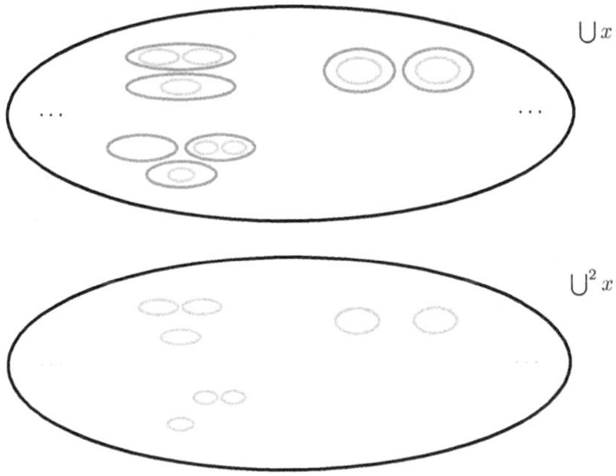

Figure 4.16 Iterated union.

Exercise 4.10. Prove that $\bigcup^m(\bigcup^n x) = \bigcup^{n+m} x$.

Exercise 4.11. (i) Let $x := \{\{\{y_1\}, \{y_1, y_2\}\}, \{\{y_1\}\}\}$, where y_1 and y_2 are two distinct sets. Compute $\bigcup x$ and $\bigcup^2 x$.

(ii) Let $x := \{\emptyset\}$ and $y := \{\{\emptyset\}\}$. Compute $\bigcup x$, $\bigcup y$, and $\bigcup^2 y$.

(iii) Prove that, if $x = \{y\}$, then $\bigcup^{n+1} x = \bigcup^n y$ for every $n \in \mathrm{N}$.

Lemma 4.12. *For every set x:*

1. *If $y \in \bigcup^n x$, then $y \subset \bigcup^{n+1} x$.*
2. *If $y \subset x$, then $\bigcup^n y \subset \bigcup^n x$ for every $n \in \mathrm{N}$.*
3. *$z \in \bigcup^{n+m+1} x$ if and only if there exists a set $y \in \bigcup^n x$ such that $z \in \bigcup^m y$.*

Exercise 4.13. Prove lemma 4.12.

4.4.2 Generated Equivalence Relation

Let us fix two sets x and y. Given a non-empty set $r \subset \mathcal{P}(x \times y)$, whose elements are by definition pre-relations between x and y, the intersection $\bigcap r$ is a pre-relation between x and y, since it belongs to $\mathcal{P}(x \times y)$. In fact, $\bigcap r \subset \bigcup r \subset \bigcup \mathcal{P}(x \times y) = x \times y$ (see exercises 1-(1, 2) and 7-1 of section 4.5).

Exercise 4.14. Let $r \subset \mathcal{P}(x \times x)$ be such that every element $\rho \in r$ is an equivalence pre-relation. Prove that, if $r \neq \emptyset$, then $\bigcap r$ is an equivalence pre-relation. Show that the analogous statement about (strict) order pre-relations holds too.

Given a pre-relation $\rho \subset x \times x$, we denote by $r_\rho \subset \mathcal{P}(x \times x)$ the set of the equivalence pre-relations in x that contain ρ, that is:

$$r_\rho := \{\eta \in \mathcal{P}(x \times x) : (\eta \text{ is an equivalence pre-relation}) \wedge (\rho \subset \eta)\}.$$

Exercise 4.15. Prove that, for every set x and every $\rho \subset x \times x$, the set r_ρ is not empty and it admits a minimum with respect to inclusion, that coincides with $\bigcap r_\rho$.

Exercise 4.15 justifies the following definition.

Definition 4.16. Given a set x and a pre-relation $\rho \subset x \times x$, we call *equivalence pre-relation generated by* ρ the equivalence pre-relation $\bar{\rho} := \bigcap r_\rho$—that is, the minimum equivalence pre-relation in x that contains ρ. ◇

Exercise 4.17. Prove that, under the hypotheses of definition 4.16, the equivalence pre-relation $\bar{\rho}$ can be characterised as follows: $y \, \bar{\rho} \, z$ if and only if there exists a finite sequence $(w_0, \ldots, w_n)$ of elements of x, for any $n \in \mathrm{N}^*$, such that:

- $w_0 = y$ and $w_n = z$;
- $w_i \, \rho \, w_{i+1}$ or $w_{i+1} \, \rho \, w_i$ or $w_i = w_{i+1}$ for every $i \in \{0, \ldots, n-1\}$.

Exercise 4.18. (i) Prove that ρ is an equivalence pre-relation if and only if $\bar{\rho} = \rho$. (ii) What is the equivalence pre-relation generated by the empty pre-relation (that is, by $\emptyset \subset x \times x$)?

Exercise 4.19. Show that, given a non-empty set x and a pre-relation $\rho \subset x \times x$, the set of (strict) order pre-relations in x, that contain ρ, can be empty.

Remark: It follows that the notion of "generated (strict) order pre-relation" is meaningless in general.

Of course, given a relation (x, x, ρ), we call *equivalence relation generated by* (x, x, ρ) the equivalence relation $(x, x, \bar{\rho})$.

4.4.3 Increasing Families of Functions

The following proposition and corollary will be useful several times later on.

Proposition 4.20. *Let us consider an indexed set of functions* $\Phi = \{(x_i, y_i, f_i)\}_{i \in I}$ *such that, for every* $i, j \in I$, *we have* $f_i \subset f_j$ *or* $f_j \subset f_i$ *(it follows that respectively* $x_i \subset x_j$ *or* $x_j \subset x_i$*). We set* $\bar{x} := \bigcup_{i \in I} x_i$, $\bar{y} := \bigcup_{i \in I} y_i$ *and* $\bar{f} := \bigcup_{i \in I} f_i$. *We have that*

1. $(\bar{x}, \bar{y}, \bar{f})$ *is a function; and*
2. *if every function* (x_i, y_i, f_i) *is injective/surjective/bijective, then the function* $(\bar{x}, \bar{y}, \bar{f})$ *is injective/surjective/bijective.*

Proof. (1) First of all we verify that $(\bar{x}, \bar{y}, \bar{f})$ is a relation. Let us fix an element $(a, b) \in \bar{f}$. By definition of union, there exists $i \in I$ such that $(a, b) \in f_i \subset x_i \times y_i$. Since $x_i \subset \bar{x}$ and $y_i \subset \bar{y}$ (see exercise 11–(5) in section 4.5), it follows that $x_i \times y_i \subset \bar{x} \times \bar{y}$ (see exercise 6-2 in section 2.8), hence $(a, b) \in \bar{x} \times \bar{y}$. Since this hold for any $(a, b) \in \bar{f}$, we conclude that $\bar{f} \subset \bar{x} \times \bar{y}$, hence $(\bar{x}, \bar{y}, \bar{f})$ is a relation. Let us verify that it is a function. We fix $a \in \bar{x}$. By definition of union, there exists $i \in I$ such that $a \in x_i$. Since f_i is a function, there exists $b \in y_i \subset \bar{y}$ such that $(a,b) \in f_i \subset \bar{f}$, hence $(a,b) \in \bar{f}$. It follows that, for every $a \in \bar{x}$, there exists $b \in \bar{y}$ such that $(a,b) \in \bar{f}$. Let us now suppose that $(a,b),(a,b') \in \bar{f}$. By definition of union, there exist $i,j \in I$ such that $(a,b) \in f_i$ and $(a,b') \in f_j$. By hypothesis, up to exchange i and j, we can suppose $f_i \subset f_j$, so that $(a,b),(a,b') \in f_j$. Since f_j is a function, we deduce that $b = b'$. Since this holds for any $(a,b),(a,b') \in \bar{f}$, we conclude that $\bar{f}$ is a function.

(2) Let us suppose that every (x_i, y_i, f_i) is injective and that $\bar{f}(a) = \bar{f}(a') = b$. This means that $(a,b),(a',b) \in \bar{f}$. By definition of union, there exist $i,j \in I$ such that $(a,b) \in f_i$ and $(a',b) \in f_j$. By hypothesis, up to exchange i and j, we can suppose $f_i \subset f_j$, so that $(a,b),(a',b) \in f_j$—that is, $f_j(a) = f_j(a') = b$. Since f_j is injective, we deduce that $a = a'$. Since this holds for every a and a' such that $\bar{f}(a) = \bar{f}(a')$, we conclude that $\bar{f}$ is injective. Let us now suppose that every (x_i, y_i, f_i) is surjective and let us fix $b \in \bar{y}$. By definition of union, there exists $i \in I$ such that $b \in y_i$. Since (x_i, y_i, f_i) is surjective, there exists $a \in x_i$ such that $f_i(a) = b$—that is, $(a,b) \in f_i \subset \bar{f}$. It follows that $\bar{f}(a) = b$. Since this holds for every $b \in \bar{y}$, we conclude that $(\bar{x}, \bar{y}, \bar{f})$ is surjective. It easily follows that, if every (x_i, y_i, f_i) is bijective, then $(\bar{x}, \bar{y}, \bar{f})$ is bijective. □

Corollary 4.21. *Let Φ be a set of pre-functions (that is, for every $f \in \Phi$ there exist x and y such that f is a pre-function from x to y) that is totally ordered by inclusion (that is, for every $f, f' \in \Phi$ we have $f \subset f'$ or $f' \subset f$). We set $\bar{f} := \bigcup \Phi$. We have that*

1. *$\bar{f}$ is a pre-function;*

2. *defining $\mathrm{dom}\Phi := \{\mathrm{dom}(f) : f \in \Phi\}$ and $\mathrm{Im}\Phi := \{\mathrm{Im}(f) : f \in \Phi\}$, we have $\mathrm{dom}(\bar{f}) = \bigcup \mathrm{dom}\Phi$ and $\mathrm{Im}(\bar{f}) = \bigcup \mathrm{Im}\Phi$; and*

3. *if every element of Φ is injective, then $\bar{f}$ is injective.*

Proof. We set $\Phi' := \{(\mathrm{dom}(f), \mathrm{Im}(f), f) : f \in \Phi\}$. We define the surjective functions $\varphi_1 \colon \Phi \to \mathrm{dom}\Phi$, $f \mapsto \mathrm{dom}(f)$, $\varphi_2 \colon \Phi \to \mathrm{Im}\Phi$, $f \mapsto \mathrm{Im}(f)$, and $\varphi_3 \colon \Phi \to \Phi'$, $f \mapsto (\mathrm{dom}(f), \mathrm{Im}(f), f)$, the third one being actually bijective. Moreover, we set $\bar{x} := \bigcup \mathrm{dom}\Phi$, $\bar{y} := \bigcup \mathrm{Im}\Phi$, and (as defined in the statement) $\bar{f} := \bigcup \Phi$. We think of the surjective functions φ_1, φ_2, and φ_3 as the indexed sets $\mathrm{dom}\Phi = \{\mathrm{dom}(f)\}_{f \in \Phi}$, $\mathrm{Im}\Phi = \{\mathrm{Im}(f)\}_{f \in \Phi}$ and $\Phi' = \{(\mathrm{dom}(f), \mathrm{Im}(f), f)\}_{f \in \Phi}$. Since the union of an indexed set coincides with the one of the underlying set, by applying proposition 4.20-1 we get that $(\bar{x}, \bar{y}, \bar{f})$ is a function, hence $\bar{f}$ is a pre-function and $\mathrm{dom}(\bar{f}) = \bar{x}$. Since every element of Φ' is surjective by construction, it follows from proposition 4.20-2 that $(\bar{x}, \bar{y}, \bar{f})$ is surjective too, hence $\mathrm{Im}(\bar{f}) = \bar{y}$. Lastly, if every element of Φ is injective, then every element of Φ' is trivially injective too, hence, again by proposition 4.20-2, the function $(\bar{x}, \bar{y}, \bar{f})$ is injective, so that $\bar{f}$ is injective. □

4.5 EXERCISES

4.5.1 Part I: Sets

1. Prove that:

1. $\bigcap x \subset \bigcup x$ for every $x \neq \emptyset$.
2. If $y \subset x$, then $\bigcup y \subset \bigcup x$ and $\bigcap y \supset \bigcap x$, assuming that $x, y \neq \emptyset$ in the case of intersection (in the case of union, cfr. lemma 4.12-2 with $n = 1$).
3. If $y \subset z$ for every $y \in x$, then $\bigcup x \subset z$.
4. If $z \subset y$ for every $y \in x$, then $z \subset \bigcap x$.
5. If $y \in x$, then $y \subset \bigcup x$ (cfr. lemma 4.12-1 with $n = 0$) and $\bigcap x \subset y$.

2. Prove that, for any sets $x_1, \ldots, x_n$:

1. $x_1 \cup \ldots \cup x_n = \bigcup\{x_1, \ldots, x_n\}$
2. $x_1 \cap \ldots \cap x_n = \bigcap\{x_1, \ldots, x_n\}$

3. Prove that, for any sets $x_1, \ldots, x_n$:

1. $\bigcup(x_1 \cup \ldots \cup x_n) = \left(\bigcup x_1\right) \cup \ldots \cup \left(\bigcup x_n\right)$
2. $\bigcap(x_1 \cup \ldots \cup x_n) = \left(\bigcap x_1\right) \cap \ldots \cap \left(\bigcap x_n\right)$
3. $\bigcup(x_1 \cap \ldots \cap x_n) \subset \left(\bigcup x_1\right) \cap \ldots \cap \left(\bigcup x_n\right)$, but the inverse inclusion does not hold in general
4. $\bigcap(x_1 \cap \ldots \cap x_n) \supset \left(\bigcap x_1\right) \cup \ldots \cup \left(\bigcap x_n\right)$, but the inverse inclusion does not hold in general

4. Given a set x, we define $\bar{\bigcup} x := \{\bigcup y : y \in x\}$. We have the natural surjective function $\mu\colon x \twoheadrightarrow \bar{\bigcup} x$, $y \mapsto \bigcup y$. Similarly, we define $\bar{\bigcap} x := \{\bigcap y : y \in x\}$. We have the natural surjective function $\nu\colon x \twoheadrightarrow \bar{\bigcap} x$, $y \mapsto \bigcap y$. Prove that, for every set x:

1. $\bigcup\bigcup x = \bigcup\bar{\bigcup} x$
2. $\bigcap\bigcup x = \bigcap\bar{\bigcap} x$
3. $\bigcup\bigcap x \subset \bigcap\bar{\bigcup} x$, but the inverse inclusion does not hold in general
4. $\bigcap\bigcap x \supset \bigcup\bar{\bigcap} x$, but the inverse inclusion does not hold in general

Remark: This exercise generalises exercise 3. In fact, exercise 2 shows that, choosing $x = \{x_1, \ldots, x_n\}$, we get exercise 3 as a particular case.

5. Prove that $\bigcap\bigcup x \subset \bigcup\bigcap x$, but the inverse inclusion does not hold in general.

6. Given two sets x and y, we define $x \mathbin{\bar{\cup}} y := \{x \cup z : z \in y\}$ and $x \mathbin{\bar{\cap}} y := \{x \cap z : z \in y\}$. Prove that, for every x and y:

1. $x \cup \bigcup y = \bigcup(x \mathbin{\bar{\cup}} y)$
2. $x \cap \bigcap y = \bigcap(x \mathbin{\bar{\cap}} y)$

3. $x \cap \bigcup y = \bigcup (x \bar{\cap} y)$
4. $x \cup \bigcap y = \bigcap (x \bar{\cup} y)$

Remark: Cfr. exercise 3 of section 2.8.

7. Given a set x, we define $\bar{\mathcal{P}}(x) := \{\mathcal{P}(y) : y \in x\}$. Prove that, for every set x:

1. $x = \bigcup \mathcal{P}(x)$
2. $x \subset \mathcal{P}\big(\bigcup x\big)$, but the inverse inclusion does not hold in general
3. $\mathcal{P}\big(\bigcap x\big) = \bigcap \bar{\mathcal{P}}(x)$
4. $\mathcal{P}\big(\bigcup x\big) \supset \bigcup \bar{\mathcal{P}}(x) \supset x$, but the inverse inclusions do not hold in general

Remark: Cfr. exercise 7 of section 2.8, items (1) and (2). We also remark that the dual statements to items (1) and (2) would be the following ones, quite meaningless: $\emptyset = \bigcap \mathcal{P}(x)$ and $\emptyset \subset \mathcal{P}\big(\bigcap x\big)$.

8. Let X be a fixed set, in which we apply the complement operation. Given $x \subset \mathcal{P}(X)$, we define $x^{\bar{\complement}} := \{y^{\complement} : y \in x\}$. Prove that, for every $x \subset \mathcal{P}(X)$, $(\bigcup x)^{\complement} = \bigcap x^{\bar{\complement}}$ and $(\bigcap x)^{\complement} = \bigcup x^{\bar{\complement}}$ (*De Morgan's laws*).

Remark: Cfr. exercise 5-7 of section 2.8.

9. Given a function $f : x \to y$, we have seen that it induces the functions $\dot{f} : \mathcal{P}(x) \to \mathcal{P}(y)$ and $\ddot{f} : \mathcal{P}(y) \to \mathcal{P}(x)$. By iterating this construction, we get $\dot{\dot{f}}, \ddot{\dot{f}} : \mathcal{P}(\mathcal{P}(x)) \to \mathcal{P}(\mathcal{P}(y))$ and $\dot{\ddot{f}}, \ddot{\ddot{f}} : \mathcal{P}(\mathcal{P}(y)) \to \mathcal{P}(\mathcal{P}(x))$. Prove that:

1. $\dot{f}\big(\bigcup a\big) = \bigcup \dot{\dot{f}}(a)$ and $\dot{f}\big(\bigcap a\big) \subset \bigcap \dot{\dot{f}}(a)$ for every $a \subset \mathcal{P}(x)$
2. $\dot{f}\big(\bigcap a\big) = \bigcap \dot{\dot{f}}(a)$ for every $a \subset \mathcal{P}(x)$ if and only if f is injective
3. $\ddot{f}\big(\bigcup b\big) = \bigcup \dot{\ddot{f}}(b)$ and $\ddot{f}\big(\bigcap b\big) = \bigcap \dot{\ddot{f}}(b)$ for every $b \subset \mathcal{P}(y)$

Remark: Cfr. exercise 8 in section 3.5.

10. (1) Prove that the function $\eta_x : \bigsqcup x \twoheadrightarrow \bigcup x$ is injective (hence bijective) if and only if $y \cap z = \emptyset$ for every $y, z \in x$ such that $y \neq z$.

(2) Assuming $x \neq \emptyset$, when is the function $\Delta_x : \bigcap x \hookrightarrow \prod x$ surjective (hence bijective)?

4.5.2 Part II: Indexed Sets

11. Let $x = \{x_i\}_{i \in I}$ be an indexed set. Prove that:

1. $\bigcap_i x_i \subset \bigcup_i x_i$, assuming $I \neq \emptyset$.
2. If $J \subset I$, then $\bigcup_j x_j \subset \bigcup_i x_i$ and $\bigcap_j x_j \supset \bigcap_i x_i$, assuming $I, J \neq \emptyset$ in the case of intersection.
3. If $x_i \subset z$ for every $i \in I$, then $\bigcup_i x_i \subset z$.
4. If $z \subset x_i$ for every $i \in I$, then $z \subset \bigcap_i x_i$.
5. If $j \in I$, then $x_j \subset \bigcup_i x_i$ and $\bigcap_i x_i \subset x_j$.

12. Given an indexed set $x = \{x_i\}_{i\in I}$, we define the indexed sets $\bar{\bigcup} x = \{\bigcup x_i\}_{i\in I}$ and $\bar{\bigcap} x = \{\bigcap x_i\}_{i\in I}$. Prove that:

1. $\bigcup \bigcup_i x_i = \bigcup_i (\bigcup x_i)$
2. $\bigcap \bigcup_i x_i = \bigcap_i (\bigcap x_i)$
3. $\bigcup \bigcap_i x_i \subset \bigcap_i (\bigcup x_i)$, but the inverse inclusion does not hold in general
4. $\bigcap \bigcap_i x_i \supset \bigcup_i (\bigcap x_i)$, but the inverse inclusion does not hold in general

Remark: Cfr. exercise 4.

13. Prove that $\bigcap \bigcup_i x_i \subset \bigcup \bigcap_i x_i$, but the inverse inclusion does not hold in general.

Remark: Cfr. exercise 5.

14. Given a set x and an indexed set $y = \{y_i\}_{i\in I}$, we get the indexed sets $x \bar{\cup} y := \{x \cup y_i\}_{i\in I}$ and $x \bar{\cap} y := \{x \cap y_i\}_{i\in I}$. Prove that:

1. $x \cup \bigcup_i y_i = \bigcup_i (x \cup y_i)$
2. $x \cap \bigcap_i y_i = \bigcap_i (x \cap y_i)$
3. $x \cap \bigcup_i y_i = \bigcup_i (x \cap y_i)$
4. $x \cup \bigcap_i y_i = \bigcap_i (x \cup y_i)$

Remark: Cfr. exercise 3 of section 2.8, and exercise 6 here.

15. Given an indexed set $x = \{x_i\}_{i\in I}$, we define $\bar{\mathcal{P}}(x) = \{\mathcal{P}(x_i)\}_{i\in I}$. Prove that:

1. $x \subset \mathcal{P}(\bigcup_i x_i)$, but the inverse inclusion does not hold in general
2. $\mathcal{P}(\bigcap_i x_i) = \bigcap_i \mathcal{P}(x_i)$
3. $\mathcal{P}(\bigcup_i x_i) \supset \bigcup_i \mathcal{P}(x_i) \supset x$, but the inverse inclusions do not hold in general

Remark: Cfr. exercise 7 of section 2.8, items (1) and (2), and exercise 7 here.

16. Let X be a fixed set, in which we apply the complement operation. Given an indexed set $x = \{x_i\}_{i\in I}$, such that $x \subset \mathcal{P}(X)$ as a set, we get the indexed set $x^{\bar{\complement}} = \{x_i^{\complement}\}_{i\in I}$. Prove that $(\bigcup_i x_i)^{\complement} = \bigcap_i x_i^{\complement}$ and $(\bigcap_i x_i)^{\complement} = \bigcup_i x_i^{\complement}$ (*De Morgan's laws*).

Remark: Cfr. exercise 5-7 of section 2.8, and exercise 8 here.

17. Given a function $f: x \to y$ and two indexed sets $a = \{a_i\}_{i\in I}$ e $b = \{b_j\}_{j\in J}$, such that $a \subset \mathcal{P}(x)$ and $b \subset \mathcal{P}(y)$ as sets, we get the indexed sets $\dot{\bar{f}}(a) = \{\dot{f}(a_i)\}_{i\in I}$ and $\ddot{\bar{f}}(b) = \{\ddot{f}(b_j)\}_{j\in J}$. Prove that:

1. $\dot{f}(\bigcup_i a_i) = \bigcup_i \dot{f}(a_i)$ and $\dot{f}(\bigcap_i a_i) \subset \bigcap_i \dot{f}(a_i)$ for every $a = \{a_i\}_{i\in I}$ such that $a \subset \mathcal{P}(x)$ as a set
2. $\dot{f}(\bigcap_i a_i) = \bigcap_i \dot{f}(a_i)$ for every $a = \{a_i\}_{i\in I}$ such that $a \subset \mathcal{P}(x)$ as a set if and only if f is injective
3. $\ddot{f}(\bigcup_j b_j) = \bigcup_j \ddot{f}(b_j)$ and $\ddot{f}(\bigcap_j b_j) = \bigcap_j \ddot{f}(b_j)$ for every $b = \{b_j\}_{j\in J}$ such that $b \subset \mathcal{P}(y)$ as a set

Remark: Cfr. exercise 8 in section 3.5 and exercise 9 in this section.

18. Let $x = \{x_i\}_{i \in I}$ be an indexed set and $p \subset \mathcal{P}(I)$ be a partition of I. Prove that:

1. $\bigcup_{i \in I} x_i = \bigcup_{a \in p} \bigcup_{i \in a} x_i$
2. $\bigcap_{i \in I} x_i = \bigcap_{a \in p} \bigcap_{i \in a} x_i$

This means the following. For every $a \in p$, we consider the indexed set $y_a = \{x_i\}_{i \in a}$. Equivalently, if $x = \{x_i\}_{i \in I}$ is the indexed set $\varphi\colon I \to x$, $i \mapsto x_i$, then $y_a = \{x_i\}_{i \in a}$ is the indexed set $\varphi|_a\|_{y_a} : a \to y_a$, $i \mapsto x_i$, where $y_a = \dot{\varphi}(a)$ as a set. Then, we consider the indexed set $z = \{z_a\}_{a \in p}$, where z_a is the union of the indexed set y_a—that is, $z_a := \bigcup_{i \in a} x_i$. We have to show that $\bigcup_{i \in I} x_i = \bigcup_{a \in p} z_a$. An analogous explication holds with respect to intersection.

Remark: This property is the *generalised associativity*, as we are going to show in Appendix B.

19. Consider the indexed set $x = \{y\}_{i \in I}$—that is, such that $x_i = y$ for every $i \in I$. Prove that $\bigsqcup_i y = y \times I$.

20. (1) Prove that the function $\bar{\eta}_x : \bigsqcup_i x_i \twoheadrightarrow \bigcup_i x_i$ is injective (hence bijetive) if and only if $x_i \cap x_j = \emptyset$ for every $i, j \in I$ such that $i \neq j$.

(2) Assuming $x \neq \emptyset$, when is the function $\bar{\Delta}_x : \bigcap_i x_i \hookrightarrow \prod_i x_i$ surjective (hence bijective)?

CHAPTER 5

Universal Properties

The Cartesian product is not strictly associative neither commutative. In fact, $((a, b), c) \in (x \times y) \times z$ and $(a, (b, c)) \in x \times (y \times z)$ are distinct elements, and similarly $(a, b) \in x \times y$ is different from $(b, a) \in y \times x$. Therefore, $(x \times y) \times z \neq x \times (y \times z)$ and $x \times y \neq y \times x$. Nevertheless, it is intuitively clear that associativity and commutativity hold up to the natural bijections $\Phi_{x,y,z} \colon (x \times y) \times z \to x \times (y \times z)$, $((a, b), c) \mapsto (a, (b, c))$, and $\Psi_{x,y} \colon x \times y \to y \times x$, $(a, b) \mapsto (b, a)$. We remark that the ternary product $x \times y \times z$ coincides with $(x \times y) \times z$, because we defined an ordered triple as $(a, b, c) := ((a, b), c)$. However, we could equivalently define $(a, b, c) := (a, (b, c))$, and, in this case, $x \times y \times z$ would coincide with $x \times (y \times z)$. A similar remark holds about the n-ary product for any $n \geq 3$.

For these reasons, it is natural to inquire if it is possible to provide a more suitable definition of Cartesian product, that allows to essentially identify $(x \times y) \times z$ with $x \times (y \times z)$ and $x \times y$ with $y \times x$. In principle, we could simply define the Cartesian product up to a bijection—that is, we could state that a Cartesian product of $x_1, \dots, x_n$ is any set z such that there exists a bijection between z and $x_1 \times \dots \times x_n$. In this way, we would trivially get associativity and commutativity. Nevertheless, this definition would be definitely too weak. For example, fixing a set x with 5 elements and a set y with 3 elements, any set of 15 elements would be a Cartesian product of x and y, losing any connection with the idea of ordered pairs. Therefore, we are in the following situation:

- If we define the Cartesian product of $x_1, \dots, x_n$ as the specific set $x_1 \times \dots \times x_n$, as we have done up to now, then we do not get associativity and commutativity. This is not reasonable, since there is no essential distinction between $(x \times y) \times z$ and $x \times (y \times z)$. Similarly, if we are not interested in fixing and order between x and y, then there is no essential distinction between $x \times y$ and $y \times x$.[1]

[1] When we apply an operation, we need to impose an order, since ab and ba are different in general. Nevertheless, when the operation is commutative, the order becomes immaterial. In the case of Cartesian product, we would like to make the order immaterial (unless there is a specific reason to fix it) because of the natural identification between $x \times y$ and $y \times x$.

DOI: 10.1201/9781003565291-5

- If we define Cartesian product up to a bijection, then we lose the specific properties of this operation, since we lose any connection with the idea of ordered n-uples.

Hence, we need to find an intermediate definition, stating in a suitable way that $(x \times y) \times z$ and $x \times (y \times z)$, both thought of as Cartesian products of x, y, and z, can be identified *only* through the bijection $((a, b), c) \mapsto (a, (b, c))$, not through any other one. In this way, two equivalent elements $((a, b), c)$ and $(a, (b, c))$ maintain their specific role of ordered triples. Similarly, $x \times y$ and $y \times x$ have to be identified *only* through the bijection $(a, b) \mapsto (b, a)$. More generally, we need a definition such that any two representatives of the Cartesian product of $x_1, \ldots, x_n$ are identified up to a *unique* bijection.

The same discussion holds about disjoint union, with no essential variations. In fact, the elements of $(x \sqcup y) \sqcup z$ are of the form $((a, \emptyset), \emptyset)$, $((b, \{\emptyset\}), \emptyset)$ or $(c, \{\emptyset\})$, with $a \in x$, $b \in y$ and $c \in z$; the elements of $x \sqcup (y \sqcup z)$ are of the form $(a, \emptyset)$, $((b, \emptyset), \{\emptyset\})$ or $((c, \{\emptyset\}), \{\emptyset\})$; and the elements of $x \sqcup y \sqcup z$ are of the form $(a, \emptyset)$, $(b, \{\emptyset\})$ or $(c, \{\{\emptyset\}\})$. Similarly, an element of $x \sqcup y$ is of the form $(a, \emptyset)$ or $(b, \{\emptyset\})$, with $a \in x$ and $b \in y$, while an element of $y \sqcup x$ is of the form $(b, \emptyset)$ or $(a, \{\emptyset\})$. Therefore, also in this case, we aim to find a definition such that two representatives of the disjoint union of $x_1, \ldots, x_n$ are identified up to a *unique* bijection, inducing associativity and commutativity. Moreover, we have a problem more, that was not present in the case of Cartesian product: in the definition of $x_1 \sqcup \ldots \sqcup x_n$, we used the sets $\emptyset, \{\emptyset\}, \ldots, \{\cdots\{\emptyset\}\cdots\}$ as "tags" to separate $x_1, \ldots, x_n$, but we could use any other family of n fixed distinct sets $a_1, \ldots, a_n$, defining the disjoint union as $(x_1 \times \{a_1\}) \cup \ldots \cup (x_n \times \{a_n\})$. It is quite clear that the choice of such sets should be immaterial. Again, the natural bijection that identifies $(y_i, \{\cdots\{\emptyset\}\cdots\})$ with (y_i, a_i) has to be the unique admissible one.

In this chapter, we provide a suitable definition of Cartesian product and disjoint union, with the features sketched above, by introducing the corresponding *universal properties*.

5.1 UNIVERSAL PROPERTY OF CARTESIAN PRODUCT

Let us begin by considering two non-empty sets x_1 and x_2. We have the surjective maps $\pi_1 \colon x_1 \times x_2 \twoheadrightarrow x_1$, $(a_1, a_2) \mapsto a_1$, and $\pi_2 \colon x_1 \times x_2 \twoheadrightarrow x_2$, $(a_1, a_2) \mapsto a_2$, called *projections*. Given a set y and two functions $f_1 \colon y \to x_1$ and $f_2 \colon y \to x_2$, it is easy to verify that there exists a unique function $f \colon y \to x_1 \times x_2$ such that $\pi_1 \circ f = f_1$ and $\pi_2 \circ f = f_2$. Such a function is defined by $f(b) := \big(f_1(b), f_2(b)\big)$ for every $b \in y$. We represent this property through the following commutative diagram:

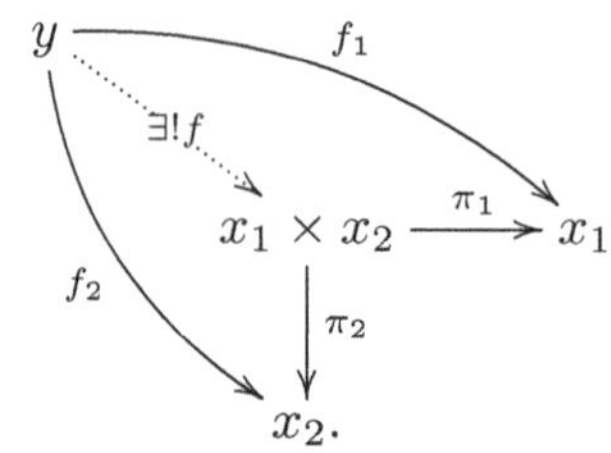

If we do not assume x_1 and x_2 to be non-empty, then $x_1 \times x_2$ may be empty, hence π_1 and π_2 may not be surjective, but the existence and uniqueness of f holds anyway. In fact, if $x_1 = \emptyset$ or $x_2 = \emptyset$, then necessarily $y = \emptyset$ and f is the empty function.

We can easily generalise this picture to the product of n sets. In fact, given $x_1, \ldots, x_n$, for each $i \in \{1, \ldots, n\}$ we have the *projection* $\pi_i \colon x_1 \times \cdots \times x_n \twoheadrightarrow x_i$, $(a_1, \ldots, a_n) \mapsto a_i$. The $(n+1)$-uple $(x_1 \times \cdots \times x_n, \pi_1, \ldots, \pi_n)$ can be characterised through the following universal property.

Proposition 5.1 (Universal property of Cartesian product). *Fixing n sets $x_1, \ldots, x_n$, for any set y and any family of functions $f_1 \colon y \to x_1, \ldots, f_n \colon y \to x_n$, there exists a unique function $f \colon y \to x_1 \times \cdots \times x_n$ such that $\pi_i \circ f = f_i$ for every i. Such a function is defined by $f(z) := (f_1(z), \ldots, f_n(z))$ and we denote it as $f =: f_1 \dot{\times} \cdots \dot{\times} f_n$.*

Exercise 5.2. Prove proposition 5.1.

This property suggests the following more general definition of Cartesian product.

Definition 5.3. A *Cartesian product* of n fixed sets $x_1, \ldots, x_n$ is an ordered $(n+1)$-uple $(x, p_1, \ldots, p_n)$ such that

- x is a set;
- $p_i \colon x \to x_i$ is a function for each i, called *projection*; and
- for any set y and any family of functions $f_1 \colon y \to x_1, \ldots, f_n \colon y \to x_n$, there exists a unique function $f \colon y \to x$ satisfying the condition $p_i \circ f = f_i$ for every i. ♢

Proposition 5.1 guarantees the existence of a Cartesian product of $x_1, \ldots, x_n$. The following proposition establishes its essential uniqueness.

Proposition 5.4. *Let us suppose that $(x, p_1, \ldots, p_n)$ and $(x', p'_1, \ldots, p'_n)$ are two Cartesian products of n fixed sets $x_1, \ldots, x_n$. Then there exists a* unique *bijection $\phi \colon x \xrightarrow{\approx} x'$ such that $p'_i \circ \phi = p_i$ for every $i \in \{1, \ldots, n\}$.*

Exercise 5.5. Prove proposition 5.4. (*Hint:* Apply the universal property to $(x, p_1, \ldots, p_n)$, with $y = x'$ and $f_i = p'_i$. Conversely, apply the universal property to $(x', p'_1, \ldots, p'_n)$, with $y = x$ and $f_i = p_i$. Consider the compositions of the (unique) induced functions and apply uniqueness as stated in the universal property.)

Exercise 5.6. Prove that, if $x_1, \ldots, x_n$ are non-empty, then the projections p_i in definition 5.3 are necessarily surjective.

Exercise 5.7. Let $(x, p_1, \ldots, p_n)$ and $(x', p'_1, \ldots, p'_n)$ be two Cartesian products of n fixed sets $x_1, \ldots, x_n$, and let $\phi\colon x \xrightarrow{\approx} x'$ be the bijection induced by the universal property. Moreover, given a set y and a family of functions $f_1\colon y \to x_1, \ldots, f_n\colon y \to x_n$, we call $f\colon y \to x$ and $f'\colon y \to x'$ the functions induced by $f_1, \ldots, f_n$ by the universal property. Prove that $f' = \phi \circ f$.

5.1.1 Remarks on Cartesian Product

It is important to observe that, while in proposition 5.1 the projections π_i are induced by $x_1, \ldots, x_n$, in definition 5.3 the projections p_i are part of the datum. Therefore, a Cartesian product of $x_1, \ldots, x_n$ is not simply a set, but it is formed by a set *and n projections*.

If $(x, p_1, \ldots, p_n)$ is a product of $x_1, \ldots, x_n$, then any set x' such that there exits a bijection $\phi\colon x \xrightarrow{\approx} x'$ is the underlying set of another product—that is, $(x', p'_1, \ldots, p'_n)$, where $p'_i := p_i \circ \phi^{-1}$. Nevertheless, the projections determine the role of each element in the set. In fact, let us fix any Cartesian product $(x, p_1, \ldots, p_n)$ and the canonical one $(x_1 \times \cdots \times x_n, \pi_1, \ldots, \pi_n)$. The unique bijection induced by proposition 5.4 is $\phi\colon x \xrightarrow{\approx} x_1 \times \cdots \times x_n$, $y \mapsto (p_1(y), \ldots, p_n(y))$. Thus, an element $y \in x$ has the same role of the ordered n-uple $(p_1(y), \ldots, p_n(y)) \in x_1 \times \cdots \times x_n$, and it cannot be "confused" with any other element of x.

5.1.2 Induced Function between Cartesian Products

In the universal property 5.1 we considered a fixed domain y for $f_1, \ldots, f_n$, but we can also fix a domain for each function. More precisely, given n functions $f_1\colon y_1 \to x_1$, …, $f_n\colon y_n \to x_n$, we get the function $g\colon y_1 \times \cdots \times y_n \to x_1 \times \cdots \times x_n$, $(b_1, \ldots, b_n) \mapsto (f_1(b_1), \ldots, f_n(b_n))$. It is the unique function such that $\pi^x_i \circ g = f_i \circ \pi^y_i$, where $\pi^x_i\colon x_1 \times \cdots \times x_n \twoheadrightarrow x_i$ and $\pi^y_i\colon y_1 \times \cdots \times y_n \twoheadrightarrow y_i$ are the corresponding projections. We use the notation $g =: f_1 \ddot{\times} \cdots \ddot{\times} f_n$. Such a property follows more generally from the universal property, as the following proposition shows.

Proposition 5.8. *Let us suppose that $(x, p_1, \ldots, p_n)$ and $(y, q_1, \ldots, q_n)$ are Cartesian products of $x_1, \ldots, x_n$ and $y_1, \ldots, y_n$ respectively. Then, given n functions $f_1\colon y_1 \to x_1$, …, $f_n\colon y_n \to x_n$, there exists a unique function $g\colon y \to x$ such that $p_i \circ g = f_i \circ q_i$ for every i.*

The condition $p_i \circ g = f_i \circ q_i$ can be expressed through the commutativity of the following diagram:

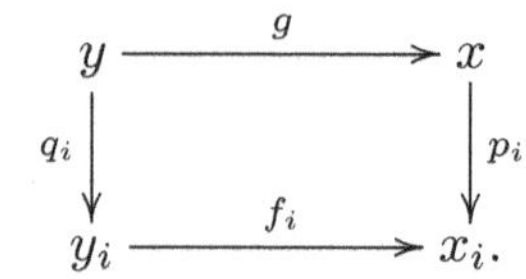

Exercise 5.9. Prove proposition 5.8.

5.2 UNIVERSAL PROPERTY OF DISJOINT UNION

The universal property of disjoint union is dual to the one of Cartesian product—that is, it is analogous to the latter, but "reversing the arrows". Let us show the details. Again, we begin by considering two sets x_1 and x_2. We have the injective maps $\iota_1 \colon x_1 \hookrightarrow x_1 \sqcup x_2$, $a_1 \mapsto (a_1, \emptyset)$, and $\iota_2 \colon x_2 \hookrightarrow x_1 \sqcup x_2$, $a_2 \mapsto (a_2, \{\emptyset\})$, called *embeddings*. Given a set y and two functions $f_1 \colon x_1 \to y$ and $f_2 \colon x_2 \to y$, it is easy to verify that there exists a unique function $f \colon x_1 \sqcup x_2 \to y$ such that $f \circ \iota_1 = f_1$ and $f \circ \iota_2 = f_2$. Such a function is defined by $f\big((a_1, \emptyset)\big) := f_1(a_1)$ and $f\big((a_2, \{\emptyset\})\big) := f_2(a_2)$. We represent this property through the following diagram:

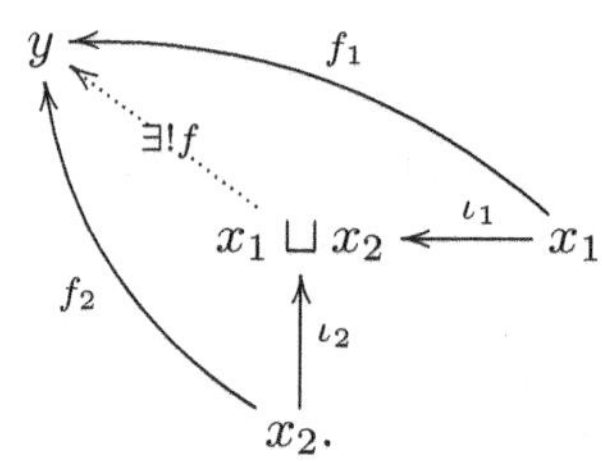

We can easily generalise this picture to the disjoint union of n sets. In fact, given $x_1, \ldots, x_n$, for each $i \in \{1, \ldots, n\}$ we have the *embedding* $\iota_i \colon x_i \hookrightarrow x_1 \sqcup \ldots \sqcup x_n$, $a_i \mapsto (a_i, \{\cdots\{\emptyset\}\cdots\})$. The $(n+1)$-uple $(x_1 \sqcup \ldots \sqcup x_n, \iota_1, \ldots, \iota_n)$ can be characterised through the following universal property.

Proposition 5.10 (Universal property of disjoint union). *Fixing n sets $x_1, \ldots, x_n$, for any set y and any family of functions $f_1 \colon x_1 \to y, \ldots, f_n \colon x_n \to y$, there exists a unique function $f \colon x_1 \sqcup \ldots \sqcup x_n \to y$ such that $f \circ \iota_i = f_i$ for every i. Such a function is defined by $f\big((a_i, \{\cdots\{\emptyset\}\cdots\})\big) := f_i(a_i)$ and we denote it as $f =: f_1 \dot{\sqcup} \cdots \dot{\sqcup} f_n$.*

Exercise 5.11. Prove proposition 5.10.

This property suggests the following more general definition of disjoint union.

Definition 5.12. A *disjoint union* of n fixed sets $x_1, \ldots, x_n$ is an ordered $(n+1)$-uple $(x, j_1, \ldots, j_n)$ such that:

- x is a set;
- $j_i : x_i \to x$ is a function for each i, called *embedding*;
- for any set y and any family of functions $f_1 : x_1 \to y, \ldots, f_n : x_n \to y$, there exists a unique function $f : x \to y$ satisfying the condition $f \circ j_i = f_i$ for every i. ◊

Proposition 5.10 guarantees the existence of a disjoint union of $x_1, \ldots, x_n$. The following proposition establishes its essential uniqueness.

Proposition 5.13. *Let us suppose that* $(x, j_1, \ldots, j_n)$ *and* $(x', j'_1, \ldots, j'_n)$ *are two disjoint unions of* n *fixed sets* $x_1, \ldots, x_n$*. Then there exists a* unique *bijection* $\phi : x \xrightarrow{\approx} x'$ *such that* $\phi \circ j_i = j'_i$.

Exercise 5.14. Prove proposition 5.13. (*Hint:* Apply the universal property to $(x, j_1, \ldots, j_n)$, with $y = x'$ and $f_i = j'_i$. Conversely, apply the universal property to $(x', j'_1, \ldots, j'_n)$, with $y = x$ and $f_i = j_i$. Consider the compositions of the (unique) induced functions and apply uniqueness as stated in the universal property.)

Exercise 5.15. Prove that the embeddings j_i in definition 5.12 are necessarily injective.

Exercise 5.16. Let $(x, j_1, \ldots, j_n)$ and $(x', j'_1, \ldots, j'_n)$ be two disjoint unions of n fixed sets $x_1, \ldots, x_n$, and let $\phi : x \xrightarrow{\approx} x'$ be the bijection induced by the universal property. Moreover, given a set y and a family of functions $f_1 : x_1 \to y, \ldots, f_n : x_n \to y$, we call $f : x \to y$ and $f' : x' \to y$ the functions induced by $f_1, \ldots, f_n$ by the universal property. Prove that $f = f' \circ \phi$.

5.2.1 Remarks on Disjoint Union

Similarly to the case of Cartesian product, a disjoint union of $x_1, \ldots, x_n$ is not simply a set, but it is formed by a set *and* n *embeddings*. Moreover, if $(x, j_1, \ldots, j_n)$ is a disjoint union of $x_1, \ldots, x_n$, then any set x' such that there exits a bijection $\phi : x \xrightarrow{\approx} x'$ is the underlying set of another disjoint union—that is, $(x', j'_1, \ldots, j'_n)$, where $j'_i := \phi \circ j_i$. Nevertheless, the embeddings determine the role of each element in the set. In fact, let us fix any disjoint union $(x, j_1, \ldots, j_n)$ and the canonical one $(x_1 \sqcup \ldots \sqcup x_n, \iota_1, \ldots, \iota_n)$. The unique bijection induced by proposition 5.13 is $\phi : x_1 \sqcup \ldots \sqcup x_n \xrightarrow{\approx} x$, $(y_i, \{\cdots\{\emptyset\}\cdots\}) \mapsto j_i(y)$. Thus, the element $j_i(y) \in x$ has the same role of the element $(y_i, \{\cdots\{\emptyset\}\cdots\}) \in x_1 \sqcup \ldots \sqcup x_n$, and it cannot be "confused" with any other element of x.

5.2.2 Induced Function between Disjoint Unions

In the universal property 5.10 we considered a fixed codomain y for $f_1, \ldots, f_n$, but we can also fix a codomain for each function. More precisely, given n functions $f_1 : x_1 \to y_1, \ldots, f_n : x_n \to y_n$, we get the function $g : x_1 \sqcup \ldots \sqcup x_n \to$

$y_1 \sqcup \ldots \sqcup y_n$, $(a_i, \{\cdots\{\emptyset\}\cdots\}) \mapsto (f_i(a_i), \{\cdots\{\emptyset\}\cdots\})$. It is the unique function such that $g \circ \iota_i^x = \iota_i^y \circ f_i$, where $\iota_i^x \colon x_i \hookrightarrow x_1 \sqcup \ldots \sqcup x_n$ and $\iota_i^y \colon y_i \hookrightarrow y_1 \sqcup \ldots \sqcup y_n$ are the corresponding embeddings. We use the notation $g =: f_1 \ddot{\sqcup} \ldots \ddot{\sqcup} f_n$. Such a property follows more generally from the universal property, as the following proposition shows.

Proposition 5.17. *Let us suppose that $(x, j_1, \ldots, j_n)$ and $(y, k_1, \ldots, k_n)$ are disjoint unions of $x_1, \ldots, x_n$ and $y_1, \ldots, y_n$ respectively. Then, given n functions $f_1 \colon x_1 \to y_1, \ldots, f_n \colon x_n \to y_n$, there exists a unique function $g \colon x \to y$ such that $g \circ j_i = k_i \circ f_i$ for every i.*

The condition $g \circ j_i = k_i \circ f_i$ can be expressed through the commutativity of the following diagram:

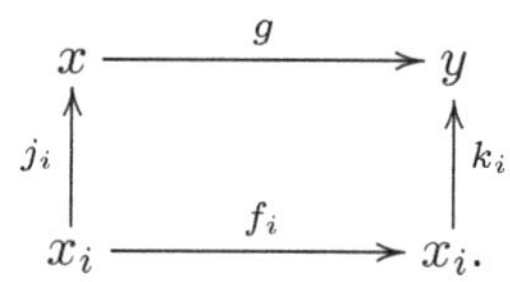

Exercise 5.18. Prove proposition 5.17.

5.2.3 Independence of the Separating Sets

Let us fix any two families of distinct sets $u_1, \ldots, u_n$ and $u'_1, \ldots, u'_n$. We consider the two following $(n+1)$-uples:

- $(x, j_1, \ldots, j_n)$, where $x := (x_1 \times \{u_1\}) \cup \ldots \cup (x_n \times \{u_n\})$ and $j_i(a_i) := (a_i, u_i)$;
- $(x', j'_1, \ldots, j'_n)$, where $x' := (x_1 \times \{u'_1\}) \cup \ldots \cup (x_n \times \{u'_n\})$ and $j'_i(a_i) := (a_i, u'_i)$.

The reader can easily verify that both $(n+1)$-uples are disjoint unions of $x_1, \ldots, x_n$ and that the unique bijection $\phi \colon x \xrightarrow{\approx} x'$ that commutes with the embeddings is $\phi((a_i, u_i)) = (a_i, u'_i)$. It follows that the choice of the sets $u_1, \ldots, u_n$ is immaterial.

5.2.4 Disjoint Union of One Set

By definition 2.9 the disjoint union of one set x_1 is $x_1 \times \{\emptyset\}$, hence it does not coincide with x_1 itself. Now we can settle this little problem through the universal property. In fact, $(x_1 \times \{\emptyset\}, j_1)$, where $j_1(a_1) = (a_1, \emptyset)$, is a disjoint union of x_1, as well as (x_1, id_{x_1}), as the reader can easily verify. The unique bijection provided by the universal property is $\phi \colon x_1 \xrightarrow{\approx} x_1 \times \{\emptyset\}$, $a_1 \mapsto (a_1, \emptyset)$. Therefore, we have an essential identification between the representative provided by definition 2.9 and (x_1, id_{x_1}).

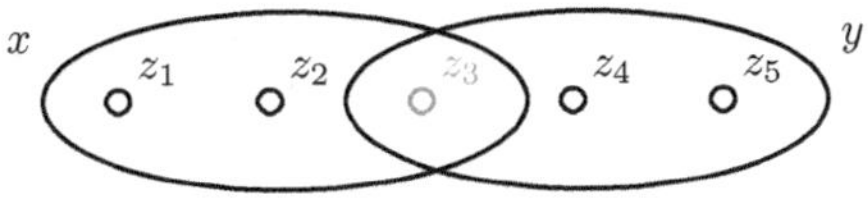

Figure 5.1 Two sets.

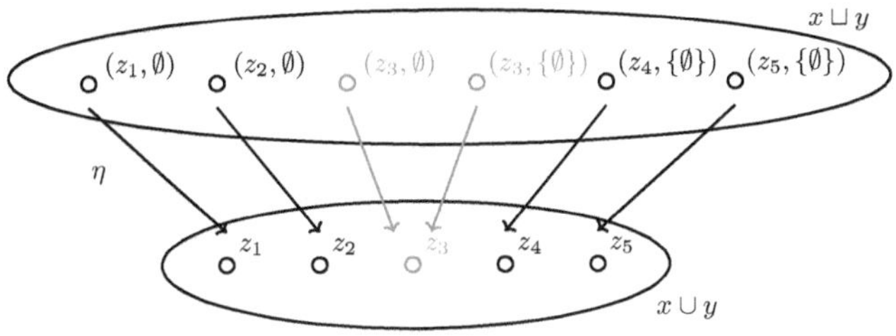

Figure 5.2 The projection induced by universal property.

5.3 TWO RELEVANT PARTICULAR CASES

Given $x_1, \dots, x_n$, we have the canonical inclusions $\mu_i \colon x_1 \cap \ldots \cap x_n \hookrightarrow x_i$. Applying the universal property 5.1, with $y = x_1 \cap \ldots \cap x_n$ and $f_i = \mu_i$, the induced function $\mu_1 \dot{\times} \cdots \dot{\times} \mu_n$ is the *diagonal embedding*, that we also call Δ:

$$\begin{aligned} \Delta \colon x_1 \cap \ldots \cap x_n &\hookrightarrow x_1 \times \cdots \times x_n \\ z &\mapsto (z, \dots, z). \end{aligned} \tag{5.1}$$

Dually, given $x_1, \dots, x_n$, we have the canonical inclusions $\kappa_i \colon x_i \hookrightarrow x_1 \cup \ldots \cup x_n$. Applying the universal property 5.10, with $y = x_1 \cup \ldots \cup x_n$ and $f_i = \kappa_i$, the induced function $\kappa_1 \dot{\sqcup} \ldots \dot{\sqcup} \kappa_n$ is the following surjective map, that we also call η:

$$\begin{aligned} \eta \colon x_1 \sqcup \ldots \sqcup x_n &\twoheadrightarrow x_1 \cup \ldots \cup x_n \\ (z, \{\cdots \{\emptyset\} \cdots\}) &\mapsto z. \end{aligned} \tag{5.2}$$

These maps are particular cases of (4.4) and (4.3), identifying the sequence $(x_1, \dots, x_n)$ with the indexed set $\varphi \colon \{1, \dots, n\} \to \{x_1, \dots, x_n\}$, $i \mapsto x_i$.[2] The map (5.2) "forgets the tags", so that it identifies the two copies of each element of $x \cap y$; it is injective (hence bijective) if and only if the sets $x_1, \dots, x_n$ are pairwise disjoint (see exercise 5 at the end of the chapter). Considering the two sets in figure 5.1, the map η is represented in figure 5.2.

Exercise 5.19. (1) Prove that the function $\eta \colon x_1 \sqcup \ldots \sqcup x_n \twoheadrightarrow x_1 \cup \ldots \cup x_n$ is injective (hence bijective) if and only if $x_i \cap x_j = \emptyset$ for every i and j such that

[2]In defining φ, we are using natural numbers as elements of the universe. We will describe how in chapter 7.

$i \neq j$. (2) When is the function $\Delta\colon x_1 \cap \ldots \cap x_n \hookrightarrow x_1 \times \cdots \times x_n$ surjective (hence bijective)?

Remark 5.20. The content of this section shows a clear duality between Cartesian product and disjoint union and between intersection and union. In fact, for the reader who knows the language of categories, the Cartesian product is the product in the category of sets, while the disjoint union is the coproduct in the same category. Moreover, the intersection is the product fibred over the union, and dually the union is the coproduct fibred over the intersection. ♢

5.4 PROPERTIES OF THE OPERATIONS

Universal properties allow to state associativity and commutativity of Cartesian product and disjoint union in a rigorous way, as well as distributivity of the former over the latter. In order to avoid unnecessarily complicated technical details, we just sketch how to achieve this aim.

5.4.1 Associativity

Associativity of Cartesian product can be stated simply by affirming that $((x \times y) \times z, \bar{\pi}_1, \bar{\pi}_2, \bar{\pi}_3)$ and $(x \times (y \times z), \bar{\pi}'_1, \bar{\pi}'_2, \bar{\pi}'_3)$, where the projections are defined in the natural way, are both Cartesian products of x, y and z, hence they are canonically identified thanks to the universal property. Nevertheless, in this statement, the operation '$\times$' is applied only to the sets x, y, and z, while the projections $\bar{\pi}_i$ and $\bar{\pi}'_i$ are defined a posteriori, in such a way that the corresponding quadruples satisfy the universal property. Hence, which is exactly the associative operation?

In order to reach a more suitable statement, we have to involve the projections in the operation '$\times$', so that $\bar{\pi}_i$ and $\bar{\pi}'_i$ are part of the result. We just outline how to proceed.

Example 5.21. Let us consider a Cartesian product (x, p_1, p_2) of x_1 and x_2 and a Cartesian product (y, q_1, q_2) of y_1 and y_2. We set $(x, p_1, p_2) \times (y, q_1, q_2) := (x \times y, p_1 \circ \pi_1, p_2 \circ \pi_1, q_1 \circ \pi_2, q_2 \circ \pi_2)$, where $\pi_1\colon x \times y \to x$ and $\pi_2\colon x \times y \to y$ are the canonical projections. The reader can easily verify that $(x, p_1, p_2) \times (y, q_1, q_2)$ is a Cartesian product of x_1, x_2, y_1, y_2.

The previous example suggests that we can define the binary operation '$\times$' between two Cartesian products, by applying it to the underlying sets and by composing the corresponding projections in the natural way. Moreover, we can embed the "old" Cartesian product of sets as a particular case of this operation. In fact, a single set x can be identified with the Cartesian product (x, id_x) of x itself and we have $(x, \mathrm{id}_x) \times (y, \mathrm{id}_y) = (x \times y, \pi_1, \pi_2)$. In this way, we can state associativity properly as follows:

> ($\star$) Given three sets x, y and z, the ordered quadruples $((x, \mathrm{id}_x) \times (y, \mathrm{id}_y)) \times (z, \mathrm{id}_z)$ and $(x, \mathrm{id}_x) \times ((y, \mathrm{id}_y) \times (z, \mathrm{id}_z))$ are both Cartesian products of x, y, and z.

Similarly, in the case of disjoint union, we can define the binary operation '$\sqcup$' between disjoint unions. Again we identify a single set x with the disjoint union (x, id_x) of x itself and we have $(x, \mathrm{id}_x) \sqcup (y, \mathrm{id}_y) = (x \sqcup y, \iota_1, \iota_2)$. In this way, we can state associativity properly as follows:

> $(*)$ Given x, y and z, the ordered quadruples $((x, \mathrm{id}_x) \sqcup (y, \mathrm{id}_y)) \sqcup (z, \mathrm{id}_z)$ and $(x, \mathrm{id}_x) \sqcup ((y, \mathrm{id}_y) \sqcup (z, \mathrm{id}_z))$ are both disjoint unions of x, y and z.

5.4.2 Commutativity

Commutativity of Cartesian product can be simply stated as follows:

> $(\star\star)$ Given x and y, the triple $(x, \mathrm{id}_x) \times (y, \mathrm{id}_y)$ is a Cartesian product of x and y and the triple $(y, \mathrm{id}_y) \times (x, \mathrm{id}_x)$ is a Cartesian product of y and x.

The only subtlety here concerns the order of the two sets, that is not the same in the two cases. However, changing the order is not a difficult task, as the following exercise shows.

Exercise 5.22. For every permutation σ of n elements, prove that $(x, p_1, \ldots, p_n)$ is a Cartesian product of $x_1, \ldots, x_n$ if and only if $(x, p_{\sigma(1)}, \ldots, p_{\sigma(n)})$ is a Cartesian product of $x_{\sigma(1)}, \ldots, x_{\sigma(n)}$. Deduce that, for every permutation σ of n elements, if $(x, p_1, \ldots, p_n)$ is a Cartesian product of $x_1, \ldots, x_n$ and $(y, q_1, \ldots, q_n)$ is a Cartesian product of $x_{\sigma(1)}, \ldots, x_{\sigma(n)}$, then there exists a unique bijection $\phi\colon x \xrightarrow{\approx} y$ such that $q_i \circ \phi = p_{\sigma(i)}$ for every $i \in \{1, \ldots, n\}$.

As a particular case of the previous exercise, considering the canonical Cartesian products $(x \times y, \pi_1, \pi_2)$ and $(y \times x, \pi_1', \pi_2')$, we have that $\phi\colon x \times y \to y \times x$, $(a, b) \mapsto (b, a)$, is the unique bijection such that $\pi_2' \circ \phi = \pi_1$ and $\pi_1' \circ \phi = \pi_2$—that is, it is the unique bijection that commutes with the projections coherently with the inversion of the order. It follows that $(\star\star)$ is a correct statement of commutativity up to a unique bijection. The same argument holds about disjoint union, hence we can formulate commutativity as follows:

> $(**)$ Given x and y, the triple $(x, \mathrm{id}_x) \sqcup (y, \mathrm{id}_y)$ is a disjoint union of x and y and the triple $(y, \mathrm{id}_y) \sqcup (x, \mathrm{id}_x)$ is a disjoint union of y and x.

5.4.3 Distributivity

It is natural to think that Cartesian product is distributive with respect to disjoint union, because of the natural bijection $\Theta_{x,y,z}\colon (x \sqcup y) \times z \to (x \times z) \sqcup (y \times z)$, $((a, \emptyset), c) \mapsto ((a, c), \emptyset)$, $((b, \{\emptyset\}), c) \mapsto ((b, c), \{\emptyset\})$. We can state this result intrinsically as follows:

- We have the natural embeddings $\iota_1\colon x \hookrightarrow x \sqcup y$ and $\iota_2\colon y \hookrightarrow x \sqcup y$.

- We get the induced functions $\iota_1 \ddot{\times} \mathrm{id}_z : x \times z \to (x \sqcup y) \times z$ and $\iota_2 \ddot{\times} \mathrm{id}_z : y \times z \to (x \sqcup y) \times z$.
- We get the induced function $(\iota_1 \ddot{\times} \mathrm{id}_z) \dot{\sqcup} (\iota_2 \ddot{\times} \mathrm{id}_z) : (x \times z) \sqcup (y \times z) \to (x \sqcup y) \times z$.

The reader can easily verify that the last function is given by $((a, c), \emptyset) \mapsto ((a, \emptyset), c)$ and $((b, c), \{\emptyset\}) \mapsto ((b, \{\emptyset\}), c)$, hence it is a bijection, its inverse being $\Theta_{x,y,z}$. This construction can be easily generalised to get the bijection $\Theta_{x_1,\ldots,x_n,z} : (x_1 \sqcup \ldots \sqcup x_n) \times z \to (x_1 \times z) \sqcup \ldots \sqcup (x_n \times z)$ for every $x_1, \ldots, x_n, z$.[3]

5.5 SETS AND INDEXED SETS

We can characterise Cartesian product and disjoint union of (indexed) sets through the corresponding universal property as well.

5.5.1 Cartesian Product

Given a set x, the pair $(\prod x, \{\pi_y : y \in x\})$ is the canonical representative of the Cartesian product of x, defined through the following universal property.[4]

Definition 5.23. A *Cartesian product* of the set x is a pair $(c, \{p_y : y \in x\})$ such that:

- c is a set;
- $p_y : c \to y$ is a function for each $y \in x$, called *projection*;
- for any set z and any set of functions $\{(f_y : z \to y) : y \in x\}$, there exists a unique function $f : z \to c$ satisfying the condition $p_y \circ f = f_y$ for every $y \in x$. ♢

Proposition 5.24. *Let us suppose that* $(c, \{p_y : y \in x\})$ *and* $(c', \{p'_y : y \in x\})$ *are two Cartesian products of* x. *Then there exists a* unique *bijection* $\phi : c \xrightarrow{\approx} c'$ *such that* $p'_y \circ \phi = p_y$ *for every* $y \in x$.

Similarly, given an indexed set $x = \{x_i\}_{i \in I}$, the pair $(\prod_i x_i, \{\pi_i\}_{i \in I})$ is the canonical representative of the corresponding Cartesian product, defined through the following universal property.[5]

[3]This formulation of distributivity is intrinsic, since it follows from universal properties, but it relies on a certain induced function, that is not part of the operations themselves. Hence, it would be more satisfactory to state distributivity as we sketched for associativity and commutativity, but we do not need to elaborate the details here.

[4]The pair $(\prod x, \{\pi_y : y \in x\})$ satisfies the universal property 5.23 even if we do not assume the axiom of choice. In fact, let us suppose that $\emptyset \notin x$ and $\prod x = \emptyset$. Then, for every $z \neq \emptyset$, there exists no set of functions of the form $\{(f_y : z \to y) : y \in x\}$, otherwise, fixing any $w \in z$, we would get the element $f \in \prod x$, $y \mapsto f_y(w)$. If we choose $z = \emptyset$, then then universal property is trivially satisfied, hence it holds in any case.

[5]Again, the pair $(\prod_i x_i, \{\pi_i\}_{i \in I})$ satisfies the universal property 5.25 even if we do not assume the axiom of choice.

Definition 5.25. A *Cartesian product* of the indexed set $x = \{x_i\}_{i \in I}$ is a pair $(c, \{p_i\}_{i \in I})$ such that:

- c is a set;
- $p_i \colon c \to x_i$ is a function for each $i \in I$, called *projection*;
- for any set z and any indexed set of functions $\{f_i \colon z \to x_i\}_{i \in I}$, there exists a unique function $f \colon z \to c$ satisfying the condition $p_i \circ f = f_i$ for every $i \in I$. ◊

Proposition 5.26. *Let us suppose that $(c, \{p_i\}_{i \in I})$ and $(c', \{p'_i\}_{i \in I})$ are two Cartesian products of $x = \{x_i\}_{i \in I}$. Then there exists a* unique *bijection $\phi \colon c \xrightarrow{\approx} c'$ such that $p'_i \circ \phi = p_i$ for every $i \in I$.*

Exercise 5.27. Prove propositions 5.24 and 5.26.

Exercise 5.28. Prove that, if $c \neq \emptyset$, then the projections p_y in definition 5.23 and the projections p_i in definition 5.25 are necessarily surjective.

Exercise 5.29. Let $(c, \{p_y : y \in x\})$ and $(c', \{p'_y : y \in x\})$ be two Cartesian products of x, and let $\phi \colon c \xrightarrow{\approx} c'$ be the bijection induced by the universal property. Moreover, given a set z and a family of functions $\{(f_y \colon z \to y) : y \in x\}$, we call $f \colon z \to c$ and $f' \colon z \to c'$ the functions induced by the universal property. Prove that $f' = \phi \circ f$. Prove the analogous statement about indexed sets.

In the universal property 5.25 we considered a fixed domain z for each f_i, but we can also consider an indexed set over the same set I on the side of the domain, as the following proposition shows.

Proposition 5.30. *Let us suppose that $(c, \{p_i\}_{i \in I})$ and $(e, \{q_i\}_{i \in I})$ are Cartesian products of $x = \{x_i\}_{i \in I}$ and $z = \{z_i\}_{i \in I}$ respectively. Then, given an indexed set of functions $\{f_i \colon z_i \to x_i\}_{i \in I}$, there exists a unique function $g \colon e \to c$ such that $p_i \circ g = f_i \circ q_i$ for every $i \in I$.*

Exercise 5.31. Prove proposition 5.30.

5.5.2 Disjoint Union

Given a set x, the pair $(\bigsqcup x, \{\iota_y : y \in x\})$ is the canonical representative of the disjoint union of x, defined through the following universal property.

Definition 5.32. A *disjoint union* of a set x is a pair $(d, \{j_y : y \in x\})$ such that:

- z is a set;
- $j_y \colon y \to d$ is a function for each $y \in x$, called *embedding*;
- for any set z and any set of functions $\{(f_y \colon y \to z) : y \in x\}$, there exists a unique function $f \colon d \to z$ satisfying the condition $f \circ j_y = f_y$ for every $y \in x$. ◊

Proposition 5.33. *Let us suppose that* $(d, \{j_y : y \in x\})$ *and* $(d', \{j'_y : y \in x\})$ *are two disjoint unions of* x. *Then there exists a* unique *bijection* $\phi\colon d \xrightarrow{\approx} d'$ *such that* $\phi \circ j_y = j'_y$ *for every* $y \in x$.

Similarly, given an indexed set $x = \{x_i\}_{i \in I}$, the pair $(\bigsqcup_i x_i, \{\iota_i\}_{i \in I})$ is the canonical representative of the corresponding disjoint union, defined through the following universal property.

Definition 5.34. A *disjoint union* of an indexed set $\{x_i\}_{i \in I}$ is a pair $(d, \{j_i\}_{i \in I})$ such that:

- z is a set;
- $j_i\colon x_i \to d$ is a function for each $i \in I$, called *embedding*;
- for any set z and any indexed set of functions $\{f_i\colon x_i \to z\}_{i \in I}$, there exists a unique function $f\colon d \to z$ satisfying the condition $f \circ j_i = f_i$ for every $i \in I$. ♢

Proposition 5.35. *Let us suppose that* $(d, \{j_i\}_{i \in I})$ *and* $(d', \{j'_i\}_{i \in I})$ *are two disjoint unions of* $\{x_i\}_{i \in I}$. *Then there exists a* unique *bijection* $\phi\colon d \xrightarrow{\approx} d'$ *such that* $\phi \circ j_i = j'_i$ *for every* $i \in I$.

Exercise 5.36. Prove propositions 5.33 and 5.35.

Exercise 5.37. Prove that the embeddings j_y in definition 5.32 and the embeddings j_i in definition 5.34 are necessarily injective.

Exercise 5.38. Let $(d, \{j_y : y \in x\})$ and $(d', \{j'_y : y \in x\})$ be two disjoint unions of x, and let $\phi\colon x \xrightarrow{\approx} x'$ be the bijection induced by the universal property. Moreover, given a set z and a family of functions $\{(f_y\colon y \to z) : y \in x\}$, we call $f\colon d \to z$ and $f'\colon d' \to z$ the functions induced by the universal property. Prove that $f = f' \circ \phi$. Prove the analogous statement about indexed sets.

In the universal property 5.34 we considered a fixed domain z for each f_i, but we can also consider an indexed set over the same set I on the side of the domain, as the following proposition shows.

Proposition 5.39. *Let us suppose that* $(d, \{j_i\}_{i \in I})$ *and* $(h, \{k_i\}_{i \in I})$ *are disjoint unions of* $x = \{x_i\}_{i \in I}$ *and* $z = \{z_i\}_{i \in I}$ *respectively. Then, given an indexed set of functions* $\{f_i\colon x_i \to z_i\}_{i \in I}$, *there exists a unique function* $g\colon d \to h$ *such that* $g \circ j_i = k_i \circ f_i$ *for every* $i \in I$.

Exercise 5.40. Prove proposition 5.39.

Given $x \neq \emptyset$, we have the canonical inclusion $j_y\colon \bigcap x \hookrightarrow y$ for every $y \in x$. Applying the universal property of Cartesian product, the induced function $\prod_{y \in x} j_y$ is (4.2). Dually, given a set x, we have the canonical inclusions $\kappa_y\colon y \hookrightarrow \dot{\bigcup} x$. Applying the universal property of disjoint union, the induced function $\bigsqcup_{y \in x} \kappa_y$ is (4.1). Similarly, in the case of indexed sets, we get (4.4) and (4.3).

5.5.3 Properties of the Operations

Up to now we introduced various operations among sets, that satisfy some important properties, like associativity, commutativity, and many others. Nevertheless, the exposition of these properties was spread along the text and the exercises, therefore we have not listed them in a unitary. We fill this gap in Appendix B, also adding some statements that did not fit in the previous chapters.

5.6 EXERCISES

1. (i) Prove that the Cartesian products of a single set x are of the form (y, ϕ), where $\phi \colon y \xrightarrow{\approx} x$ is a bijection. (ii) Prove that the disjoint unions of a single set x are of the form (y, ϕ), where $\phi \colon x \xrightarrow{\approx} y$ is a bijection.

2. Prove that, given $x_1, \ldots, x_k$, such that x_i contains n_i elements for every $i \in \{1, \ldots, k\}$, the set $x_1 \times \cdots \times x_k$ contains $n_1 \cdots n_k$ elements and the set $x_1 \sqcup \ldots \sqcup x_k$ contains $n_1 + \cdots + n_k$ elements.

3. Prove that, if at least one of the sets $x_1, \ldots, x_n$ is empty, the universal property of Cartesian product keeps on holding without variations, $(\emptyset, \emptyset, \ldots, \emptyset)$ being the unique Cartesian product in this case.

4. Prove that the disjoint unions of $x_1, \ldots, x_n, \emptyset, \ldots, \emptyset$ are of the form $(x, j_1, \ldots, j_n, \emptyset, \ldots, \emptyset)$, where $(x, j_1, \ldots, j_n)$ is a disjoint union of $x_1, \ldots, x_n$.

5. Given n functions $f_1 \colon x_1 \to y_1, \ldots, f_n \colon x_n \to y_n$, since a (pre-)function is in particular a set, we get the sets $f_1 \times \cdots \times f_n$ and $f_1 \sqcup \ldots \sqcup f_n$.

1. Given $f_1 \colon y \to x_1, \ldots, f_n \colon y \to x_n$, which is the relation between the sets $f_1 \times \cdots \times f_n$ and $f_1 \dot{\times} \cdots \dot{\times} f_n$?
2. Given $f_1 \colon y_1 \to x_1, \ldots, f_n \colon y_n \to x_n$, which is the relation between the sets $f_1 \times \cdots \times f_n$ and $f_1 \ddot{\times} \cdots \ddot{\times} f_n$?
3. Given $f_1 \colon x_1 \to y, \ldots, f_n \colon x_n \to y$, which is the relation between the sets $f_1 \sqcup \ldots \sqcup f_n$ and $f_1 \dot{\sqcup} \ldots \dot{\sqcup} f_n$?
4. Given $f_1 \colon x_1 \to y_1, \ldots, f_n \colon x_n \to y_n$, which is the relation between the sets $f_1 \sqcup \ldots \sqcup f_n$ and $f_1 \ddot{\sqcup} \ldots \ddot{\sqcup} f_n$?

CHAPTER 6

Sets and Classes

We know that, in general, we cannot consider $x := \{y : \varphi(y)\}$ a set in our universe, even if φ is a wf (see section 2.3). Nevertheless, it is not forbidden to think of all the sets y that satisfy $\varphi(y)$, obtaining a meta-theoretical collection of sets. Since such a collection is not necessarily a set, we call it *class*. It follows that, for any set-theoretical property φ, the class $\mathsf{c} := \{y : \varphi(y)\}$ is well-defined in the meta-theory, but it can be a set or not. If it is not, we say that it is a *proper class*. Intuitively, a proper class is "too big" to form a set. For example, the class of all sets $\mathsf{u} := \{x : x = x\}$ is proper and it coincides with the whole universe. On the contrary, $\emptyset$ and $\{\emptyset\}$ are sets, hence classes too, but not proper classes. It is important to remark that a set can belong to a fixed class or not, but a class *cannot* belong to another class, since a class only contains sets.

This brief chapter is dedicated to the notion of (proper) class and to admissible operations among classes.

6.1 DEFINITIONS AND OPERATIONS AMONG CLASSES

In the meta-theory, we can extend to classes most of the definitions and operations we introduced up to now among sets.

6.1.1 Inclusion and Equality

Given two classes c and d, we say that $\mathsf{c} \subset \mathsf{d}$ if and only if every element of c belongs to d too. In this case, c is a *subclass* of d. Hence, if $\mathsf{c} = \{x : \varphi(x)\}$ and $\mathsf{d} = \{x : \psi(x)\}$, then $\mathsf{c} \subset \mathsf{d}$ if and only if $\forall x\big(\varphi(x) \Rightarrow \psi(x)\big)$. It follows that $\mathsf{c} = \mathsf{d}$ if and only if $\mathsf{c} \subset \mathsf{d}$ and $\mathsf{d} \subset \mathsf{c}$, that is equivalent to $\forall x\big(\varphi(x) \Leftrightarrow \psi(x)\big)$.

6.1.2 Unary Operations

When $\mathsf{c} \subset \mathsf{d}$, where d is a fixed class, the *complement* of c in d is formed by the elements of d that do not belong to c. Hence, if $\mathsf{c} = \{x : \varphi(x)\}$ and

DOI: 10.1201/9781003565291-6

$\mathsf{d} = \{x : \psi(x)\}$, by hypothesis $\forall x(\varphi(x) \Rightarrow \psi(x))$ and by definition $\mathsf{c}^{\complement} = \{x : \psi(x) \wedge \sim \varphi(x)\}$. As a particular case, the bigger class d can be the whole universe u. In this way, we get an *absolute* definition of complement (that was not possible among sets): given a class c, the class $\mathsf{c}^{\complement}$ is formed by all of the sets that do not belong to c. Equivalently, if $\mathsf{c} = \{x : \varphi(x)\}$, then $\mathsf{c}^{\complement} = \{x : \sim\varphi(x)\}$.

Moreover, given a class c, we define the *power class* $\mathcal{P}(\mathsf{c}) := \{x : x \subset \mathsf{c}\}$. It is important to remark that $\mathcal{P}(\mathsf{c})$ only contains the subclasses of c *that are sets*, since a proper class cannot belong to another class.

6.1.3 n-Ary Operations

Given a finite family of classes $\mathsf{c}_1, \dots, \mathsf{c}_n$, we can apply the operations defined in section 2.6. In particular, we have the *union* $\mathsf{c}_1 \cup \dots \cup \mathsf{c}_n$, the *intersection* $\mathsf{c}_1 \cap \dots \cap \mathsf{c}_n$, the *Cartesian productintersection* $\mathsf{c}_1 \times \cdots \times \mathsf{c}_n$ and the *disjoint union* $\mathsf{c}_1 \sqcup \dots \sqcup \mathsf{c}_n$. If $n = 2$, we have the *class difference* $\mathsf{c}_1 \setminus \mathsf{c}_2$ and the *symmetric difference* $\mathsf{c}_1 \,\Delta\, \mathsf{c}_2$. It follows that, if $\mathsf{c}_i = \{x : \varphi_i(x)\}$ for every $i \in \{1, \dots, n\}$, then:

- $\mathsf{c}_1 \cup \dots \cup \mathsf{c}_n = \{x : \varphi_1(x) \vee \dots \vee \varphi_n(x)\}$
- $\mathsf{c}_1 \cap \dots \cap \mathsf{c}_n = \{x : \varphi_1(x) \wedge \dots \wedge \varphi_n(x)\}$
- $\mathsf{c}_1 \times \cdots \times \mathsf{c}_n = \{(x_1, \dots, x_n) : \varphi_1(x_1) \wedge \dots \wedge \varphi_n(x_n)\}$
- $\mathsf{c}_1 \sqcup \dots \sqcup \mathsf{c}_n = (\mathsf{c}_1 \times \{\emptyset\}) \cup \dots \cup (\mathsf{c}_n \times \{\cdots\{\emptyset\}\cdots\})$
- $\mathsf{c}_1 \setminus \mathsf{c}_2 = \{x : \varphi_1(x) \wedge \sim\varphi_2(x)\}$
- $\mathsf{c}_1 \,\Delta\, \mathsf{c}_2 = (\mathsf{c}_1 \cup \mathsf{c}_2) \setminus (\mathsf{c}_1 \cap \mathsf{c}_2) = (\mathsf{c}_1 \setminus \mathsf{c}_2) \cup (\mathsf{c}_2 \setminus \mathsf{c}_1)$

6.1.4 Relations and Functions

Given two classes c and d, we define a *binary pre-relation* between them as a subclass $\rho \subset \mathsf{c} \times \mathsf{d}$. If $\mathsf{c} = \mathsf{d}$, we say that ρ is *internal* to c. The only difference with respect to sets is that we cannot define the notion of relation, since, if at least one of the two classes c and d is proper, then we cannot form the triple $(\mathsf{c}, \mathsf{d}, \rho)$. Nevertheless, by abuse of language, we will frequently use the word "relation" to indicate a pre-relation. The domain and the image are defined as among sets, but they are classes. More generally, fixing $\mathsf{c}_1, \dots, \mathsf{c}_n$ for any $n \in \mathrm{N}^*$, an *n-ary pre-relation* between them is a subclass $\rho \subset \mathsf{c}_1 \times \cdots \times \mathsf{c}_n$. As a particular case of binary pre-relation, we define a *pre-function* $f \colon \mathsf{c} \to \mathsf{d}$, that can be injective, surjective, or bijective, as between sets. Moreover, we have the *identity pre-function* $\mathrm{id}_{\mathsf{c}} \colon \mathsf{c} \to \mathsf{c}$ and, when f is bijective, the *inverse pre-function* $f^{-1} \colon \mathsf{d} \to \mathsf{c}$. Also in this case, we will frequently use the word "function" to indicate a pre-function.

We observe that the class $\mathcal{P}(\mathsf{c}_1 \times \cdots \times \mathsf{c}_n)$ contains the n-ary pre-relations between $\mathsf{c}_1, \dots, \mathsf{c}_n$ *that are sets*, therefore not all of them. About the possibility of constructing the class Func($\mathtt{c}, \mathtt{d}$), see exercise 7 at the end of the chapter.

A binary pre-relation, internal to a fixed class $\mathtt{c}$, can be *reflexive*, *anti-reflexive*, *symmetric*, *anti-symmetric*, or *transitive*. In particular, it can be an *equivalence relation* or a *(strict) (total) order relation*. In the former case, the *quotient class* and the corresponding projection are defined only provided that each equivalence class is a set.[1] Considering the equivalence relations in $\mathtt{c}$ with this property, we cannot state the bijection $\Theta_{\mathtt{c}}$ in proposition 3.22 as a function, since the domain and the codomain would contain classes, but we can affirm in the meta-theoretical language that each equivalence relation corresponds to a unique partition and conversely. A similar consideration holds about proposition 3.28.

6.1.5 Indexed Classes

We can consider an indexed class $\mathtt{x} = \{x_i\}_{i \in \mathtt{I}}$, where $\mathtt{I}$ and $\mathtt{x}$ are classes (thus, their elements are sets). This is equivalent to consider a surjective function between classes $\varphi \colon \mathtt{I} \to \mathtt{x}$. For a fixed (indexed) class, the operations of *union*, *intersection*, *separation*, and *disjoint union* are defined, with the same properties discussed in sections 4.1–4.3. In the case of intersection, the empty family is allowed too, since $\bigcap \emptyset = \mathtt{u}$, where $\mathtt{u}$ is the class of all sets. On the contrary, we *cannot* define Cartesian product. In fact, if $\mathtt{c}$ is a proper class such that $\emptyset \notin \mathtt{c}$, then each element $f \in \prod \mathtt{c}$ would be a proper class too (see exercise 8 in section 6.5).

6.2 STATING COMPREHENSION WITH CLASSES

The comprehension hypothesis states that, if a is a set and φ is a set-theoretical property, then $x := \{y \in a : \varphi(y)\}$ is a set too. This means that, if we consider a class $\mathtt{c} := \{y : \varphi(y)\}$ and a set a, then the class $x := \mathtt{c} \cap a$ is a set. Hence, we can state the comprehension hypothesis in the following equivalent form:

(A4') *The intersection between a class and a set is a set.*

Equivalently:

(A4") *Every subclass of a set is a set.*

The intuitive idea is that a proper class is "too big" to be a set, hence sets are "sufficiently small" classes. Of course a class, that is smaller or equal to a small one, is small too.

Exercise 6.1. Show that (A4') and (A4") are equivalent.

Remark 6.2. It follows from (A4") that, *if a class* $\mathtt{c}$ *is not empty, then* $\bigcap \mathtt{c}$ *is a set.* In fact, fixing any $x \in \mathtt{c}$, we have $\left(\bigcap \mathtt{c}\right) \subset x$. ◇

[1] The expression "equivalence *class*", used in the framework of sets as well (as we did in section 3.3), might be confusing, but it is so common in literature that we cannot avoid it.

6.3 REPLACEMENT HYPOTHESIS

We have seen that, informally speaking, a proper class is "too big" to form a set. Therefore, any class that is smaller or equal to a set is reasonably a set. We applied this principle to subclasses of sets in (A4"), but we can also apply it as follows. Let us consider a *surjective* function between classes $f\colon \mathtt{c} \to \mathtt{d}$. Because of surjectivity, it is natural to think that the codomain $\mathtt{d}$ is smaller or equal to the domain $\mathtt{c}$, since we can associate to every element $y \in \mathtt{d}$ at least one element $x \in \mathtt{c}$ such that $f(x) = y$.[2] Therefore, we make the following assumption:

(A6) *Given a surjective function $f\colon \mathtt{c} \to \mathtt{d}$ between classes, if $\mathtt{c}$ is a set, then $\mathtt{d}$ is a set too.*

Equivalently, we can formulate this assumption as follows:

(A6') *Given a function $f\colon x \to \mathtt{u}$, where x is a set and $\mathtt{u}$ is the universe, the image of f is a set.*

Exercise 6.3. Show that (A6) and (A6') are equivalent.

We call (A6) or (A6') *replacement hypothesis*. It will be applied several times from the next chapter on.[3]

6.4 WHAT IS A CLASS?

Up to now we thought of a class as a meta-theoretical collection of objects, that is not an object of the universe of set theory. This is a possible interpretation, but one can also think of the language of classes as an alternative formulation of the language of set-theoretical properties—that is, one can identify the class $\mathtt{c} := \{x : \varphi(x)\}$ with the wf φ itself. With this approach, the statement $x \in \mathtt{c}$ is just an equivalent formulation of $\varphi(x)$ and the operations between classes are equivalent ways to construct wfs from other ones. For example, given two classes $\mathtt{c} := \{x : \varphi(x)\}$ and $\mathtt{d} := \{x : \psi(x)\}$, the statement $x \in \mathtt{c} \cap \mathtt{d}$ is just an alternative way to express $\varphi(x) \wedge \psi(x)$. Similarly, the statement $x \in \mathtt{c} \times \mathtt{d}$ is an alternative way to state $\exists y, z\big(\varphi(y) \wedge \psi(z) \wedge x = (y, z)\big)$. The inclusion $\mathtt{c} \subset \mathtt{d}$, where $\mathtt{c} = \{x : \varphi(x)\}$ and $\mathtt{d} := \{x : \psi(x)\}$, is equivalent to $\forall x\big(\varphi(x) \Rightarrow \psi(x)\big)$ and so on.

One is free to choose her or his preferred approach, but we think that interpreting classes as meta-theoretical collections of objects is a clearer and more efficient point of view, since it is easier to visualise and to handle. The fact that the collection itself is not a member of the universe does not constitute a problem for our intuition, but only a fact to remember when performing formal constructions and proofs.

[2]Here we ignore the axiom of choice, since it is only an intuitive argument.

[3]In the axiomatic framework, (A6) is used to show that binary Cartesian product is well-defined in the universe without assuming this fact explicitly, as we did in (A3). Nevertheless, this is not particularly relevant for us.

Moreover, in principle it is also possible to enrich the universe with classes themselves, thinking of a proper class as a higher-level object with respect to sets. In this case, we could deal with 2-classes (also called *hyperclasses*) as meta-theoretical collections of classes. We could keep on in this way, including 2-classes as higher-level objects in the universe and dealing with meta-theoretical 3-classes. Actually, one can also include n-classes for every $n \in \mathrm{N}$ in the universe, a 0-class being a set and a 1-class being a class, imposing that a k-class can belong to an n-class if and only if $k < n$ or $k = n = 0$. Nevertheless, this is not the approach we are considering here, therefore we ignore it and we keep on working in the universe of sets.

6.5 EXERCISES

1. Let $\mathtt{u} := \{x : x = x\}$ be the class that coincides with the whole universe. Prove that:

1. $\mathcal{P}(\mathtt{u}) = \mathtt{u}$
2. $\left(\bigcup \mathtt{u}\right) = \mathtt{u}$
3. $\left(\bigcap \mathtt{u}\right) = \emptyset$
4. $\mathtt{u} \times \mathtt{u} \subsetneq \mathtt{u}$
5. $\mathtt{u} \sqcup \mathtt{u} \subsetneq \mathtt{u}$

2. Let $\mathtt{s}$ be the class of singletons, that is:

$$\mathtt{s} := \Big\{x : \exists y\big((y \in x) \wedge \forall z((z \in x) \Rightarrow (z = y))\big)\Big\}$$

Prove that $\mathtt{s}$ is a proper class.

3. Prove that, if $f\colon \mathtt{c} \hookrightarrow x$ is an injective function between classes and x is a set, then $\mathtt{c}$ is a set too.

4. Let $\mathtt{c}$ be a class. Prove that:

1. If $\mathcal{P}(\mathtt{c})$ is a set, then $\mathtt{c}$ is a set.
2. If $\bigcup \mathtt{c}$ is a set, then $\mathtt{c}$ is a set.
3. If $\hat{\mathtt{c}}$ is a set, then $\mathtt{c}$ is a set.
4. If $\bigsqcup \mathtt{c}$ is a set, then $\mathtt{c}$ is a set.

5. Let $\mathtt{c}_1, \ldots, \mathtt{c}_n$ be classes. Prove that:

1. If $\mathtt{c}_1 \times \cdots \times \mathtt{c}_n$ is a set and $\mathtt{c}_1, \ldots, \mathtt{c}_n \neq \emptyset$, then $\mathtt{c}_1, \ldots, \mathtt{c}_n$ are sets.
2. If $\mathtt{c}_1 \sqcup \ldots \sqcup \mathtt{c}_n$ is a set, then $\mathtt{c}_1, \ldots, \mathtt{c}_n$ are sets.
3. If $\mathtt{c}_1 \cup \ldots \cup \mathtt{c}_n$ is a set, then $\mathtt{c}_1, \ldots, \mathtt{c}_n$ are sets.

6. We recall the absolute definition of complement: if $\mathtt{c} = \{x : \varphi(x)\}$, then $\mathtt{c}^{\mathtt{C}} := \{x : \sim\varphi(x)\}$.

- Prove that, if $\mathtt{c}$ is a set, then $\mathtt{c}^{\mathtt{C}}$ is a proper class.
- Show with an example that it is possible that both $\mathtt{c}$ and $\mathtt{c}^{\mathtt{C}}$ are proper classes.

7. (i) Prove that, if $f\colon \mathtt{c} \to \mathtt{d}$ is a pre-function between classes, then f is set if and only if $\mathtt{c}$ is a set (hence, f is a proper class if and only if $\mathtt{c}$ is a proper class).

(ii) Deduce that, if x is a set and $\mathtt{d}$ is a class, then the class $\mathrm{Func}(x, \mathtt{d})$, defined as the class of pre-functions from x to $\mathtt{d}$, is well-defined. On the contrary, if $\mathtt{c}$ is a proper class, then $\mathrm{Func}(\mathtt{c}, \mathtt{d})$ is not a class (it is a hyper-class), unless $\mathtt{d}$ is empty.

(iii) Prove that, if x is a non-empty set and $\mathtt{d}$ is a class, then $\mathrm{Func}(x, \mathtt{d})$ is a set if and only if $\mathtt{d}$ is a set.

8. Show that, if $\mathtt{c}$ is a proper class such that $\emptyset \notin \mathtt{c}$, then each element $f \in \prod \mathtt{c}$ is a proper class too. Therefore, $\prod \mathtt{c}$ cannot be defined as a class (it is a hyper-class).

9. Given two sets x and y and a class $\mathtt{d}$, construct a (natural) bijection $P\colon \mathrm{Func}(x \times y, \mathtt{d}) \xrightarrow{\approx} \mathrm{Func}(x, \mathrm{Func}(y, \mathtt{d}))$.

10. Prove that, by applying assumptions `(A1)`–`(A6)`, we can construct all of the finite sets described in section 2.4 and only them (see exercise 14 in section 2.8 and the corresponding remark).

CHAPTER 7

Embedding Natural Numbers in Set Theory

We are going to *define* natural numbers as sets in the universe, but we will see that this definition does not really replace the corresponding meta-theoretical notion. However, the set-theoretical construction allows to use natural numbers in ordinary mathematics within the present foundational setting. Moreover, it will be the starting point to construct integral, rational, real, and complex numbers with set-theoretical tools.

7.1 FIRST ATTEMPT

Notation 7.1. We denote by $\mathbf{0}, \mathbf{1}, \mathbf{2}, \ldots$ the set-theoretical natural numbers we are going to construct, while we keep on denoting by $0, 1, 2, \ldots$ the meta-theoretical ones. Coherently, we call '$\mathbb{N}$' the set of natural numbers within the universe, while 'N' is the set formed by the meta-theoretical ones. Moreover, we denote by '$\boldsymbol{s}$' the successor operation in $\mathbb{N}$, while 's' is the one in N.

The basic idea we start from is the following one: we set $\mathbf{0} := \emptyset$, $\mathbf{1} := \{\mathbf{0}\}$, $\mathbf{2} := \{\mathbf{0}, \mathbf{1}\}$, $\mathbf{3} := \{\mathbf{0}, \mathbf{1}, \mathbf{2}\}$, ..., defining recursively $\boldsymbol{s}(\boldsymbol{n}) := \{\mathbf{0}, \ldots, \boldsymbol{n}\}$. Explicitly:

$$\mathbf{0} = \emptyset \qquad \mathbf{1} = \{\emptyset\} \qquad \mathbf{2} = \{\emptyset, \{\emptyset\}\} \qquad \mathbf{3} = \{\emptyset, \{\emptyset\}, \{\emptyset, \{\emptyset\}\}\} \qquad \cdots$$

In this way, the number $\boldsymbol{n} \in \mathbb{N}$ is a finite (hereditary) set of n elements. We can think of it as the canonical representative of the class formed by the sets of n elements.

This idea is coherent, but it cannot be realised by recursion as we did, since induction and recursion in $\mathbb{N}$ will be stated and proved (not assumed intuitively) after having defined $\mathbb{N}$ itself. In principle, we could apply the meta-theoretical recursion as follows. We define the sequence of sets $f(n)$ by $f(0) := \emptyset$ and $f(s(n)) := \{f(0), \ldots, f(n)\}$. Then, we set $\boldsymbol{n} := f(n)$ and we assume that

DOI: 10.1201/9781003565291-7

the class $\mathbb{N} := \{\boldsymbol{n} : n \in \mathrm{N}\}$ is a set in our universe. This approach is coherent, but we are just creating a "copy" of the meta-theoretical numbers within the universe, that is quite useless. For this reason, in the following section we will introduce another approach, that is internal to set theory and does not rely on the meta-theoretical knowledge of natural numbers. Afterwards, we will discuss whether this approach makes such a meta-theoretical knowledge unnecessary (basically we think it does not, but this is just our opinion).

7.2 THE CONSTRUCTION

We start by defining what is the successor of any set.

Definition 7.2. Given a set x, the *successor* of x the set $\boldsymbol{s}(x) := x \cup \{x\}$. ♢

We get the function $s\colon \mathtt{u} \to \mathtt{u}$, $x \mapsto \boldsymbol{s}(x)$, where $\mathtt{u}$ is the class of all sets. We observe that $y \in \boldsymbol{s}(x)$ if and only if $y \in x$ or $y = x$.

Definition 7.3. A set x is *closed under successor* if the two following conditions hold:

- $\emptyset \in x$;
- for every $y \in x$, we have $\boldsymbol{s}(y) \in x$.

We call $\mathtt{cs}$ the class of sets that are closed under successor. ♢

It is immediate to verify that $\mathtt{cs}$ is closed by intersection (that is, the intersection of a non-empty subclass of $\mathtt{cs}$ is an element of $\mathtt{cs}$). It follows that, if $\mathtt{cs}$ is not empty, then the intersection of $\mathtt{cs}$ belongs to $\mathtt{cs}$ itself, and it is its minimum with respect to inclusion.

We aim to define $\mathbb{N}$ as the minimum of $\mathtt{cs}$, but, up to now, we cannot exclude that $\mathtt{cs}$ is empty. In fact, an element of $\mathtt{cs}$ is necessarily infinite,[1] and exercise 10 of section 6.5 shows that, if we reduce our universe to the class of finite sets constructed in section 2.4, then the assumptions made up to now remain coherent. Therefore, there is no way to prove that $\mathtt{cs}$ is not empty on the basis of (A1)–(A6). This makes the following new hypothesis necessary:

(A7) $\mathtt{cs} \neq \emptyset$.

We call this assumption *infinity hypothesis*. It immediately follows from (A7) and remark 6.2 that the intersection of $\mathtt{cs}$ is a set, hence we define

$$\mathbb{N} := \bigcap \mathtt{cs}. \tag{7.1}$$

[1] We will formalise this statement later on, but the reader can guess that it holds, since an element of $\mathtt{cs}$ has to contain $\emptyset, \boldsymbol{s}(\emptyset), \boldsymbol{s}(\boldsymbol{s}(\emptyset)), \ldots$

7.2.1 Main Properties of Natural Numbers

We set $\mathbf{0} := \emptyset$ (that is, '$\mathbf{0}$' is just another symbol to denote the empty set). Definition 7.3 trivially implies that $\mathbf{0} \in \mathbb{N}$ and $\boldsymbol{s}(\boldsymbol{n}) \in \mathbb{N}$ for every $\boldsymbol{n} \in \mathbb{N}$. We aim to show that every element of $\mathbb{N}$ is of the form $\mathbf{0}, \boldsymbol{s}(\mathbf{0}), \boldsymbol{s}(\boldsymbol{s}(\mathbf{0})), \ldots$ We start from the following *proposition* (not assumption).

Proposition 7.4 (Peano's axioms). *The following properties of the set* $\mathbb{N}$ *hold:*

(i) $\boldsymbol{s}(\boldsymbol{n}) \neq \mathbf{0}$ *for every* $\boldsymbol{n} \in \mathbb{N}$.

(ii) $\boldsymbol{s}(\boldsymbol{n}) = \boldsymbol{s}(\boldsymbol{m})$ *if and only if* $\boldsymbol{n} = \boldsymbol{m}$.

(iii) Induction Principle: *if* $A \subset \mathbb{N}$ *satisfies* $\mathbf{0} \in A$ *and* $\boldsymbol{n} \in A \Rightarrow \boldsymbol{s}(\boldsymbol{n}) \in A$ *for every* $\boldsymbol{n} \in \mathbb{N}$*, then* $A = \mathbb{N}$ *(that is,* $\boldsymbol{n} \in A$ *for every* $\boldsymbol{n} \in \mathbb{N}$*).*

Proof of (i) and (iii). (i) Since $\boldsymbol{s}(\boldsymbol{n}) = \boldsymbol{n} \cup \{\boldsymbol{n}\}$ and $\boldsymbol{n} \in \{\boldsymbol{n}\}$, it follows that $\boldsymbol{n} \in \boldsymbol{s}(\boldsymbol{n})$, hence $\boldsymbol{s}(\boldsymbol{n}) \neq \emptyset = \mathbf{0}$. (iii) Since $A \subset \mathbb{N}$ by hypothesis, the condition $\boldsymbol{n} \in A \Rightarrow \boldsymbol{s}(\boldsymbol{n}) \in A$ for every $\boldsymbol{n} \in \mathbb{N}$ is equivalent to the condition $a \in A \Rightarrow \boldsymbol{s}(a) \in A$ for every set a. This menas that A is closed under successor—that is, $A \in \mathsf{cs}$. Hence, definition (7.1) implies that $\mathbb{N} \subset A$, therefore $A = \mathbb{N}$. □

We postponed the proof of (ii), since we need to apply (iii) to get it. We remark that the induction principle (iii), stated through a subset $A \subset \mathbb{N}$, is equivalent to the one stated through a set-theoretical property φ in $\mathbb{N}$. In fact, given φ, we get the subset $A := \{\boldsymbol{n} \in \mathbb{N} : \varphi(\boldsymbol{n})\}$ and, given a subset $A \subset \mathbb{N}$, we get the property $\varphi(\boldsymbol{n}) := (\boldsymbol{n} \in A)$.[2]

Lemma 7.5. $\bigcup \boldsymbol{s}(\boldsymbol{n}) = \boldsymbol{n}$ *for every* $\boldsymbol{n} \in \mathbb{N}$.

Proof. We prove the result by induction. We have $\bigcup \boldsymbol{s}(\mathbf{0}) = \bigcup\{\emptyset\} = \emptyset = \mathbf{0}$. If $\bigcup \boldsymbol{s}(\boldsymbol{n}) = \boldsymbol{n}$, then

$$\begin{aligned}\bigcup \boldsymbol{s}(\boldsymbol{s}(\boldsymbol{n})) &= \bigcup(\boldsymbol{s}(\boldsymbol{n}) \cup \{\boldsymbol{s}(\boldsymbol{n})\}) \overset{(\star)}{=} \left(\bigcup \boldsymbol{s}(\boldsymbol{n})\right) \cup \boldsymbol{s}(\boldsymbol{n}) \\ &\overset{(\star\star)}{=} \boldsymbol{n} \cup (\boldsymbol{n} \cup \{\boldsymbol{n}\}) = \boldsymbol{n} \cup \{\boldsymbol{n}\} = \boldsymbol{s}(\boldsymbol{n}).\end{aligned}$$

The equality $(\star)$ follows from exercise 3-1 of section 4.5, while $(\star\star)$ follows from the inductive hypothesis and the definition of successor. □

*Proof of 7.4-*ii. ($\Leftarrow$) Obvious. ($\Rightarrow$) If $\boldsymbol{s}(\boldsymbol{n}) = \boldsymbol{s}(\boldsymbol{m})$, then $\bigcup \boldsymbol{s}(\boldsymbol{n}) = \bigcup \boldsymbol{s}(\boldsymbol{m})$, hence lemma 7.5 implies that $\boldsymbol{n} = \boldsymbol{m}$. □

Remark 7.6. The statement of ex. 1.4 holds in $\mathbb{N}$ too, that is:

- every $\boldsymbol{n} \in \mathbb{N}^*$ is the successor of another unique natural number
- $\boldsymbol{s}(\boldsymbol{n}) \neq \boldsymbol{n}$ for every $\boldsymbol{n} \in \mathbb{N}$

---

[2] Of course the property $\varphi(\boldsymbol{n}) := (\boldsymbol{n} \in A)$ cannot be used to *define* A as a subset of $\mathbb{N}$ through assumption `(A4)`, but, once that A has been already given, the property φ trivially characterises it.

The proof is the same, applying the induction in $\mathbb{N}$ instead of the meta-theoretical induction in N. ◇

The following lemma states two simple properties of natural numbers that will be useful below. We set $\mathbb{N}^* := \mathbb{N} \setminus \{\mathbf{0}\}$.

Lemma 7.7. *We have that:*

(i) $\mathbf{0} \in \boldsymbol{n}$ *for every* $\boldsymbol{n} \in \mathbb{N}^*$.

(ii) *If* $\boldsymbol{m} \in \boldsymbol{n}$, *then* $\boldsymbol{s}(\boldsymbol{m}) \in \boldsymbol{s}(\boldsymbol{n})$.

Proof. We prove the two items by induction. (i) The statement is equivalent to $\mathbf{0} \in \boldsymbol{s}(\boldsymbol{n})$ for every $\boldsymbol{n} \in \mathbb{N}$. Since $\boldsymbol{s}(\mathbf{0}) = \{\mathbf{0}\}$, trivially $\mathbf{0} \in \boldsymbol{s}(\mathbf{0})$. If $\mathbf{0} \in \boldsymbol{s}(\boldsymbol{n})$, then, since $\boldsymbol{s}(\boldsymbol{s}(\boldsymbol{n})) = \boldsymbol{s}(\boldsymbol{n}) \cup \{\boldsymbol{s}(\boldsymbol{n})\}$, we have $\mathbf{0} \in \boldsymbol{s}(\boldsymbol{n}) \subset \boldsymbol{s}(\boldsymbol{s}(\boldsymbol{n}))$. (ii) Since $\boldsymbol{m} \in \mathbf{0}$ is never verified, the statement trivially holds for $\boldsymbol{n} = \mathbf{0}$. Let us suppose that the statement holds for $\boldsymbol{n}$ and that $\boldsymbol{m} \in \boldsymbol{s}(\boldsymbol{n})$. If $\boldsymbol{m} \neq \boldsymbol{n}$, then, by the inductive hypothesis, $\boldsymbol{s}(\boldsymbol{m}) \in \boldsymbol{s}(\boldsymbol{n}) \subset \boldsymbol{s}(\boldsymbol{s}(\boldsymbol{n}))$. If $\boldsymbol{m} = \boldsymbol{n}$, then $\boldsymbol{s}(\boldsymbol{m}) = \boldsymbol{s}(\boldsymbol{n}) \in \boldsymbol{s}(\boldsymbol{s}(\boldsymbol{n}))$. Hence, in any case, $\boldsymbol{s}(\boldsymbol{m}) \in \boldsymbol{s}(\boldsymbol{s}(\boldsymbol{n}))$. □

7.2.2 Recursive Definitions

The validity of recursive definitions must be proved as well, not assumed. A recursive definition of $f\colon \mathbb{N} \to \mathsf{c}$, where c is a fixed class, is constructed as follows:

- We establish what is $f(\mathbf{0})$.
- Assuming to know $f(\boldsymbol{n})$, we define $f(\boldsymbol{s}(\boldsymbol{n}))$. This can be formalised as follows: we fix a function $h\colon \mathsf{c} \to \mathsf{c}$ and we set $f(\boldsymbol{s}(\boldsymbol{n})) := h(f(\boldsymbol{n}))$.

This leads to the following proposition.

Proposition 7.8. *Given*

- *a class* c;
- *a fixed element* $a \in \mathsf{c}$; *and*
- *a function* $h\colon \mathsf{c} \to \mathsf{c}$,

there exists a unique function $f\colon \mathbb{N} \to \mathsf{c}$ *such that*

I. $f(\mathbf{0}) = a$; *and*

II. $f(\boldsymbol{s}(\boldsymbol{n})) = h(f(\boldsymbol{n}))$ *for every* $\boldsymbol{n} \in \mathbb{N}$.

Proof. `Uniqueness`: Let us suppose that f and f' verify I and II and let us prove by induction that $f(\boldsymbol{n}) = f'(\boldsymbol{n})$ for every $\boldsymbol{n} \in \mathbb{N}$. If $\boldsymbol{n} = \mathbf{0}$, then $f(\mathbf{0}) = f'(\mathbf{0}) = a$ by I. If $f(\boldsymbol{n}) = f'(\boldsymbol{n})$, then $f(\boldsymbol{s}(\boldsymbol{n})) \overset{\text{II}}{=} h(f(\boldsymbol{n})) = h(f'(\boldsymbol{n})) \overset{\text{II}}{=} f'(\boldsymbol{s}(\boldsymbol{n}))$.

`Existence`: Let us show by induction that, for every $\boldsymbol{n} \in \mathbb{N}$, there exists a unique function $f_{[\boldsymbol{n}]}\colon \boldsymbol{s}(\boldsymbol{n}) \to \mathsf{c}$ such that:

$\mathrm{I}_{\boldsymbol{n}}$. $f_{[\boldsymbol{n}]}(\mathbf{0}) = a$, where $\mathbf{0} \in \boldsymbol{s}(\boldsymbol{n})$ by lemma 7.7-i;

$\mathrm{II}_{\boldsymbol{n}}$. for every $\boldsymbol{m} \in \boldsymbol{n}$:

$$f_{[\boldsymbol{n}]}(\boldsymbol{s}(\boldsymbol{m})) = h(f_{[\boldsymbol{n}]}(\boldsymbol{m})). \qquad (\#)$$

In the l.h.s. of (#), we have $\boldsymbol{s}(\boldsymbol{m}) \in \boldsymbol{s}(\boldsymbol{n})$ by lemma 7.7-ii. In the r.h.s., we have $\boldsymbol{m} \in \boldsymbol{s}(\boldsymbol{n})$, since $\boldsymbol{m} \in \boldsymbol{n} \subset \boldsymbol{s}(\boldsymbol{n})$.

For $\boldsymbol{n} = \mathbf{0}$, there exists the unique function $f_{[\mathbf{0}]} : \{\mathbf{0}\} \to \mathsf{c}$, defined by $f_{[\mathbf{0}]}(\mathbf{0}) := a$, trivially satisfying $\mathrm{I}_{\mathbf{0}}$ and $\mathrm{II}_{\mathbf{0}}$. Assuming existence and uniqueness of $f_{[\boldsymbol{n}]}$, we set:

$$f_{[\boldsymbol{s}(\boldsymbol{n})]}(\boldsymbol{m}) := \begin{cases} f_{[\boldsymbol{n}]}(\boldsymbol{m}) & \boldsymbol{m} \in \boldsymbol{s}(\boldsymbol{n}) \\ h(f_{[\boldsymbol{n}]}(\boldsymbol{n})) & \boldsymbol{m} = \boldsymbol{s}(\boldsymbol{n}). \end{cases} \qquad (*)$$

In $(*)$, we have $\boldsymbol{m} \in \boldsymbol{s}(\boldsymbol{s}(\boldsymbol{m})) = \boldsymbol{s}(\boldsymbol{m}) \cup \{\boldsymbol{s}(\boldsymbol{m})\}$, hence $\boldsymbol{m} \in \boldsymbol{s}(\boldsymbol{n})$ or $\boldsymbol{m} = \boldsymbol{s}(\boldsymbol{n})$, that are the two cases considered there. It follows from $(*)$ that $f_{[\boldsymbol{s}(\boldsymbol{n})]}(\mathbf{0}) = f_{[\boldsymbol{n}]}(\mathbf{0}) = a$, the latter equality being due to the inductive hypothesis. Moreover, for every $\boldsymbol{m} \in \boldsymbol{n}$, that implies $\boldsymbol{s}(\boldsymbol{m}) \in \boldsymbol{s}(\boldsymbol{n})$ by lemma 7.7-ii, we have

$$f_{[\boldsymbol{s}(\boldsymbol{n})]}(\boldsymbol{s}(\boldsymbol{m})) \overset{(*)}{=} f_{[\boldsymbol{n}]}(\boldsymbol{s}(\boldsymbol{m})) \overset{(\#)}{=} h(f_{[\boldsymbol{n}]}(\boldsymbol{m})) \overset{(*)}{=} h(f_{[\boldsymbol{s}(\boldsymbol{n})]}(\boldsymbol{m})).$$

When $\boldsymbol{m} = \boldsymbol{n}$, we have

$$f_{[\boldsymbol{s}(\boldsymbol{n})]}(\boldsymbol{s}(\boldsymbol{n})) \overset{(*)}{=} h(f_{[\boldsymbol{n}]}(\boldsymbol{n})) \overset{(*)}{=} h(f_{[\boldsymbol{s}(\boldsymbol{n})]}(\boldsymbol{n})).$$

Therefore, $f_{[\boldsymbol{s}(\boldsymbol{n})]}$ satisfies $\mathrm{I}_{\boldsymbol{s}(\boldsymbol{n})}$ and $\mathrm{II}_{\boldsymbol{s}(\boldsymbol{n})}$, proving existence. About uniqueness, let us suppose that $f'_{[\boldsymbol{s}(\boldsymbol{n})]} : \boldsymbol{s}(\boldsymbol{s}(\boldsymbol{n})) \to \mathsf{c}$ satisfies $\mathrm{I}_{\boldsymbol{s}(\boldsymbol{n})}$ and $\mathrm{II}_{\boldsymbol{s}(\boldsymbol{n})}$ too. It is immediate to verify that $f'_{[\boldsymbol{s}(\boldsymbol{n})]}|_{\boldsymbol{s}(\boldsymbol{n})} : \boldsymbol{s}(\boldsymbol{n}) \to \mathsf{c}$ satisfies $\mathrm{I}_{\boldsymbol{n}}$ and $\mathrm{II}_{\boldsymbol{n}}$, hence, by the inductive hypothesis, $f'_{[\boldsymbol{s}(\boldsymbol{n})]}|_{\boldsymbol{s}(\boldsymbol{n})} = f_{[\boldsymbol{n}]}$. It follows from $(*)$ that

$$f'_{[\boldsymbol{s}(\boldsymbol{n})]}(\boldsymbol{m}) = f_{[\boldsymbol{s}(\boldsymbol{n})]}(\boldsymbol{m}) \qquad \forall \boldsymbol{m} \in \boldsymbol{s}(\boldsymbol{n}). \qquad (!)$$

It remains to consider $\boldsymbol{m} = \boldsymbol{s}(\boldsymbol{n})$. We have

$$f'_{[\boldsymbol{s}(\boldsymbol{n})]}(\boldsymbol{s}(\boldsymbol{n})) \overset{(\#)}{=} h(f'_{[\boldsymbol{s}(\boldsymbol{n})]}(\boldsymbol{n})) \overset{(!)}{=} h(f_{[\boldsymbol{s}(\boldsymbol{n})]}(\boldsymbol{n})) \overset{(\#)}{=} f_{[\boldsymbol{s}(\boldsymbol{n})]}(\boldsymbol{s}(\boldsymbol{n})).$$

This proves that $f'_{[\boldsymbol{s}(\boldsymbol{n})]} = f_{[\boldsymbol{s}(\boldsymbol{n})]}$.

We have shown by induction that $f_{[\boldsymbol{n}]}$ exists unique for every $\boldsymbol{n} \in \mathbb{N}$. Now we set $f(\boldsymbol{n}) := f_{[\boldsymbol{n}]}(\boldsymbol{n})$ for every $\boldsymbol{n} \in \mathbb{N}$. It follows that $f(\mathbf{0}) = f_{[\mathbf{0}]}(\mathbf{0}) = a$ and

$$f(\boldsymbol{s}(\boldsymbol{n})) = f_{[\boldsymbol{s}(\boldsymbol{n})]}(\boldsymbol{s}(\boldsymbol{n})) \overset{(*)}{=} h(f_{[\boldsymbol{n}]}(\boldsymbol{n})) = h(f(\boldsymbol{n}))$$

as required. □

Remark 7.9. The proof of proposition 7.8 may seem circular at a first glance, since the reader may think that $f_{[\boldsymbol{n}]}$ has been defined recursively on $\boldsymbol{n}$. Actually, this is not the case. In fact, we proved by induction existence and uniqueness of $f_{[\boldsymbol{n}]}$ for every $\boldsymbol{n}$ (more precisely, we considered the subset $A \subset \mathbb{N}$, defined by $A := \{\boldsymbol{n} \in \mathbb{N} : \exists! f_{[\boldsymbol{n}]} : \mathrm{I}_{\boldsymbol{n}} \wedge \mathrm{II}_{\boldsymbol{n}}\}$, and we proved by induction that $A = \mathbb{N}$). In the inductive proof, we assumed that $\boldsymbol{n} \in A$ to construct $f_{[\boldsymbol{s}(\boldsymbol{n})]}$, without relying on recursion. Afterwards, we used the unique $f_{[\boldsymbol{n}]}$ for every $\boldsymbol{n}$ to define f, again without relying on recursion. Because of uniqueness, neither the axiom of choice has been applied with respect to the assignment $\boldsymbol{n} \mapsto f_{[\boldsymbol{n}]}$. ◇

It may happen that the recursive definition of $f(\boldsymbol{s}(\boldsymbol{n}))$ depends on $\boldsymbol{n}$ too, not only of $f(\boldsymbol{n})$. It is also possible to define recursively a function $f(z, \boldsymbol{n})$, where z ranges within a fixed class d (as in the definition of sum, product, and power, see section 7.2.3 below). In this case, the function h may depend on z too, or, more generally, on the whole function $z \mapsto f(z, \boldsymbol{n})$. All of these cases can be deduced from proposition 7.8. The reader can find the detailed statements and proofs in appendix C.

7.2.3 Operations and Order

Now all the tools used in section 1.4.1 are available in the set-theoretical framework. Therefore, we can define the same operations and relations and prove the corresponding properties within the universe, with the same techniques. For example, we define the sum $+ : \mathbb{N} \times \mathbb{N} \to \mathbb{N}$ by $\boldsymbol{n} + \boldsymbol{0} := \boldsymbol{n}$ and $\boldsymbol{n} + \boldsymbol{s}(\boldsymbol{m}) := \boldsymbol{s}(\boldsymbol{n} + \boldsymbol{m})$. The relations '$\leq$' and '$<$' are total order relations in $\mathbb{N}$, the second one being strict.

Remark: All of the exercises of section 1.4.1 can be stated and solved in the same way within the universe, replacing N by $\mathbb{N}$.

Exercise 7.10. Prove that:

(i) $\boldsymbol{n} \subset \mathbb{N}$ for every $\boldsymbol{n} \in \mathbb{N}$

(ii) $\mathbb{N} = \bigcup \mathbb{N}$

Exercise 7.11. Prove that, for every $\boldsymbol{n}, \boldsymbol{m} \in \mathbb{N}$:

(i) $\boldsymbol{m} < \boldsymbol{n}$ if and only if $\boldsymbol{m} \in \boldsymbol{n}$

(ii) $\boldsymbol{m} \leq \boldsymbol{n}$ if and only if $\boldsymbol{m} \subset \boldsymbol{n}$

Remark: It follows from exercise 7.10-i and exercise 7.11-i that $\boldsymbol{n} = \{\boldsymbol{m} \in \mathbb{N} : \boldsymbol{m} < \boldsymbol{n}\}$, that is the idea we started from in section 7.1. Hence, $\boldsymbol{s}(\boldsymbol{n}) = \{\boldsymbol{m} \in \mathbb{N} : \boldsymbol{m} \leq \boldsymbol{n}\}$. Moreover, it follows from exercise 7.11 that $(\boldsymbol{m} \in \boldsymbol{n}) \Rightarrow (\boldsymbol{m} \subset \boldsymbol{n})$.

7.2.4 Sequences

We formalise the notion of (finite) sequence of sets within the universe (cfr. section 1.4.3).

Definition 7.12. Let c be a class (in particular, it can be a set). A *sequence in* c is a function $f : \mathbb{N} \to \mathsf{c}$. A *finite sequence of length* $\boldsymbol{n}$ in c is a function $f : \boldsymbol{n} \to \mathsf{c}$, where $\boldsymbol{n} \in \mathbb{N}^*$. If $\mathsf{c} = \mathsf{u}$, where u is the class of all sets, then f is a *(finite) sequence of sets* or simply a *(finite) sequence.* ◇

If a sequence $f\colon \mathbb{N} \to x$ or $f\colon \boldsymbol{n} \to x$ is surjective, then it coincides with the indexed set $x = \{x_{\boldsymbol{n}}\}_{\boldsymbol{n}\in\mathbb{N}}$ or $x = \{x_{\boldsymbol{m}}\}_{\boldsymbol{m}\in\boldsymbol{n}}$. That's because, since $\mathbb{N}$ has been defined in our universe, the natural numbers and $\mathbb{N}$ itself are possible sets of indices.

Notation 7.13. We denote by $\mathrm{Seq}(\mathsf{c})$, $\mathrm{FSeq}_{\boldsymbol{n}}(\mathsf{c})$, and $\mathrm{FSeq}(\mathsf{c})$ the classes respectively of sequences, finite sequences of length $\boldsymbol{n}$, and finite sequences of any length in c. Moreover, we usually set $x_{\boldsymbol{n}} := f(\boldsymbol{n})$ and we denote the whole sequence by $\boldsymbol{(}x_{\boldsymbol{0}}, x_{\boldsymbol{1}}, x_{\boldsymbol{2}}, \dots\boldsymbol{)}$ or by $\boldsymbol{(}x_{\boldsymbol{0}}, \dots, x_{\boldsymbol{n-1}}\boldsymbol{)}$ (with bold parentheses). We use the notation $\mathrm{FSeq}^*(\mathsf{c})$ or $\mathrm{FSeq}^*_{\boldsymbol{n}}(\mathsf{c})$ when considering finite sequences that start from the index $\mathbf{1}$—that is, of the form $\boldsymbol{(}x_{\boldsymbol{1}}, \dots, x_{\boldsymbol{n}}\boldsymbol{)}$.

Exercise 7.14. Show that $\mathrm{Seq}(\mathsf{c})$, $\mathrm{FSeq}_{\boldsymbol{n}}(\mathsf{c})$, and $\mathrm{FSeq}(\mathsf{c})$ are sets if and only if c is a set.

We can use finite sequences to formalise n-ary operations and iterations of operations in the universe—for example, we define $\cup\colon \mathrm{FSeq}^*(\mathsf{u}) \to \mathsf{u}$, $\boldsymbol{(}x_{\boldsymbol{1}}, \dots, x_{\boldsymbol{n}}\boldsymbol{)} \mapsto x_{\boldsymbol{1}} \cup \dots \cup x_{\boldsymbol{n}} := \{y : \exists \boldsymbol{i} \in \boldsymbol{s}(\boldsymbol{n})^* \, (y \in x_{\boldsymbol{i}})\}$; similarly, the recursive definition $\bigcup^{\mathbf{0}} x := x$ and $\bigcup^{\boldsymbol{s}(\boldsymbol{n})} x := \bigcup(\bigcup^{\boldsymbol{n}} x)$ can be performed in $\mathbb{N}$ instead of N. For every $\boldsymbol{n} \in \mathbb{N}$, we have the canonical bijection

$$\begin{aligned} \Psi_{\boldsymbol{n},\mathsf{c}}\colon \mathsf{c} \times \overset{\boldsymbol{n}\text{ times}}{\cdots} \times \mathsf{c} &\xrightarrow{\approx} \mathrm{FSeq}^*_{\boldsymbol{n}}(\mathsf{c}) \\ (y_{\boldsymbol{1}}, \dots, y_{\boldsymbol{n}}) &\mapsto \boldsymbol{(}y_{\boldsymbol{1}}, \dots, y_{\boldsymbol{n}}\boldsymbol{)}, \end{aligned} \tag{7.2}$$

showing that a finite sequence of length $\boldsymbol{n}$ can be essentially identified with an element of $\mathsf{c} \times \dots \times \mathsf{c}$.

Exercise 7.15. Formalise the recursive definition of the ordered $\boldsymbol{n}$-uple $(y_{\boldsymbol{1}}, \dots, y_{\boldsymbol{n}})$.

7.2.5 Second Form of Induction and Recursion

We state the second form of induction within the universe of sets.

Proposition 7.16. *If $A \subset \mathbb{N}$ satisfies $\mathbf{0} \in A$ and $(\forall \boldsymbol{m} < \boldsymbol{n}(\boldsymbol{m} \in A)) \Rightarrow (\boldsymbol{n} \in A)$ for every $\boldsymbol{n} \in \mathbb{N}$, then $A = \mathbb{N}$.*

Proposition 7.4-iii, proposition 7.16, and the well-ordering principle for natural numbers (cfr. exercise 1.16) are equivalent, since the resolutions of exercises 1.15 and 1.17 hold within the universe of sets with no essential variations. This proves proposition 7.16 and the well-ordering principle for $\mathbb{N}$ within set theory. In the case of recursive definitions, we have the following statement.

Proposition 7.17. *Given*

- *a class* c*;*
- *a fixed element* $a \in \mathsf{c}$*; and*
- *a function* $h\colon \mathrm{FSeq}(\mathsf{c}) \to \mathsf{c}$*,*

there exists a unique function $f\colon \mathbb{N} \to \mathsf{c}$ *such that*

I. $f(\boldsymbol{0}) = a$; *and*

II. $f(\boldsymbol{s}(\boldsymbol{n})) = h(f|_{\boldsymbol{s}(\boldsymbol{n})}) = h((f(\boldsymbol{0}), \ldots, f(\boldsymbol{n})))$ *for every* $\boldsymbol{n} \in \mathbb{N}$.

Exercise 7.18. Prove proposition 7.17.

Exercise 7.19. Show that proposition 7.17 is equivalent to proposition 7.8. *Hint:* Adapt the resolution of exercise 1.15 about recursive definitions.

Remark: Exercises 1.18 and 1.19 hold within the universe of sets, the resolutions being analogous.

The comments after remark 7.9 hold about proposition 7.17 too. Again, the reader can find the details in appendix C.

Since we formalised both forms of induction and recursion, all of the properties of natural numbers considered in chapter 1 can be proven set-theoretically. In particular, the decimal representation, described in section 1.4.4, holds in $\mathbb{N}$ without variations. The same consideration applies to the properties stated in section 1.5 (the notion of divisor, existence and uniqueness of quotient and reminder, existence and uniqueness of the decomposition in prime factors, and existence and uniqueness of GCD and LCM).

7.2.6 Finite and Infinite Sets

Now we can formalise within set theory the following notions.

Definition 7.20. A set x

- *contains* $\boldsymbol{n}$ *elements* if there exists a bijection $f\colon \boldsymbol{n} \xrightarrow{\approx} x$;
- is *finite* it there exists $\boldsymbol{n} \in \mathbb{N}$ such that x contains $\boldsymbol{n}$ elements;
- is *infinite* if it is not finite;
- is *denumerable* if there exists a bijection $f\colon \mathbb{N} \xrightarrow{\approx} x$; and
- is *countable* or *at most denumerable* if it is finite or denumerable. ◇

Among the previous definitions, the notions of denumerable and countable set are new, while the first three were introduced meta-theoretically in section 1.4.3. Actually, we could introduce the last two meta-theoretically as well, but it was not necessary. We will see in the next section that the meta-theoretical definitions agree with the ones given in 7.20, as expected (see exercise 7.28 below).

Exercise 7.21. Prove that, for every $\boldsymbol{n}, \boldsymbol{m} \in \mathbb{N}$:

(i) there exists an injective function $f\colon \boldsymbol{n} \hookrightarrow \boldsymbol{m}$ if and only if $\boldsymbol{n} \leq \boldsymbol{m}$;

(ii) there exists a surjective function $f\colon \boldsymbol{n} \twoheadrightarrow \boldsymbol{m}$ if and only if $\boldsymbol{n} \geq \boldsymbol{m}$;

(iii) an injective function $f\colon \boldsymbol{n} \hookrightarrow \boldsymbol{m}$, with $\boldsymbol{n} \leq \boldsymbol{m}$, is bijective if and only if $\boldsymbol{n} = \boldsymbol{m}$;

(iv) a surjective function $f\colon \boldsymbol{n} \twoheadrightarrow \boldsymbol{m}$, with $\boldsymbol{n} \geq \boldsymbol{m}$, is bijective if and only if $\boldsymbol{n} = \boldsymbol{m}$;

(v) there exists no injective function $f\colon \mathbb{N} \hookrightarrow \boldsymbol{n}$;

(vi) there exists no surjective function $f\colon \boldsymbol{n} \twoheadrightarrow \mathbb{N}$.

It follows that:

(vii) there exists a bijection $f\colon \boldsymbol{n} \xrightarrow{\approx} \boldsymbol{m}$ if and only if $\boldsymbol{n} = \boldsymbol{m}$;

(viii) there exists no bijection $f\colon \boldsymbol{n} \xrightarrow{\approx} \mathbb{N}$;

(ix) if x contains $\boldsymbol{n}$ elements and $\boldsymbol{m}$ elements at the same time, then $\boldsymbol{n} = \boldsymbol{m}$;

(x) if a set x is infinite, then, for every $\boldsymbol{n} \in \mathbb{N}$, there exists a finite subset $y \subset x$ that contains $\boldsymbol{n}$ elements.

Remark: This exercise is the set-theoretical version of exercise 4 of section 1.5, the latter being meta-theoretical. We only skipped the converse of item (x) and items (xi) and (xii), since they will be stated and proved more naturally in items (vii), (viii), (ix) and (x) of the next exercise.

Exercise 7.22. Let x be a finite set with $\boldsymbol{n}$ elements, y be a finite set with $\boldsymbol{m}$ elements and z be an infinite set. We suppose $\boldsymbol{n}, \boldsymbol{m} > \boldsymbol{0}$. Prove that:

(i) there exists an injective function $f\colon x \hookrightarrow y$ if and only if $\boldsymbol{n} \leq \boldsymbol{m}$; such a function is bijective if and only if $\boldsymbol{n} = \boldsymbol{m}$;

(ii) there exists a surjective function $f\colon x \twoheadrightarrow y$ if and only if $\boldsymbol{n} \geq \boldsymbol{m}$; such a function is bijective if and only if $\boldsymbol{n} = \boldsymbol{m}$;

(iii) there exists an injective function $f\colon x \hookrightarrow z$, but no injective function $f\colon z \hookrightarrow x$;

(iv) there exists a surjective function $f\colon z \twoheadrightarrow x$, but no surjective function $f\colon x \twoheadrightarrow z$.

Deduce that:

(v) if x and y are finite, then there exists a bijection between them if and only if they contain the same number of elements;

(vi) if x is finite and y is infinite, then there exists no bijection between them;

(vii) if x is a finite set with $\boldsymbol{n}$ elements, y is a finite set with $\boldsymbol{m}$ elements and $y \subset x$, then $\boldsymbol{m} \leq \boldsymbol{n}$; in this case, $x = y$ if and only if $\boldsymbol{n} = \boldsymbol{m}$;

(viii) a set x is infinite if and only if, for every $\boldsymbol{n} \in \mathbb{N}$, there exists a finite subset $y \subset x$ that contains $\boldsymbol{n}$ elements (equivalently, there exists an injective function $\iota_{\boldsymbol{n}}\colon \boldsymbol{n} \hookrightarrow x$ for every $\boldsymbol{n} \in \mathbb{N}$);

(ix) if x is finite and $y \subset x$, then y is finite;

(x) if y is infinite and $y \subset x$, then x is infinite;

(xi) a denumerable set is infinite.

Remark: This exercise is the set-theoretical version of exercise 2 of section 3.5, the latter being meta-theoretical. We only added item (xi), since in chapter 3 we had not introduced the notion of denumerability yet.

In item (viii) of exercise 7.22, it would be more natural to affirm that a set x is infinite if and only if it contains a denumerable subset (equivalently, there exists an injective function $\iota\colon \mathbb{N} \hookrightarrow x$), but this is true only assuming the axiom of choice, that we will introduce in chapter 9. Even assuming this stronger version, exercise 7.22 does not settle the following important problem: is every infinite set denumerable? Equivalently: given any two infinite sets x and y, is there a bijection between them? We will see in chapter 10 that the answer is definitely negative (independently of the axiom of choice).

7.2.7 Alternative Approach

In chapters 11 and 12 we will consider another approach to natural numbers, defining them as particular cases of ordinals or cardinals. In that setting, we will define the strict order relation '$<$' in $\mathbb{N}$ as the belonging relation '$\in$' (cfr. ex. 7.11), therefore there will be no need to use the operation of sum to define it. Moreover, we will define the operations of sum, product, and power through elementary set-theoretical operations, instead of using induction. In principle, we could adopt only that point of view to deal with natural numbers within set theory, but we think that the approach of the present chapter is interesting anyway, since it leads directly to the definition of $\mathbb{N}$, without relying on the theory of ordinals and cardinals.

7.3 META-THEORETICAL CONSIDERATIONS

In section 7.2, we defined and studied the set $\mathbb{N}$ without mentioning the meta-theoretical set N. Of course, we used the basic tools of naive set theory, that have been constructing assuming the meta-theoretical knowledge of natural numbers, but, if we "take" those tools without minding the corresponding constructions, the definition of $\mathbb{N}$ is self-contained. In this section, we resume the explicit intuitive knowledge of natural numbers, and we use it to show that $\mathbb{N}$ is a good set-theoretical model of N.

7.3.1 N *vs* $\mathbb{N}$

We prove that there exits a meta-theoretical bijection between N and $\mathbb{N}$ that respects the operations and the order.

Notation 7.23. For any $n \in \mathrm{N}$, we set $\boldsymbol{s}^n(x) := \boldsymbol{s}(\cdots \boldsymbol{s}(x)\cdots)$, where '$\boldsymbol{s}$' is applied n times. More precisely, by recursion in the meta-theory, we set $\boldsymbol{s}^0(x) := x$ and $\boldsymbol{s}^{s(n)}(x) := \boldsymbol{s}(\boldsymbol{s}^n(x))$.

Lemma 7.24. *The following two properties hold:*

(i) $\boldsymbol{s}^n(\emptyset) \in \mathbb{N}$ *for every* $n \in \mathrm{N}$

(ii) $\boldsymbol{s}^n(\emptyset) = \boldsymbol{s}^m(\emptyset)$ *if and only if* $n = m$

Proof. (i) Since $\mathbb{N} \in \mathtt{cs}$, it immediately follows by meta-theoretical induction. (ii) ($\Leftarrow$) Obvious. ($\Rightarrow$) Up to exchange n and m, we suppose $n \leq m$—that is, $m = n + k$. We have to prove that necessarily $k = 0$. Hence, we prove by induction on n that $\boldsymbol{s}^n(\emptyset) \neq \boldsymbol{s}^{n+s(h)}(\emptyset)$ for every fixed $h \in \mathrm{N}$. If $n = 0$, then we get $\emptyset \neq \boldsymbol{s}^{s(h)}(\emptyset)$—that is, $\emptyset \neq \boldsymbol{s}(\boldsymbol{s}^h(\emptyset))$. Defining $\boldsymbol{h} := \boldsymbol{s}^h(\emptyset)$, this means that $\mathbf{0} \neq \boldsymbol{s}(\boldsymbol{h})$, that is the statement of proposition 7.4-i. Let us assume that the result holds for n and, by contradiction, that $\boldsymbol{s}^{s(n)}(\emptyset) = \boldsymbol{s}^{s(n)+s(h)}(\emptyset)$, that is, $\boldsymbol{s}^{s(n)}(\emptyset) = \boldsymbol{s}^{s(n+s(h))}(\emptyset)$. This means that $\boldsymbol{s}(\boldsymbol{s}^n(\emptyset)) = \boldsymbol{s}(\boldsymbol{s}^{n+s(h)}(\emptyset))$. From proposition 7.4-ii we get $\boldsymbol{s}^n(\emptyset) = \boldsymbol{s}^{n+s(h)}(\emptyset)$, against the inductive hypothesis. □

Proposition 7.25. *For every $\boldsymbol{n} \in \mathbb{N}$, there exists a unique $n \in \mathrm{N}$ such that $\boldsymbol{n} = \boldsymbol{s}^n(\emptyset)$.*

Proof. `Existence`: We set $A := \{\boldsymbol{n} \in \mathbb{N} : \exists n \in \mathrm{N} : \boldsymbol{n} = \boldsymbol{s}^n(\emptyset)\}$. Let us show that A is closed under successor. Obviously $\mathbf{0} \in A$, since $\mathbf{0} = \emptyset = \boldsymbol{s}^0(\emptyset)$. If $\boldsymbol{n} := \boldsymbol{s}^n(\emptyset) \in A$, then $\boldsymbol{s}(\boldsymbol{n}) = \boldsymbol{s}(\boldsymbol{s}^n(\emptyset)) = \boldsymbol{s}^{s(n)}(\emptyset) \in A$. Definition (7.1) immediately implies that $\mathbb{N} \subset A$, hence $A = \mathbb{N}$. `Uniqueness`: It immediately follows from lemma 7.24-ii. □

Thanks to proposition 7.25, the meta-theoretical sequence $n \mapsto \boldsymbol{s}^n(\emptyset)$ identifies N with $\mathbb{N}$—that is, it is a meta-theoretical bijection between them. From now on, we use the notational convention that, for any character that denotes a meta-theoretical natural number, its bold version denotes the corresponding set-theoretical number and vice-versa—that is, $\boldsymbol{n} = \boldsymbol{s}^n(\emptyset)$, $\boldsymbol{m} = \boldsymbol{s}^m(\emptyset)$, etc.[3] In particular, we simply denote by $n \mapsto \boldsymbol{n}$ the identification between N and $\mathbb{N}$.

Proposition 7.26. *Under the correspondence $n \mapsto \boldsymbol{n}$, we have:*

- $0 \mapsto \mathbf{0}$ *and* $s(n) \mapsto \boldsymbol{s}(\boldsymbol{n})$
- $n + m \mapsto \boldsymbol{n} + \boldsymbol{m}$, $nm \mapsto \boldsymbol{nm}$, *and* $n^m \mapsto \boldsymbol{n}^{\boldsymbol{m}}$
- $n \leq m$ *(respectively, $n < m$) if and only if $\boldsymbol{n} \leq \boldsymbol{m}$ (respectively, $\boldsymbol{n} < \boldsymbol{m}$)*

Exercise 7.27. Prove proposition 7.26.

Exercise 7.28. Show that the notions formalised in definition 7.20 agree with the corresponding meta-theoretical ones stated in section 1.4.3.

Thanks to the bijection $n \mapsto \boldsymbol{n}$, we can identify a meta-theoretical sequence $x_0, \ldots, x_n$ or $x_0, x_1, \ldots$ with the set-theoretical one $(x_{\mathbf{0}}, \ldots, x_{\boldsymbol{n}})$ or $(x_{\mathbf{0}}, x_{\mathbf{1}}, \ldots)$, where $x_{\boldsymbol{i}} := x_i$. It follows that, given a meta-theoretical sequence $x_0, x_1, \ldots$, the set $\{x_0, x_1, \ldots\}$ is well-defined in our universe by assumption `(A6)`, since it is the image of the function $\boldsymbol{i} \mapsto x_{\boldsymbol{i}} = x_i$. This remark extends assumption `(A2)` to infinite sequences.

[3]This is just a notational shortcut, since formally 'n' and '$\boldsymbol{n}$' are two distinct symbols (the notion of "boldness" is not formalisable). Thus, for each $n \in \mathrm{N}$ fixed, we should write each time "we set $\boldsymbol{n} := \boldsymbol{s}^n(\emptyset)$" and, for each $\boldsymbol{n} \in \mathbb{N}$ fixed, we should write "we denote by n the unique element of N such that $\boldsymbol{n} := \boldsymbol{s}^n(\emptyset)$".

7.3.2 What is a Natural Number?

We have seen that natural numbers can be embedded in the set-theoretical universe by constructing a suitable model. Now it is natural to inquire whether the meta-theoretical knowledge of natural numbers is actually necessary or not. In other words, couldn't we avoid to introduce N intuitively?

We do not have a clear answer to this question, but we can make some comments about it. Let us suppose not to assume the meta-theoretical knowledge of N. In this case, we should review the exposition of a big part of the theory discussed above because of the following reasons:

1. We could not state that $\alpha_1 \wedge \ldots \wedge \alpha_n$ and $\alpha_1 \vee \ldots \vee \alpha_n$ are well-defined and do not depend on the order because of associativity and commutativity; similarly, we could not state that the composition of functions $f_n \circ \cdots \circ f_1$ is well-defined because of associativity. We could not affirm that a proposition α depends on the variables $x_1, \ldots, x_n$ either.

2. We could not state that $x := \{y_1, \ldots, y_n\}$ is well-defined in our universe (as a direct assumption or as a consequence of the existence of pairs) before introducing natural numbers.

3. We could not define the concept of well-founded formula in section 2.2.1, since we used the notion of *finite sequence* of formulas, that relies on the notion of natural number; hence, we could not formulate comprehension hypothesis `(A4)` properly (see section 2.7).

4. We could not define the n-ary operations (like $x_1 \cup \ldots \cup x_n$ or $x_1 \times \cdots \times x_n$) before introducing natural numbers.

5. ...

About items 2 and 4, it would be not so problematic to deal with them after introducing natural numbers. In the former case, we should state the axiom of pairs and show that the image of any set-theoretical finite sequence is a set. Similarly, in the latter case, we should consider the union or the Cartesian product only of a set-theoretical finite sequence; nevertheless, it is quite unnatural not to state these definitions and properties immediately after introducing the corresponding binary operations. The situation is more delicate about items 1 and 3. In these cases, we should first embed propositional calculus and quantifiers within set theory, carefully verifying that we are not assuming (explicitly or implicitly) any meta-theoretical knowledge of natural numbers. Afterwards, we would be free to deal with $\boldsymbol{n}$ variables or propositions and we could provide the correct definition of well-founded formula in set-theoretical terms. We are not able to establish definitely whether this is possible or not (we feel it is not), but, in any case, it seems quite cumbersome to apply such a construction just to deal with $x_1, \ldots, x_n$ or $\alpha_1 \wedge \ldots \wedge \alpha_n$, without having this possibility since the beginning. In principle, we could settle item 3 by stating `(A4)` through the intuitive notion of property, as we did for the meta-theoretical induction (and recursion) principle. Nevertheless, such a formulation would not be precise enough; since (A4) is a statement internal to set theory, a non-precise formulation should be better avoided. Summarising, even if we are not able to decide whether the meta-theoretical knowledge of natural numbers is an intrinsic necessity or not, we think it is fundamental for the sake of clarity and naturality of the exposition. That's way we chose the present approach and we waited until this chapter to embed natural numbers in set theory.

Therefore, if someone asks us "What is a natural number?", our personal answer is the following one: "It is a meta-theoretical concept, based on the intuitive notions of *zero* and *successor*, that can be faithfully and autonomously reproduced within set theory, leading to a well-defined model of each natural number as a (hereditary) set and of the whole meta-theoretical set of natural numbers as a (hereditary) set too."

7.4 EXERCISES

Remark: The following five exercises are the set-theoretical versions of exercises 8–10 of section 2.8, exercises 3-2 and 6 of section 3.5, and exercise 2 of section 5.6. The remaining items of exercise 3 of section 3.5 easily follow with the same proof.

1. Prove that, if x contains $\boldsymbol{n}$ elements, y contains $\boldsymbol{m}$ elements and $x \cap y$ contains $\boldsymbol{k}$ elements, then:

- $x \sqcup y$ contains $\boldsymbol{n} + \boldsymbol{m}$ elements;
- $x \setminus y$ contains $\boldsymbol{n} - \boldsymbol{k}$ elements;
- $x \,\Delta\, y$ contains $\boldsymbol{n} + \boldsymbol{m} - \boldsymbol{2k}$ elements;
- $x \cup y$ contains $\boldsymbol{n} + \boldsymbol{m} - \boldsymbol{k}$ elements;
- $x \times y$ contains $\boldsymbol{nm}$ elements.

2. Prove that, if x contains $\boldsymbol{n}$ elements and y contains $\boldsymbol{m}$ elements, then $\mathrm{Func}(x, y)$ contains $\boldsymbol{m}^{\boldsymbol{n}}$ elements.

3. Prove that, given a set x, there exists a natural bijection between $\mathcal{P}(x)$ and $\mathrm{Func}(x, \boldsymbol{2})$.

Remark: It immediately follows from exercises 2 and 3 that, if x contains $\boldsymbol{n}$ elements, then $\mathcal{P}(x)$ contains $\boldsymbol{2}^{\boldsymbol{n}}$ elements.

4. Prove that, if x contains $\boldsymbol{n}$ elements and $y \subset x$ contains $\boldsymbol{k}$ elements, then there exists a bijection $\varphi\colon \boldsymbol{n} \xrightarrow{\approx} x$ such that $y = \dot{\varphi}(\boldsymbol{k})$.

5. Prove that, given a finite sequence of sets $(x_{\boldsymbol{1}}, \ldots, x_{\boldsymbol{k}})$, such that $x_{\boldsymbol{i}}$ contains $\boldsymbol{n}_{\boldsymbol{i}}$ elements for every $\boldsymbol{i} \in \{\boldsymbol{1}, \ldots, \boldsymbol{k}\}$, the set $x_{\boldsymbol{1}} \times \cdots \times x_{\boldsymbol{k}}$ contains $\boldsymbol{n}_{\boldsymbol{1}} \cdots \boldsymbol{n}_{\boldsymbol{k}}$ elements and the set $x_{\boldsymbol{1}} \sqcup \ldots \sqcup x_{\boldsymbol{k}}$ contains $\boldsymbol{n}_{\boldsymbol{1}} + \cdots + \boldsymbol{n}_{\boldsymbol{k}}$ elements.

Remark: Items (iv) and (v) of the following two exercises require some knowledge of basic algebra.

6. Consider the following equivalence relation in $\mathbb{N} \times \mathbb{N}$:

$$(\boldsymbol{n}, \boldsymbol{m}) \simeq (\boldsymbol{n}', \boldsymbol{m}') \Leftrightarrow \boldsymbol{n} + \boldsymbol{m}' = \boldsymbol{m} + \boldsymbol{n}'.$$

(i) Prove that it is actually an equivalence relation.

(ii) Prove that each equivalence class can be uniquely represented in the form $[(\boldsymbol{n}, \boldsymbol{0})]$ or $[(\boldsymbol{0}, \boldsymbol{n})]$.

We set $\mathbb{Z} := (\mathbb{N} \times \mathbb{N})/\simeq$.

(iii) Prove that the following operations are well-defined in $\mathbb{Z}$ (that is, they do not depend on the representatives chosen):

- $[(\boldsymbol{n}, \boldsymbol{m})] + [(\boldsymbol{n}', \boldsymbol{m}')] := [(\boldsymbol{n} + \boldsymbol{n}', \boldsymbol{m} + \boldsymbol{m}')]$;
- $[(\boldsymbol{n}, \boldsymbol{m})] \cdot [(\boldsymbol{n}', \boldsymbol{m}')] := [(\boldsymbol{n}\boldsymbol{n}' + \boldsymbol{m}\boldsymbol{m}', \boldsymbol{n}\boldsymbol{m}' + \boldsymbol{m}\boldsymbol{n}')]$.

(iv) Prove that $(\mathbb{Z} + \cdot)$ is a unitary integral domain.

(v) Prove that we have the injective unitary semi-ring morphism $\iota \colon \mathbb{N} \hookrightarrow \mathbb{Z}$, $\boldsymbol{n} \mapsto [(\boldsymbol{n}, 0)]$.

For every $\boldsymbol{n} \in \mathbb{N}$, we identify $\boldsymbol{n}$ with its image $\iota(\boldsymbol{n})$, hence we think that $\mathbb{N} \subset \mathbb{Z}$. It follows from item (ii) that every element of $\mathbb{Z}$ is of the form $\boldsymbol{n}$ or $-\boldsymbol{n}$, with $\boldsymbol{n} \in \mathbb{N}$.

7. Consider the following equivalence relation in $\mathbb{Z} \times \mathbb{Z}^*$, where $\mathbb{Z}^* := \mathbb{Z} \setminus \{\boldsymbol{0}\}$:

$$(a, b) \simeq (a', b') \Leftrightarrow ab' = a'b.$$

(i) Prove that it is actually an equivalence relation.

(ii) Prove that each equivalence class can be uniquely represented in the form $[(\boldsymbol{0}, \boldsymbol{1})]$, $[(\boldsymbol{n}, \boldsymbol{m})]$ or $[(-\boldsymbol{n}, \boldsymbol{m})]$, where $\boldsymbol{n}, \boldsymbol{m} \in \mathbb{N}^*$ are coprime.

We set $\mathbb{Q} := (\mathbb{Z} \times \mathbb{Z}^*)/\simeq$ and we denote $[(a, b)]$ by $\frac{a}{b}$.

(iii) Prove that the following operations are well-defined in $\mathbb{Q}$ (that is, they do not depend on the representatives chosen):

- $\frac{a}{b} + \frac{a'}{b'} := \frac{ab' + a'b}{bb'}$;
- $\frac{a}{b} \cdot \frac{a'}{b'} := \frac{aa'}{bb'}$.

(iv) Prove that $(\mathbb{Q} + \cdot)$ is a field.

(v) Prove that we have the injective unitary ring morphism $j \colon \mathbb{Z} \hookrightarrow \mathbb{Q}$, $a \mapsto \frac{a}{\mathbf{1}}$.

Remark: We have seen how to construct the ring $\mathbb{Z}$ and the field $\mathbb{Q}$ within our universe, starting from $\mathbb{N}$. Now, starting from $\mathbb{Q}$, it is possible to define the fields $\mathbb{R}$ and $\mathbb{C}$. The corresponding constructions are usually shown in basic calculus courses, therefore we do not deal with them here.

8. Show how to construct the field $\mathbb{Q}$ from the semi-ring $\mathbb{N}$ by extending first the product and then the sum.

9. Let us consider an indexed set $x = \{x_n\}_{n\in\mathbb{N}}$. Prove that:

(i) $y \in \bigcup_n \bigcap_{k\geq n} x_k$ if and only if the set $\{n \in \mathbb{N} : y \notin x_n\}$ is finite.

(ii) $y \in \bigcap_n \bigcup_{k\geq n} x_k$ if and only if the set $\{n \in \mathbb{N} : y \in x_n\}$ is infinite.

(iii) $\bigcup_n \bigcap_{k\geq n} x_k \subset \bigcap_n \bigcup_{k\geq n} x_k$, but in general the converse does not hold.

Remark: Formally, the indexed sets involved are the following ones. For every $n \in \mathbb{N}$, we define $\boldsymbol{x_n} := \{x_{n+k}\}_{k\in\mathbb{N}}$. We set $x'_n := \bigcup \boldsymbol{x_n} = \bigcup_k x_{n+k}$ and $x''_n := \bigcap \boldsymbol{x_n} = \bigcap_k x_{n+k}$. We use the following notation: $x'_n = \bigcup_{k\geq n} x_k$ and $x''_n = \bigcap_{k\geq n} x_k$. We obtain the indexed sets $x' = \{x'_n\}_{n\in\mathbb{N}}$ and $x'' = \{x''_n\}_{n\in\mathbb{N}}$. We set $\boldsymbol{x}' := \bigcap x' = \bigcap_n x'_n$ and $\boldsymbol{x}'' := \bigcup x' = \bigcup_n x'_n$. We use the following notation: $\boldsymbol{x}' = \bigcap_n \bigcup_{k\geq n} x_k$ and $\boldsymbol{x}'' = \bigcup_n \bigcap_{k\geq n} x_k$.

10. Formalise rigorously in set-theoretical terms the statement and the proof of exercise 1.21.

Remark: It is not necessary to formalise exercise 1.23, since the resolution of the latter holds within set-theory too, just replacing meta-theoretical natural numbers, functions and sequences with the corresponding set-theoretical ones.

CHAPTER 8

Well-Orders

The notion of *well-order* will be fundamental to define ordinals and cardinals, and the existence of a well-order in each set is a statement equivalent to the axiom of choice and to Zorn's lemma. In this chapter we introduce this concept and its basic properties.

8.1 DEFINITION AND FIRST PROPERTIES

Definition 8.1. A *well-order* in a set x is an order relation '$\preceq$' such that every $a \in \mathcal{P}^*(x)$ admits a minimum. A *well-ordered set* is a set endowed with a well-order. ◇

A well-order is a total order, since, for every $y, z \in x$, the subset $\{y, z\}$ admits a minimum. Moreover, if $(x, \preceq)$ is a well-ordered set and we endow any subset $a \subset x$ with the order induced by restriction, then $(a, \preceq|_{a\times a})$ is well-ordered too (since a subset of a is also a subset of x, therefore it admits a minimum).

Definition 8.2. Given two well-ordered sets $(x, \preceq_x)$ and $(y, \preceq_y)$, we define the well-ordered set $(x \sqcup y, \preceq_\sqcup)$ as follows:

- $(z, \emptyset) \preceq_\sqcup (z', \emptyset)$ if and only if $z \preceq_x z'$
- $(w, \{\emptyset\}) \preceq_\sqcup (w', \{\emptyset\})$ if and only if $w \preceq_y w'$
- $(z, \emptyset) \prec_\sqcup (w, \{\emptyset\})$ for every $z \in x$ and $w \in y$

We set $(x, \preceq_x) \sqcup (y, \preceq_y) := (x \sqcup y, \preceq_\sqcup)$. ◇

Informally, the well-ordered set $(x \sqcup y, \preceq_\sqcup)$ consists of the juxtaposition of x and y, in such a way that y strictly follows x.

Definition 8.3. Given two well-ordered sets $(x, \preceq_x)$ and $(y, \preceq_y)$, we define the well-ordered set $(x \times y, \preceq_\times)$ by $(z, w) \preceq_\times (z', w')$ if and only if one of the two following alternatives holds:

DOI: 10.1201/9781003565291-8

- $z \prec_x z'$
- $z = z'$ and $w \preceq_y w'$

We set $(x, \preceq_x) \times (y, \preceq_y) := (x \times y, \preceq_\times)$. The relation '$\preceq_\times$' is called *lexicographical order* in $x \times y$. ◇

Exercise 8.4. Prove that the relations '$\preceq_\sqcup$' and '$\preceq_\times$', introduced in definitions 8.2 and 8.3, are actually well-orders.

We treat the operations '$\sqcup$' and '$\times$' as associative. This issue will be settled rigorously in exercise 8.10 below.

It immediately follows from exercise 3.33 that a *finite* ordered set is well-ordered if and only if it is totally ordered. In particular, every natural number is a well-ordered set. Moreover, exercise 1.16 (within set theory) shows that $\mathbb{N}$ is well-ordered. Thus, for every natural number $n \in \mathbb{N}$, we get $\mathbb{N} \sqcup n$. Similarly, we get $\mathbb{N} \sqcup \mathbb{N}$, $\mathbb{N} \sqcup \mathbb{N} \sqcup n$, $\mathbb{N} \sqcup \mathbb{N} \sqcup \mathbb{N}$, and so on. We represented these well-orders graphically in figure 8.1. Afterwards, we have $\mathbb{N} \times \mathbb{N}$, and we can keep on again with disjoint union, getting $(\mathbb{N} \times \mathbb{N}) \sqcup n$, $(\mathbb{N} \times \mathbb{N}) \sqcup \mathbb{N}$, $(\mathbb{N} \times \mathbb{N}) \sqcup (\mathbb{N} \sqcup n)$, etc., as figure 8.2 shows. Then, we have $\mathbb{N} \times \mathbb{N} \times \mathbb{N}$, and we can start again. We could consider the limit of these products and start again, but we defer this discussion to chapter 11, where we will define ordinal numbers. In particular, we will see that studying well-orders is essentially equivalent to studying the class of ordinal numbers (see proposition 11.26).

We remark that $\mathbb{N} \sqcup \mathbb{N}$ can be equivalently described as $2 \times \mathbb{N}$; in general, $\mathbb{N} \sqcup \overset{n \text{ times}}{\cdots} \sqcup \mathbb{N}$ is equivalent to $n \times \mathbb{N}$. More precisely, the well-ordered sets

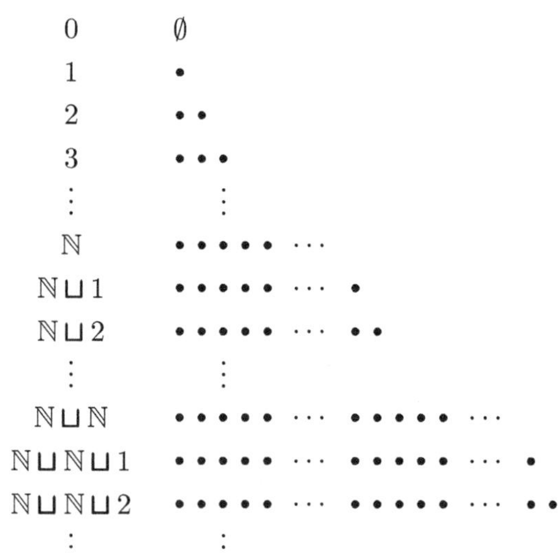

Figure 8.1 First well-orders.

Figure 8.2 Other well-orders.

$\mathbb{N} \sqcup \overset{n \text{ times}}{\cdots} \sqcup \mathbb{N}$ and $n \times \mathbb{N}$ are *isomorphic*, as we will see in the next section (see exercise 8.12 below).

8.2 WELL-ORDER ISOMORPHISMS

Intuitively, an *isomorphism* between two ordered sets is a bijection that respects the order. This means that, if there exists an isomorphism, then the two ordered sets have the same structure. The following definition formalises this concept.

Definition 8.5. Given two ordered sets $(x, \preceq)$ and $(y, \preceq)$, an *ordered-set isomorphism* from x to y is a bijection $\varphi\colon x \to y$ such that, for every $z_1, z_2 \in x$, we have $z_1 \preceq z_2 \Leftrightarrow \varphi(z_1) \preceq \varphi(z_2)$.[1] Two ordered sets are *isomorphic* if there exists an isomorphism between them. Moreover, an isomorphism between $(x, \preceq)$ and itself is called *automorphism*. ♢

Exercise 8.6. Given three ordered sets $(x, \preceq)$, $(y, \preceq)$ and $(z, \preceq)$, show that:

- The identity of x is an ordered-set automorphism (trivial).
- If $\varphi\colon x \to y$ is an ordered-set isomorphism, then its inverse $\varphi^{-1}\colon y \to x$ is an ordered-set isomorphism too.
- If $\varphi\colon x \to y$ and $\psi\colon y \to z$ are ordered-set isomorphisms, then the composition $\psi \circ \varphi\colon x \to z$ is an ordered-set isomorphism too.

It follows that the existence of an isomorphism is an equivalence relation in the class of ordered sets.

Exercise 8.7. Show that, given an ordered set $(x, \preceq)$, a set y and a bijection $\varphi\colon x \to y$, there exists a unique order relation in y, that we denote by $\varphi_*(\preceq)$, that makes φ an ordered-set isomorphism.

[1] We should use distinct symbols for the two order relations—for example, $(x, \preceq_x)$ and $(y, \preceq_y)$—but we use the same one for simplicity.

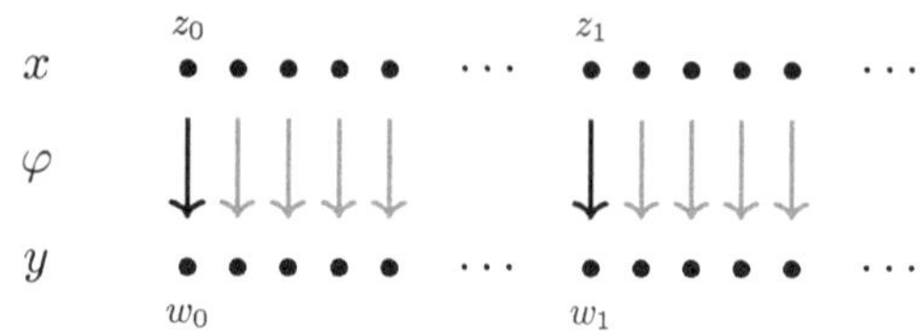

Figure 8.3 A well-ordered set isomorphism.

One of the main properties of well-orders is that, if two well-ordered sets are isomorphic, then there exists a *unique* isomorphism between them. Let us try to guess why. Figure 8.3 represents an isomorphism between well-orders.

The idea is the following one. Since φ respects the order by definition, it necessarily sends $z_0 := \min(x)$ to $w_0 := \min(y)$, thus the arrow on the left in figure 8.3 is uniquely determined. Therefore, φ restricts to a bijection from $x' := x \setminus \{z_0\}$ to $y' := y \setminus \{w_0\}$. Again, since φ respects the order, it necessarily sends $\min(x')$ to $\min(y')$, and so on. Hence, the left block of arrows in figure 8.3 is uniquely determined. We call $a \subset x$ and $b \subset y$ the sets formed respectively by the sources and the targets of these arrows. It follows that φ restricts to bijection from $x \setminus a$ to $y \setminus b$. Since φ respects the order, it necessarily sends $z_1 := \min(x \setminus a)$ to $w_1 := \min(y \setminus b)$, hence the left arrow of the right block is uniquely determined. Repeating the same procedure, we deduce that the whole φ is unique. The following lemma and proposition provide a formal proof of this result.

Lemma 8.8. *Let $(x, \preceq)$ be a well-ordered set. The only ordered-set automorphism of $(x, \preceq)$ is the identity.*

Proof. We can assume that $x \neq \emptyset$, otherwise the empty function, that coincides with the identity, is trivially the unique automorphism. Let us suppose that $\varphi\colon x \to x$ is an ordered set automorphism different from the identity. It follows that the subset $a := \{y \in x : \varphi(y) \neq y\}$ is not empty, hence it admits a minimum m. We have two possibilities: (i) $\varphi(m) \prec m$; (ii) $\varphi(m) \succ m$. If (ii) holds, then, replacing φ by φ^{-1}, we get (i), hence we can assume that (i) holds without loss of generality.[2] Since $m = \min(a)$, it follows from (i) that $\varphi(m) \notin a$, hence $\varphi(\varphi(m)) = \varphi(m)$. Since φ is a bijection, we deduce that $\varphi(m) = m$, a contradiction. □

Proposition 8.9. *Let $(x, \preceq)$ and $(y, \preceq)$ be well-orders. If there exists an ordered-set isomorphism $\varphi\colon x \to y$, then it is unique.*

Proof. Let $\varphi, \psi\colon x \to y$ be two ordered-set isomorphisms. It follows that $\psi^{-1} \circ \varphi\colon x \to x$ is an automorphism, hence lemma 8.8 implies that $\psi^{-1} \circ \varphi = \mathrm{id}_x$. Therefore, $\psi \circ \psi^{-1} \circ \varphi = \psi \circ \mathrm{id}_x$, that is, $\varphi = \psi$. □

[2]We remark that the set a does not change by replacing φ with φ^{-1}, since $\varphi(y) \neq y$ iff $\varphi^{-1}(\varphi(y)) \neq \varphi^{-1}(y)$ iff $y \neq \varphi^{-1}(y)$. In particular, the minimum m does not change.

Exercise 8.10. Prove that the two operations introduced in definitions 8.2 and 8.3 are associative, but *not* commutative, up to a unique bijection. Prove that the former operation is distributive over the latter on the left—that is, $(x, \preceq_x) \times \big((y, \preceq_y) \sqcup (z, \preceq_z)\big)$ coincides with $\big((x, \preceq_x) \times (y, \preceq_y)\big) \sqcup \big((x, \preceq_x) \times (z, \preceq_z)\big)$ up to a unique isomorphism—but *not* on the right.

Exercise 8.11. Show that, because of associativity, we can represent the well-ordered sets $(x_1, \preceq_1) \sqcup \ldots \sqcup (x_n, \preceq_n)$ and $(x_1, \preceq_1) \times \cdots \times (x_n, \preceq_n)$, for any $n \in \mathbb{N}^*$, by choosing the canonical n-ary disjoint union and Cartesian product as the underlying sets—that is, in the form respectively $(x_1 \sqcup \ldots \sqcup x_n, \preceq_\sqcup)$ and $(x_1 \times \cdots \times x_n, \preceq_\times)$.

Exercise 8.12. Show that the well-ordered sets $\mathbb{N} \sqcup \overset{n \text{ times}}{\cdots} \sqcup \mathbb{N}$ and $n \times \mathbb{N}$ are isomorphic.

8.3 INITIAL SEGMENTS

We recall and introduce some definitions about total orders, that hold in particular for well-orders.

Definition 8.13. Let $(x, \preceq)$ be a *totally* ordered set and $a \subset x$ be a subset ordered by restriction.

- $y \in x$ is an *upper bound* of a if $z \preceq y$ for every $z \in a$. We say that z is a *strict* upper bound if the strict inequality $z \prec y$ holds for every $z \in a$ (cfr. definition 3.34).
- a is *(strictly) bounded above* if it admits a (strict) upper bound (cfr. definition 3.34).
- a is an *initial segment* of $(x, \preceq)$ if $\big((y \in a) \wedge (z \in x) \wedge (z \preceq y)\big) \Rightarrow (z \in a)$. This means that, if a contains an element y, then it contains all of the elements $z \in x$ such that $z \preceq y$.
- a is a *proper initial segment* if it is an initial segment and $a \neq x$. ◇

Proposition 8.14. *Let $(x, \preceq)$ be a totally ordered set.*

(i) An element $y \in x$ induces the proper initial segment $\mathrm{prec}_x(y) := \{z \in x : z \prec y\}$.

(ii) Any proper initial segment $a \subset x$ is strictly bounded above. In particular, any element of $x \setminus a$ is a strict upper bound of a.

(iii) If $a, b \subset x$ are initial segments, then $a \subset b$ (hence, a is an initial segment of b) or $b \subset a$ (hence, b is an initial segment of a).

Proof. (i) If $z \in \mathrm{prec}_x(y)$—that is, $z \prec y$—and $w \preceq z$, then $w \preceq z \prec y$ implies $w \prec y$ (see exercise 3.30), hence $w \in \mathrm{prec}_x(y)$. This shows that $\mathrm{prec}_x(y)$ is an initial segment. It is proper, since $y \notin \mathrm{prec}_x(y)$. (ii) We fix any $y \in x \setminus a$. If $z \in a$, we have to prove that $z \prec y$. Since the order is total, $z \prec y$ or $y \preceq z$.

Let us suppose by contradiction that $y \preceq z$. Since a is an initial segment and $z \in a$, it follows that $y \in a$, a contradiction. (iii) We can assume that $a, b \neq \emptyset$ and $a \neq b$, otherwise the result is trivial. Up to exchange a and b, we assume that there exists $y \in a \setminus b$. It follows from item ii that y is a strict upper bound of b—that is, $z \prec y$ for every $z \in b$. Thus, for every $z \in b$, we have $z \prec y$ and $y \in a$, hence, a being an initial segment, $z \in a$. It follows that $b \subset a$. It is straightforward to prove that b is an initial segment of a, since it is an initial segment of x. □

Exercise 8.15. Let $(x, \preceq)$ be a totally ordered set.

(i) Let $a, b \subset x$ be subsets such that a is an initial segment of b (the latter being endowed with the restriction of the order of x) and b is an initial segment of x. Prove that a is an initial segment of x. Show that, if a is proper in b or b is proper in x, then a is proper in x.

(ii) Let $b \subset x$ be any subset and $a \subset x$ be an initial segment. Prove that, if $a \subset b$, then a is an initial segment of b (the latter being endowed with the restriction of the order of x).

(iii) Let $a = \{a_i\}_{i \in I}$ be a family of initial segments of x. Prove that $\bigcup_i a_i$ is an initial segment of x too.

Exercise 8.16. Let $(x, \preceq)$ and $(y, \preceq)$ be totally ordered sets and $\varphi\colon x \to y$ be an ordered-set isomorphism. Prove that $a \subset x$ is a (proper) initial segment if and only if $\dot{\varphi}(a) \subset y$ is a (proper) initial segment.

In general, a proper initial segment is not of the form $\mathrm{prec}_x(y)$. For example, $[0, 1] \subset [0, +\infty)$ and $[0, \sqrt{2}) \cap \mathbb{Q}^{\geq 0} \subset \mathbb{Q}^{\geq 0}$, with the order induced by restriction form $\mathbb{R}$, are proper initial segments not of that form. The following proposition shows that this is not possible in a well-ordered set.

Proposition 8.17. *If $(x, \preceq)$ is a well-ordered set and $a \subset x$ is a proper initial segment, then $a = \mathrm{prec}_x(y)$, where $y := \min(x \setminus a)$.*

Proof. By hypothesis $a \neq x$, hence y is well-defined. Let us prove that $a = \mathrm{prec}_x(y)$. If $z \in a$, let us suppose that $z \notin \mathrm{prec}_x(y)$—that is, $y \preceq z$. Since a is as initial segment, it follows that $y \in a$, a contradiction. Conversely, if $z \in \mathrm{prec}_x(y)$—that is, $z \prec y$—let us suppose that $z \notin a$. Then $z \in x \setminus a$, hence $y \preceq z$ (by definition of y), a contradiction. □

Exercise 8.18. Let $(x, \preceq)$ be a well-ordered set and $a \subset x$ be a proper initial segment. We set $y := \min(x \setminus a)$. Prove that $a \cup \{y\}$ is an initial segment of x.

In general, a totally ordered set can be isomorphic to a proper initial segment of itself, hence to a strictly bounded above subset. For example, $[0, \infty)$ and $[0, 1)$, with the order induced by restriction form $\mathbb{R}$, are isomorphic, an isomorphism being $\varphi\colon [0, \infty) \to [0, 1)$, $x \mapsto \frac{2}{\pi}\arctan(x)$. The following proposition shows that this is not possible with well-orders.

Proposition 8.19. *Let $(x, \preceq)$ be a well-ordered set. If $a \subset x$ is strictly bounded above (in particular, if it is a proper initial segment), then there are no ordered-set isomorphisms between x and a.*

Proof. We can assume that $x \neq \emptyset$ (otherwise it admits no strictly bounded above subsets) and that $a \neq \emptyset$ (otherwise it cannot be isomorphic to a non-empty set). By hypothesis there exists $y \in x$ such that $z \prec y$ for every $z \in a$. Let us suppose that $\varphi \colon x \to a$ is an ordered-set isomorphism. It follows that $\varphi(y) \in a$, hence $\varphi(y) \prec y$, therefore the subset $b := \{w \in x : \varphi(w) \prec w\}$ is not empty. We set $m := \min(b)$. Then, $\varphi(m) \prec m$, hence $\varphi(m) \notin b$ (since m is the minimum of b and $\varphi(m) \prec m$), that is, $\varphi(\varphi(m)) \succeq \varphi(m)$. Since φ is an ordered-set isomorphism, it follows that $\varphi(m) \succeq m$, a contradiction. □

We remark that a well-ordered set can be isomorphic to a proper subset of itself, if the latter is not strictly bounded above. For example, $(\mathbb{N}, \leq)$ is well-ordered. We set $\mathbb{N}^{\geq 2} := \{n \in \mathbb{N} : n \geq 2\}$. We have the ordered-set isomorphism $\varphi \colon \mathbb{N} \to \mathbb{N}^{\geq 2}$, $n \mapsto n+2$. Similarly, calling $2\mathbb{N}$ the subset of even numbers, we have the ordered-set isomorphism $\psi \colon \mathbb{N} \to 2\mathbb{N}$, $n \mapsto 2n$.

8.4 COMPARING WELL-ORDERS

The theorem we are going to state is the main one about well-orders, since it shows that any two well-orders can be compared in a very clear way. Let us introduce it intuitively. The theorem states that, given any two well-orders, we have the following possibilities:

(i) the two well-orders are isomorphic

(ii) the first one is isomorphic to a proper initial segment of the second one

(iii) the second one is isomorphic to a proper initial segment of the first one

Figure 8.4 represents these possibilities. Of course, cases (ii) and (iii) are equivalent up to exchange the two ordered sets.

This result can be equivalently formulated by affirming that, given any two well-orders, one of the two can be embedded as an initial segment of the other. Hence, if we think of isomorphism classes of well-orders, we can imagine them as the initial segments of a class that is (well-)ordered by inclusion; this class will be constructed in chapter 11 by selecting a canonical representative for each equivalence class of well-orders. Now we state and prove the theorem formally, starting from the following lemma.

Lemma 8.20. *Let us consider the following data:*

- *two well-ordered sets* $(x, \preceq)$ *and* $(y, \preceq)$
- *four initial segments* $a, a' \subset x$ *and* $b, b' \subset y$
- *two ordered set isomorphisms* $\varphi \colon a \to b$ *and* $\varphi' \colon a' \to b'$

If $a \subset a'$*, then necessarily* $b \subset b'$ *and* $\varphi = \varphi'|_a\|_b$*—that is,* $\varphi \subset \varphi'$ *as pre-functions.*

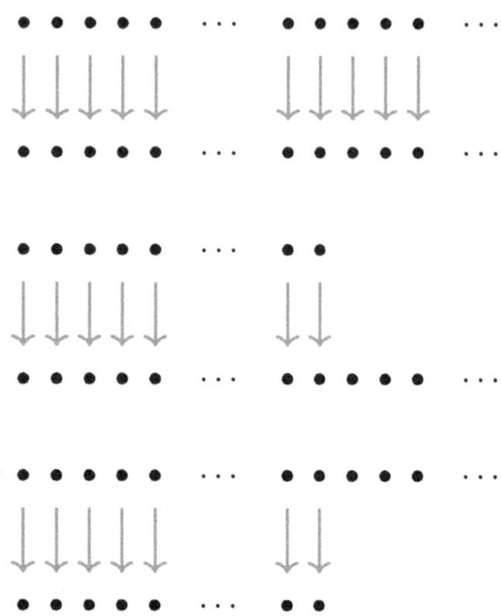

Figure 8.4 Comparison of well-orders.

Proof. We consider the restriction $\varphi'|_a||_{\dot\varphi'(a)}\colon a \to \dot\varphi'(a)$, where $\dot\varphi'(a) \subset \mathrm{Im}(\varphi') = b'$. By applying exercise 8.16 to φ', we deduce that $\dot\varphi'(a)$ is an initial segment of b'. By exercise 8.15-i, $\dot\varphi'(a)$ is an initial segment of y, therefore, by proposition 8.14-iii, we have $b \subset \dot\varphi'(a)$ or $\dot\varphi'(a) \subset b$. Since $\varphi'|_a||_{\dot\varphi'(a)} \circ \varphi^{-1}\colon b \to \dot\varphi'(a)$ is an isomorphism, by proposition 8.19 necessarily $b = \dot\varphi'(a)$ and, by proposition 8.9, $\varphi'|_a||_{\dot\varphi'(a)} = \varphi$. □

Theorem 8.21. *Let $(x, \preceq)$ and $(y, \preceq)$ be well-ordered sets. One of the following alternatives holds:*

(i) There exists an isomorphism between x and y.

(ii) There exists an isomorphism between x and a proper initial segment of y.

(iii) There exists an isomorphism between y and a proper initial segment of x.

Proof. We can assume that $x \neq \emptyset$ and $y \neq \emptyset$, otherwise the result is trivial. We consider the following set:

$$\Phi := \left\{ \varphi\colon a \xrightarrow{\approx} b : \begin{array}{l} \bullet\ a \text{ is an initial segment of } x \\ \bullet\ b \text{ is an initial segment of } y \\ \bullet\ \varphi \text{ is an ordered-set isomorphism} \end{array} \right\}$$

We think of the elements of Φ as *pre*-functions (since they are surjective by definition, we do not miss any piece of information). The set Φ is not empty. In fact, if $x_0 := \min(x)$ and $y_0 := \min(y)$, then the only pre-function $\varphi_0 \colon \{x_0\} \to \{y_0\}$ belongs to Φ.

`Step I.` The set Φ is *totally* ordered by inclusion—that is, if $\varphi, \varphi' \in \Phi$, then $\varphi \subset \varphi'$ or $\varphi' \subset \varphi$. In fact, let us fix $\varphi \colon a \to b$ and $\varphi' \colon a' \to b'$, both belonging to Φ. By proposition 8.14-iii, up to exchange φ and φ', we can assume that $a \subset a'$. By applying lemma 8.20, we deduce that $\varphi \subset \varphi'$.

`Step II.` We set $\bar{\varphi} := \bigcup \Phi$ and we claim that $\bar{\varphi} \in \Phi$,[3] therefore $\bar{\varphi}$ is the maximum of Φ. In fact, by corollary 4.21, $\bar{\varphi}$ is an injective pre-function. We call $\bar{a}$ and $\bar{b}$ respectively the domain and the image of $\bar{\varphi}$, hence $\bar{\varphi} \colon \bar{a} \xrightarrow{\approx} \bar{b}$. Again by corollary 4.21, we have $\bar{a} = \bigcup\{a : \exists \varphi \in \Phi : \mathrm{dom}(\varphi) = a\}$ and $\bar{b} = \bigcup\{b : \exists \varphi \in \Phi : \mathrm{Im}(\varphi) = b\}$, hence it follows from exercise 8.15-iii that $\bar{a}$ is an initial segment of x and $\bar{b}$ is an initial segment of y.

It remains to show that $\bar{\varphi}$ respects the order. If $z, w \in \bar{a}$ and $z \preceq w$, then, since $\bar{a}$ is the domain of $\bar{\varphi}$, there exist $u, v \in \bar{b}$ such that $(z, u), (w, v) \in \bar{\varphi}$. By definition of $\bar{\varphi}$, there exist $\varphi, \varphi' \in \Phi$ such that $(z, u) \in \varphi$ and $(w, v) \in \varphi'$. Since Φ is totally ordered, we can suppose that $\varphi \subset \varphi'$. It follows that $(z, u), (w, v) \in \varphi'$—that is, $\varphi'(z) = u$ and $\varphi'(w) = v$. Sine φ' respects the order and $z \preceq w$ by hypothesis, it follows that $u \preceq v$—that is, $\bar{\varphi}(z) \preceq \bar{\varphi}(w)$.

`Step III.` We consider the maximum $\bar{\varphi} \colon \bar{a} \to \bar{b}$ of Φ, defined in step II. If $\bar{a} = x$ and $\bar{b} = y$, then alternative (i) of the statement holds. If $\bar{a} = x$ and $\bar{b} \subsetneq y$, then (ii) holds. If $\bar{a} \subsetneq x$ and $\bar{b} = y$, then (iii) holds. We have to show that it is not possible that $\bar{a} \subsetneq x$ and $\bar{b} \subsetneq y$ at the same time. In fact, if this is the case, then there exist $z := \min(x \setminus \bar{a})$ and $w := \min(y \setminus \bar{b})$. We define $\bar{\varphi}' \colon \bar{a} \cup \{z\} \to \bar{b} \cup \{w\}$ by $\bar{\varphi}'|_{\bar{a}}\|_{\bar{b}} = \bar{\varphi}$ and $\bar{\varphi}'(z) = w$. The subsets $\bar{a} \cup \{z\}$ and $\bar{b} \cup \{w\}$ are initial segments respectively of x and y because of exercise 8.18 and it is straightforward to verify that $\bar{\varphi}'$ is an ordered-set isomorphism. It follows that $\bar{\varphi}' \in \Phi$. By construction, $\bar{\varphi} \subsetneq \bar{\varphi}'$, that is a contradiction, since $\bar{\varphi}$ is the maximum of Φ. □

The following corollary will be useful in chapter 10.

Corollary 8.22. *If $(x, \preceq)$ is an infinite well-order, such that every proper initial segment is finite, then there exists a unique isomorphism between $(x, \preceq)$ and $(\mathbb{N}, \leq)$.*

Proof. Since both $(x, \preceq)$ and $(\mathbb{N}, \leq)$ are well-orders, one of the alternatives of theorem 8.21 holds. Nevertheless, alternatives (ii) and (iii) cannot hold, since x and $\mathbb{N}$ are both infinite, while any proper initial segment of them is finite (in the case of $\mathbb{N}$, that's because every proper initial segment is of the form $\mathrm{prec}_{\mathbb{N}}(n) = n$ by proposition 8.17). Therefore, alternative (i) holds. Uniqueness immediately follows from proposition 8.9. □

---

[3]The fact that a set x satisfies $\left(\bigcup x\right) \in x$ is not pathological. For example, $x = \{\{y\}, \{z\}, \{y, z\}\}$ has this property, since $\bigcup x = \{y, z\}$. It simply means that there exists an element of x that contains all of the elements of the other elements of x.

CHAPTER 9

Axioms of Choice

We introduce four equivalent assumptions: the axiom of choice, Zorn's lemma, the Hausdorff maximal principle, and the well-ordering principle. We refer to these assumptions globally as "axioms of choice". In this chapter, we analyse each of them and we prove their equivalence.

9.1 THE AXIOM OF CHOICE

The axiom of choice may seem self-evident at a first glance. Therefore, a beginner may have some difficulties in understanding why it does not immediately follow from the other assumptions made up to now. Nevertheless, it is possible to develop the intuition in order to understand the non-triviality of this axiom, that is a cornerstone of contemporary mathematics.

The most intuitive formulation of the axiom of choice is the following one:

($\star$) For any set x, such that $\emptyset \notin x$, there exists a function $\xi \colon x \to \bigcup x$ such that $\xi(y) \in y$ for every $y \in x$.

Statement ($\star$) affirms intuitively that, given a set x, it is possible to select at the same time an element of each element of x, provided that no element of x is empty. In fact, this is the task of the function ξ in the statement, called *choice function*. We observe that $\xi \in \prod x$ by definition of Cartesian product, hence ($\star$) is equivalent to state that, if $\emptyset \notin x$, then $\prod x \neq \emptyset$.

If the reader is a beginner in set theory and foundations of mathematics, probably (s)he will be asking herself or himself the following questions: "isn't it obvious? if every element of x is not empty, what forbids me to pick up an element of each element of x?" In order to reply to these (non-trivial) questions, let us imagine to have a dialogue with this reader. We ask her or him: "how do you pick up concretely an element of each element of x"? The answer will probably be: "I choose an element $y \in x$ by chance and, since it is not empty, I pick up any element $z \in y$; afterwards, I choose $y' \in x$, different from y, and I pick up $z' \in y'$, and so on." This procedure seems quite natural, but it is endless if the set x is infinite, thus it is not acceptable in general.

DOI: 10.1201/9781003565291-9

If we were working only with finite sets, then the axiom of choice would be redundant, but this is not the case.

Therefore, considering an infinite set, how do we pick up an element of each element? Let us consider the following example by Bertrand Russell (the same of Russell's paradox). We have the two following (non-mathematical) sets: x_1 is an infinite collection of pairs of shoes, while x_2 is an infinite collection of pairs of socks. If we want to pick up an element of each element of x_1, we have a simple criterion: we select the right (or the left) shoe in each pair. What about x_2? There is no distinction between left and right socks, and we cannot pick up one sock by chance for each pair, since the collection of pairs is infinite. Hence, how do we proceed? Actually, if we imagine these pairs embedded in a 3D space, then the reader may think that there exist some ways to fix a reference frame and decide how to select a sock in each pair (for example, by fixing some base-points in the shape of a sock and measuring their coordinates). It would be quite difficult to write down exactly such a criterion, but, in order to eliminate any doubt, we can think of an infinite collection of pairs of concrete objects of any shape. In this case, it is quite evident that there is no way to select concretely one element for each pair in a finite time, neither by performing this operation pair by pair (it would be endless), nor by establishing a criterion that can be expressed through a finite sentence or formula.

This example shows that, if we believe that a choice function always exists, then we must state its existence as an independent axiom. This is precisely the task of the axiom of choice.[1]

9.1.1 Formulation of the Axiom

We state the axiom of choice in eight equivalent formulations.

AC1. For any set x, if $\emptyset \notin x$, then $\prod x \neq \emptyset$—that is, there exists a function $\xi \colon x \to \bigcup x$ such that $\xi(y) \in y$ for every $y \in x$.

AC2. For any indexed set $x = \{x_i\}_{i \in I}$, such that $x_i \neq \emptyset$ for every $i \in I$, the Cartesian product $\prod_{i \in I} x_i$ is not empty—that is, there exists a function $\xi \colon I \to \bigcup x$, $i \mapsto y_i$, such that $y_i \in x_i$ for every $i \in I$.

AC3. For any function $f \colon y \to x$, such that $f(w) \neq \emptyset$ for every $w \in y$, there exists a function $\xi \colon y \to \bigcup x$ such that $\xi(w) \in f(w)$ for every $w \in y$.

AC4. For any set x, there exists a function $\xi \colon \mathcal{P}^*(x) \to x$ such that $\xi(a) \in a$ for every $a \in \mathcal{P}^*(x)$.

[1] See the conclusions for a more detailed discussion about the independence of the axiom of choice.

AC5. For any set x, with the following two properties:

(i) $\emptyset \notin x$

(ii) for any $y, z \in x$, if $y \neq z$, then $y \cap z = \emptyset$

there exists a set c such that, for every $y \in x$, the intersection $c \cap y$ contains exactly one element.

AC6. For any set x and any partition p of x, there exists a set c such that, for every $a \in p$, the intersection $c \cap a$ contains exactly one element.

AC7. If $f\colon x \to y$ is a surjective function between sets, then it admits a right inverse—that is, there exists a function $g\colon y \to x$ such that $f \circ g = \mathrm{id}_y$.

AC8. For any binary relation (x, y, ρ), there exists a function $f\colon \mathrm{dom}(\rho) \to \mathrm{Im}(\rho)$ such that $f \subset \rho$—that is, $z \,\rho\, f(z)$ for every $z \in \mathrm{dom}(\rho)$.

Exercise 9.1. Prove that these formulations of the axiom of choice are equivalent.

9.1.2 Partition Principle

Let us consider the following statement:

> *Partition principle:* If there exists a surjective function $f\colon x \to y$, then there exists an injective function $g\colon y \to x$.

We will justify the name "partition principle" in section 10.1.1. Obviously, the axiom of choice implies the partition principle, since, in formulation AC7, the right-inverse function $g\colon y \to x$ in injective. What about the converse? Surprisingly, it is still an open problem.

9.1.3 The Axiom of Choice in Mathematics

Among the axioms of choice, Zorn's lemma is the most used in ordinary mathematics, therefore it will be easier to show concrete applications when introducing that formulation. However, the name "Zorn's lemma", and not "Zorn's axiom", is due to the fact that the axiom of choice is commonly stated as an hypothesis and the other equivalent principles as consequences.

Independently of the formulation chosen, the applications of this axiom are extremely relevant. A huge part of modern mathematics would fail without it. Nevertheless, there are some mathematicians that do not accept such assumption. That's because any proof relying on it is not constructive, since this axiom is used precisely to assume the existence of functions that we cannot construct explicitly (see section 9.1). Therefore, it usually leads to abstract existence proofs of mathematical objects that are not concretely computable.

For this reason, it is common in the mathematical literature to indicate explicitly when the axiom of choice is used in a proof, even if the authors assume it in general.

Each mathematician is free to choose her or his preferred point of view, but honestly we see no advantages in refusing the axiom of choice, while there are plenty of disadvantages. Moreover, the fact of not knowing an object concretely does not mean that its existence has no practical consequences. For example, many results in functional analysis, that are based on the axiom of choice, can be concretely applied in quantum mechanics. Similar considerations hold about plenty of statements in geometry, algebra, analysis, mathematical physics, and so on. In order to clarify this concept, let us make a non-mathematical example. Let us suppose that the police, through suitable investigations, manage to prove that someone put a bomb in a building, but without knowing which kind of bomb it is and where it is exactly. This is an existence result: there is a bomb, but we cannot describe it concretely (if it were a mathematical object, it would be non-computable). Is this a result without concrete applications? Of course not! The authorities can evacuate the building and save tens of human lives.

Therefore, from now on we make the following assumption:

(A8) *The axiom of choice holds.*

9.2 ZORN'S LEMMA AND THE HMP

As we anticipated above, among the axioms of choice, Zorn's lemma is the most applied one in every area of ordinary mathematics. Since it is strictly related to the Hausdorff maximal principle, we discuss both in this section and we show their equivalence.

9.2.1 Zorn's Lemma

We recall that an *ordered set* is a pair $(x, \preceq)$, where x is a set and $\preceq$ is an order relation in x. We define the corresponding strict order relation as follows: $y \prec z$ if and only if $y \preceq z$ and $y \neq z$ (cfr. proposition 3.28). Let us recall and state some definitions about ordered sets.

Definition 9.2. Given an ordered set $(x, \preceq)$:

- a *chain* in $(x, \preceq)$ is a subset $a \subset x$ such that the restriction $\preceq|_{a\times a}$ is a *total* order relation;[2]
- for any subset $a \subset x$, an *upper bound* of a is an element $y \in x$ such that $z \preceq y$ for every $z \in a$;

[2]Intuitively, the word "chain" may induce the reader to think of a discrete set, but this is not the case in general. For example, since $(\mathbb{R}, \leq)$ is totally ordered, $\mathbb{R}$ itself (or any other subset) is a chain.

- a *maximal element* of $(x, \preceq)$ is an element $y \in x$ such that there not exists any $z \in x$ such that $y \prec z$. ◇

ZORN'S LEMMA. Let $(x, \preceq)$ be an ordered set such that every chain admits an upper bound. Then, there exists a maximal element in x.

Assumption (A8) can be equivalently stated as follows:

(A8') *Zorn's lemma holds.*

In order to really understand the intuitive meaning of Zorn's lemma, it is necessary to know the basic notions about ordinals and cardinals, that will be introduced respectively in chapters 11 and 12. Anyway, we try to sketch such an intuitive explanation here. Let us suppose that $(x, \preceq)$ satisfies the hypothesis but not the thesis of Zorn's lemma. Trivially $x \neq \emptyset$, otherwise $\emptyset \subset x$ would be a chain with no upper bound. Hence, let us start from any element $y_0 \in x$. Since x has no maximal elements, y_0 is not maximal, hence there exists $y_1 \in x$ such that $y_0 \prec y_1$. Keeping on in this way, we construct a chain $y_0 \prec y_1 \prec y_2 \prec \cdots$. Since the hypothesis of Zorn's lemma holds, the chain $\{y_n\}_{n \in \mathbb{N}}$ admits an upper bound z_0, that is not a maximal element. Hence we can keep on and construct the chain $y_0 \prec y_1 \prec y_2 \prec \cdots \prec z_0 \prec z_1 \prec z_2 \prec \cdots$. Again this chain admits an upper bound w_0 and we can keep on again. We get an infinite sequence of chains $\{y_n\}_{n \in \mathbb{N}}, \{z_n\}_{n \in \mathbb{N}}, \{w_n\}_{n \in \mathbb{N}}, \ldots$, that is a chain itself, hence it admits an upper bound and we can start again. The idea is that, by iterating this procedure, we can construct a bigger and bigger chain, so that we manage to embed in x a proper class, that is a contradiction by assumption (A4").[3]

Zorn's lemma just states the existence of a maximal element, but a stronger condition follows as a corollary: if x satisfies the hypothesis of the lemma (and the lemma holds), then, for every $y \in x$, there exists a maximal element of x bigger or equal to y. Let us prove this fact.

Corollary 9.3 (of Zorn's lemma). *Let $(x, \preceq)$ be an ordered set such that every chain admits an upper bound. Then, for every $y \in x$, there exists a (not necessarily unique) maximal element $m \in x$ such that $y \preceq m$.*

Proof. We fix any $y \in x$ and we set $a := \{z \in x : y \preceq z\} \subset x$. Surely $a \neq \emptyset$, since $y \in a$. We endow a with the order induced by restriction from x.

Step I: the ordered set a satisfies the hypothesis of Zorn's lemma. Let $c \subset a$ be a chain and let us show that it admits an upper bound in a. In fact, if $c = \emptyset$, then y is an upper bound of c and $y \in a$. If $c \neq \emptyset$, then, since $c \subset x$, there exists an upper bound $u \in x$ of c. Taking any element $z \in c$, we have that $y \preceq z \preceq u$ (the first inequality being due to the fact that $z \in c \subset a$ and the second one to the definition of upper bound), hence $y \preceq u$, thus $u \in a$.

[3]This intuitive idea seems to be the sketch of a proof that does not require the axiom of choice, but this is not the case. In fact, for every partial chain we constructed in x, we had to choose an upper bound. Here we used AC.

Step II: there exists a maximal element m of $(x, \preceq)$ such that $m \in a$. Because of step I, we can apply Zorn's lemma to a and obtain a maximal element m of a. It remains to show that m is maximal also in x. In fact, if there exists $w \in x$ such that $m \prec w$, then $y \preceq m \prec w$, thus $y \prec w$, hence $w \in a$. This implies that m is not maximal in a, a contradiction. □

Remark: The following two exercises require some knowledge of basic algebra.

Exercise 9.4. Let $(A + \cdot)$ be a commutative ring with unit, different from the zero-ring. We recall that an ideal $I \lhd A$ is *proper* if $I \neq A$ and it is *maximal* if it is proper and it is not properly contained in any other proper ideal. Prove that, assuming Zorn's lemma:

- there exists a maximal ideal of A;
- if $I \lhd A$ is a proper ideal, then there exists a maximal ideal of A that contains I;
- if $a \in A$ is not invertible, then there exists a maximal ideal of A that contains a.

Exercise 9.5. Let V be a vector space over the field $\mathbb{K}$ (not necessarily finite-dimensional) different from $\{0\}$. Prove that, assuming Zorn's lemma:

- there exists a basis of V;
- if $\mathcal{A} \subset V$ is linearly independent, then there exists a basis of V that contains $\mathcal{A}$.

9.2.2 Hausdorff Maximal Principle

HAUSDORFF MAXIMALITY PRINCIPLE. Let $(x, \preceq)$ be an ordered set. There exists a maximal chain in x.

Let us analyse the statement. Given an ordered set $(x, \preceq)$, we call $\boldsymbol{c}_x$ the set of chains of x. It follows that $\boldsymbol{c}_x \subset \mathcal{P}(x)$ or, equivalently, $\boldsymbol{c}_x \in \mathcal{P}(\mathcal{P}(x))$. Such a set is ordered too by inclusion, therefore the ordered set $(x, \preceq)$ induces the ordered set $(\boldsymbol{c}_x, \subset)$. The Hausdorff maximal principle states that $(\boldsymbol{c}_x, \subset)$ admits a maximal element, with no hypotheses on $(x, \preceq)$. Let us show the equivalence with Zorn's lemma.

Proposition 9.6. *Zorn's lemma (ZL) and the Hausdorff maximal principle (HMP) are equivalent.*

Proof. (ZL $\Rightarrow$ HMP) Let us show that, given any ordered set $(x, \preceq)$, the ordered set $(\boldsymbol{c}_x, \subset)$ satisfies the hypothesis of Zorn's lemma, so that the thesis immediately follows.

Step I: If $c \subset \boldsymbol{c}_x$—that is, c is a set of chains of x—then $\bigcup c \subset x$. In fact, if $y \in \bigcup c$, then there exists $a \in c$ such that $y \in a$. Since $a \in c$ and $c \subset \boldsymbol{c}_x$, we deduce that $a \in \boldsymbol{c}_x$—that is, a is a chain of x. In particular, $a \subset x$. Since $y \in a$ and $a \subset x$, it follows that $y \in x$.

Step II: If $c \subset \mathfrak{c}_x$ is a chain, then $\bigcup c \subset x$ is a chain. By step I, we know that $\bigcup c \subset x$, hence we have to prove that it is totally ordered. If $y, z \in \bigcup c$, by definition there exist $a, b \in c$ such that $y \in a$ and $z \in b$. Since c is a chain, $a \subset b$ or $b \subset a$. Up to exchange them, we suppose $a \subset b$. It follows that $y, z \in b$. Since b is a chain, we have that $y \preceq z$ or $z \preceq y$, hence $\bigcup c$ is totally ordered.

Step III: If $c \subset \mathfrak{c}_x$ is a chain, then $\bigcup c \in \mathfrak{c}_x$ is an upper bound of c. By step II, we know that $\bigcup c \in \mathfrak{c}_x$. For every $a \in c$, we have $a \subset \bigcup c$ by exercise 1-5 in section 4.5. Since '$\subset$' is the order relation in $\mathfrak{c}_x$, we get the result.

(HMP $\Rightarrow$ ZL) Let $(x, \preceq)$ be an ordered set satisfying the hypothesis of Zorn's lemma. By hypothesis HMP holds, hence there exists a maximal chain $a \subset x$. Since the hypothesis of ZL holds, there exists an upper bound $y \in x$ of the chain a. We show that y is a maximal element of x. In fact, otherwise, there exists $z \in x$ such that $y \prec z$. The reader can easily verify that the subset $a' := a \cup \{z\} \subset x$ is a chain and $a \subsetneq a'$, contradicting the maximality of a. □

Assumption (A8) or (A8') can be equivalently stated as follows:

(A8") The Hausdorff maximal principle holds.

The Haudorff maximal principle just states the existence of a maximal chain. Actually, a stronger condition follows as a corollary: any chain in an ordered set is contained in a (not necessarily unique) maximal chain.

Corollary 9.7 (of HMP). *Let $(x, \preceq)$ be an ordered set. Every chain of x is contained in a (not necessarily unique) maximal chain.*

Proof. Let $(x, \preceq)$ be an ordered set and $c \subset x$ be a chain. We have seen in the proof of proposition 9.6 that the ordered set $(\mathfrak{c}_x, \subset)$ satisfies the hypothesis of Zorn's lemma, hence, knowing that HMP is equivalent to ZL, we can apply corollary 9.3 and find a maximal chain m such that $c \subset m$. □

Exercise 9.8. We have seen that, given an ordered set $(x, \preceq)$, we get the ordered set $(\mathfrak{c}_x, \subset)$. We can iterate this construction, getting the ordered set $(\mathfrak{c}_{\mathfrak{c}_x}, \subset)$ and so on. Let $\mathfrak{m} \in \mathfrak{c}_{\mathfrak{c}_x}$ be a maximal element. Prove that $m := \bigcup \mathfrak{m} \in \mathfrak{c}_x$ is a maximal element. *Remark:* This result holds independently of HMP.

Exercise 9.9. Prove corollary 9.7 directly from HMP, without passing through ZL. *Hint:* Let $c \subset x$ be a chain—that is, $c \in \mathfrak{c}_x$. We set $\mathfrak{c}_{x,c} := \{c' \in \mathfrak{c}_x : c \subset c'\} \subset \mathfrak{c}_x$. Since $(\mathfrak{c}_{x,c}, \subset)$ is an ordered set, by the HMP there exists a maximal chain $\mathfrak{m} \in \mathfrak{c}_{\mathfrak{c}_{x,c}}$. Show that $m := \bigcup \mathfrak{m} \in \mathfrak{c}_x$ is a maximal element containing c.

9.3 WELL-ORDERING PRINCIPLE

We introduced well-ordered sets in chapter 8.

ZERMELO'S WELL-ORDERING PRINCIPLE. Any set can be endowed with a well-order.

Assumption (A8), (A8') or (A8") can be equivalently formulated as follows:

(A8"') The well-ordering principle holds.

9.4 EQUIVALENCE

Let us prove the equivalence between the axiom of choice, Zorn's lemma and the well-ordering principle, the proof relying on assumptions (A1)–(A6) (we do not need (A7) here). We recall that we already proved the equivalence between Zorn's lemma and the Hausdorff maximal principle in section 9.2.

Theorem 9.10. *The axiom of choice (AC), Zorn's lemma (ZL) and the well-ordering principle (WO) are equivalent.*

Proof. (ZL $\Rightarrow$ WO) Let x be a set. We can suppose that $x \neq \emptyset$, since WO trivially holds for the empty set. We define the following set:

$$w_x := \{\preceq_a : a \subset x \text{ and } \preceq_a \text{ is a well-order in } a\}.$$

We are thinking of $\preceq_a$ as a pre-relation in a; since $a = \mathrm{dom}(\preceq_a) = \mathrm{Im}(\preceq_a)$ by reflexivity, we do not miss any piece of information. The set w_x is not empty, since, fixing any $y \in x$, we have $\preceq_{\{y\}} := \{(y, y)\} \in w_x$. We define the following order relation in w_x:

$$\preceq_a \preceq \preceq_b \quad \overset{\text{def.}}{\Longleftrightarrow} \quad \begin{cases} (1) & a \subset b \text{ and } \preceq_a = \preceq_b |_{a \times a} \\ (2) & a \text{ is an initial segment of } (b, \preceq_b) \end{cases}$$

Exercise 9.11. Show that condition (1) is equivalent to $\preceq_a \subset \preceq_b$.

It is straightforward to verify that $\preceq$ is actually an order relation (see exercise 8.15-i for transitivity).

Step I: The ordered set $(w_x, \preceq)$ satisfies the hypothesis of ZL. In fact, let $c \subset w_x$ be a non-empty chain. We set $\preceq_{\bar{a}} := \bigcup c$.

Exercise 9.12. Verify that $\preceq_{\bar{a}}$ is a well-order pre-relation in $\bar{a} = \bigcup\{a \subset x : \exists \preceq_a \in c\}$.

It follows from exercise 9.12 that $\preceq_{\bar{a}} \in w_x$. Let us prove that it is an upper bound of c. For every $\preceq_a \in c$, by definition of union we have $\preceq_a \subset \preceq_{\bar{a}}$. We have to verify that a is an initial segment of $(\bar{a}, \preceq_{\bar{a}})$. If $y \in a$, $z \in \bar{a}$ and $z \preceq_{\bar{a}} y$, then there exists $\preceq_b \in c$ such that $z \preceq_b y$ (thus $y, z \in b$). It follows that $y \in a \cap b$, hence, since c is a chain, we have two possibilities: I. a is an initial segment of $(b, \preceq_b)$; in this case, $z \in a$ by definition of initial segment, since $y \in a$, $z \in b$ and $z \preceq_b y$; II. b is an initial segment of $(a, \preceq_a)$; in this case, $z \in b \subset a$, hence $z \in a$. In both cases $z \in a$, as required.

Since ZL holds by hypothesis, by step I there exists a maximal element in w_x, that we denote by $(m, \preceq_m)$.

Step II: Necessarily $m = x$, hence $\preceq_m$ is a well-order in x, proving WO. In fact, let us suppose by contradiction that there exists $y_0 \in x \setminus m$. We set $m' := m \cup \{y_0\}$ and we define the pre-relation $\preceq_{m'} := \preceq_m \cup \{(z, y_0) : z \in m'\}$. It follows that $\preceq_{m'} |_{m \times m} = \preceq_m$ and $z \prec_{m'} y_0$ for every $z \in m$ (that is, y_0 is the maximum of m').

Exercise 9.13. Verify that $(m', \preceq_{m'})$ is a well-ordered set and that m is an initial segment of $(m', \preceq_{m'})$.

It follows from exercise 9.13 that $\preceq_{m'} \in w_x$ and $\preceq_m \prec \preceq_{m'}$, contradicting the maximality of $\preceq_m$.

(WO $\Rightarrow$ AC6) Given a non-empty set x, by hypothesis we can endow it with a well-order. We get the choice function $\xi \colon \mathcal{P}^*(x) \to x$, $a \mapsto \min(a)$.

(AC6 $\Rightarrow$ ZL) We follow [1].[4] By contradiction, let $(x, \preceq)$ be a non-empty ordered set that satisfies the hypothesis but not the thesis of ZL. Let $\xi \colon \mathcal{P}^*(x) \to x$ be a choice function.

Step I: If $c \subset x$ is a chain, then it admits a strict *upper bound.* In fact, since the hypothesis of ZL holds, there exists an upper bound $u \in x$ of c. This means that $y \preceq u$ for every $y \in c$. We have two possibilities: I. $u \notin c$; II. $u \in c$. In case I, necessarily $y \prec u$ for every $y \in c$, hence u itself is a strict upper bound. In case II, since x has no maximal elements by hypothesis, in particular u is not maximal, hence there exists $v \in x$ such that $u \prec v$. It follows that $y \preceq u \prec v$ for every $y \in c$, hence v is a strict upper bound of c.

Idea of the proof: Considering step I, we can use the choice function ξ to associate a strict upper bound to each chain of x. We call η such a function—that is, $\eta(c)$ is a strict upper bound of the chain c. Hence, we start from $y_0 := \eta(\emptyset) \in x$. Then, since $\{y_0\}$ is a chain, we set $y_1 := \eta(\{y_0\})$. It follows that $y_0 \prec y_1$, hence $\{y_0, y_1\}$ is a chain. We set $y_2 := \eta(\{y_0, y_1\})$ and we keep on in this way. We get the chain $\{y_0, y_1, y_2, \ldots\}$ and we set $z_0 := \eta(\{y_0, y_1, y_2, \ldots\})$. Again we keep on in this way, obtaining the chain $\{y_0, y_1, \ldots, z_0, z_1, \ldots\}$. Iterating this procedure, we get a chain formed by a denumerable family of sub-chains, and we keep on again. Since it is always possible to choose a strict upper bound, we can construct a bigger and bigger chain. We have to show that the union of all these chains, each one being an initial segment of the bigger ones, is a chain that is "too big" to admit a strict upper bound. This is a contradiction, since step I affirms that every chain in x admits at least one.

We prove this statement as follows. Every chain c, constructed as above, satisfies the following property: ($\star$) *For every proper initial segment a of c, we have* $\min(c \setminus a) = \eta(a)$. That's because each new element has been chosen by applying η to the previous chain. We call $\bar{c}$ the union of all the chains that satisfy ($\star$), and we verify that $\bar{c}$ is a chain that satisfies ($\star$) too, hence it is the maximum of the set of chains with this property. Nevertheless, $\bar{c}$ admits a strict upper bound (as every other chain), hence we can construct $\bar{\bar{c}} := \bar{c} \cup \{\eta(\bar{c})\}$, that satisfies ($\star$) and is strictly bigger than $\bar{c}$, a contradiction.

---

[4]The author claims that this proof was suggested to him by Prof. James Cummings.

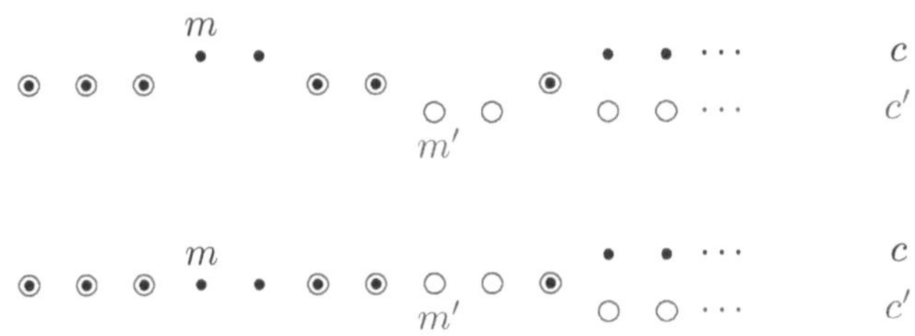

Figure 9.1 Two possible situations.

Step II: Construction of suitable well-ordered chains. Let $\boldsymbol{c}_x$ be the set of chains of x. Given $c \in \boldsymbol{c}_x$, we call $\mathrm{subs}(c) \subset x$ the set of strict upper bounds of c. Step I affirms that $\mathrm{subs}(c) \neq \emptyset$. We define the function $\eta\colon \boldsymbol{c}_x \to x$, $c \mapsto \xi(\mathrm{subs}(c))$. Since ξ is a choice function, it follows that $\eta(c) \in \mathrm{subs}(c)$ for every chain c—that is, η associates to each chain of x a corresponding strict upper bound. Moreover, we set $y_0 := \xi(x)$. We define the set $\boldsymbol{c}_{x,\eta} \subset \boldsymbol{c}_x$ as follows:

$$\boldsymbol{c}_{x,\eta} := \left\{ c \in \boldsymbol{c}_x : \begin{array}{l} \bullet\ c \text{ is well-ordered} \\ \bullet\ \min(c) = y_0 \\ \bullet\ \text{if } a \subset c \text{ is a pr. in. segment, then } \min(c \setminus a) = \eta(a) \end{array} \right\}.$$

The set $\boldsymbol{c}_{x,\eta}$ is not empty, since $\{y_0\} \in \boldsymbol{c}_{x,\eta}$. In fact, obviously $\{y_0\}$ is well-ordered and its minimum is y_0. Moreover, the only proper initial segment of $\{y_0\}$ is $\emptyset$, and $\min(\{y_0\} \setminus \emptyset) = y_0 = \xi(x) = \xi(\mathrm{subs}(\emptyset)) = \eta(\emptyset)$.

Step III: $\boldsymbol{c}_{x,\eta}$ is a chain of $(\boldsymbol{c}_x, \subset)$. In fact, let us fix $c, c' \in \boldsymbol{c}_{x,\eta}$ and suppose by contradiction that $c \not\subset c'$ and $c' \not\subset c$. Since c and c' are well-ordered, it follows that there exist $m := \min(c \setminus c')$ and $m' := \min(c' \setminus c)$. Since $m \in c$ and $m' \notin c$ (equivalently, since $m' \in c'$ and $m \notin c'$), necessarily $m \neq m'$, hence it cannot happen that $m \preceq m'$ and $m' \preceq m$. Up to exchange c and c', we suppose that $m' \not\preceq m$. Figure 9.1 represents two possible situations (in the first case m and m' are not comparable, in the second case we have $m \prec m'$).

We set $d := \mathrm{prec}_c(m) = \{y \in c : y \prec m\}$. Since $y_0 \in d$, the set d is not empty. We have that:

(i) $d \subset c \cap c'$. In fact, $d \subset c$ by definition. Moreover, if $y \in d$, then $y \notin c \setminus c'$ by definition of m, hence $y \in c'$.

(ii) d is a proper initial segment of c and $m = \min(c \setminus d)$. This is obvious by definition of d, since $d := \mathrm{prec}_c(m)$.

(iii) d is a proper initial segment of c'. In fact, let us suppose that $y \in d$, $z \in c'$ and $z \prec y$. If $z \notin c$, then $z \in c' \setminus c$, hence $m' \preceq z$. It follows that $m' \preceq z \prec y \prec m$, hence $m' \prec m$, against the hypothesis $m' \not\preceq m$. It follows that $y \in d$, $z \in c$ and $z \prec y$, hence $z \in d$ by item (ii). This proves that d is an initial segment of c'. It is proper, since $m' \in c' \setminus c \subset c' \setminus d$.

Figure 9.2 The contradictory hypothesis.

Since $c \in \boldsymbol{c}_{x,\eta}$ by construction, it follows from item (ii) and the definition of $\boldsymbol{c}_{x,\eta}$ that $\eta(d) = \min(c \setminus d) = m$. Similarly, since $c' \in \boldsymbol{c}_{x,\eta}$ by construction, it follows from item (iii) and the definition of $\boldsymbol{c}_{x,\eta}$ that $\eta(d) = \min(c' \setminus d) \in c'$. This is a contradiction, since $m \notin c'$.

Step IV: If $c, c' \in \boldsymbol{c}_{x,\eta}$, then c is an initial segment of c' or vice-versa. The result is trivial if $c = c'$, hence we suppose that $c \neq c'$. By step III we have $c \subsetneq c'$ or $c' \subsetneq c$. We assume that $c' \subsetneq c$ and we set again $m := \min(c \setminus c')$ and $d := \operatorname{prec}_c(m)$. Trivially $d \subset c'$ by definition of m. Let us prove that $d = c'$, hence the result immediately follows. Figure 9.2 represents the contradictory hypothesis $d \subsetneq c'$.

By construction $\eta(d) = \min(c \setminus d) = m$. Moreover, $d \subset c'$ by definition of m, $c' \subset c$ by hypothesis and d is an initial segment of c by construction, hence, by exercise 8.15-ii, d is an initial segment of c'. Therefore, if $d \neq c'$, then $\eta(d) = \min(c' \setminus d) \in c'$, a contradiction, since by definition $m \in c \setminus c'$.

Step V: We set $\bar{c} := \bigcup \boldsymbol{c}_{x,\eta}$. We have that $\bar{c} \in \boldsymbol{c}_{x,\eta}$ and every $c \in \boldsymbol{c}_{x,\eta}$ is an initial segment of $\bar{c}$. In fact, we prove the following statements:

(a) $\bar{c}$ is well-ordered.

(b) $y_0 = \min(\bar{c})$;

(c) every $c \in \boldsymbol{c}_{x,\eta}$ is an initial segment of $\bar{c}$

(d) $\bar{c} \in \boldsymbol{c}_{x,\eta}$

(a) If $a \subset \bar{c}$ and $a \neq \emptyset$, let us show that a admits a minimum. We fix $z_0 \in a$. By definition of union, there exists $c \in \boldsymbol{c}_{x,\eta}$ such that $z_0 \in c$, hence $a \cap c \neq \emptyset$. We set $m := \min(a \cap c)$, the latter being well-defined since $a \cap c \subset c$ and c is well-ordered. Let us prove that $m = \min(a)$. In fact, fixing any $y \in a$, there exists $c' \in \boldsymbol{c}_{x,\eta}$ such that $y \in c'$. If by contradiction $y \prec m$, then $y \in c' \setminus c$, since $m = \min(a \cap c)$. It follows from step IV that c is an initial segment of c'. Since $m \in c$, $y \in c'$ and $y \prec m$, by definition of initial segment we have $y \in c$, a contradiction. This shows that $\bar{c}$ is well-ordered.

(b) Trivially $y_0 \in \bar{c}$ by definition of union, since $y_0 \in c$ for any $c \in \boldsymbol{c}_{x,\eta}$. If $y \in \bar{c}$, by definition of union there exists $c \in \boldsymbol{c}_{x,\eta}$ such that $y \in c$. Since $y_0 = \min(c)$, it follows that $y_0 \preceq y$.

(c) If $c \in \boldsymbol{c}_{x,\eta}$, then $c \subset \bar{c}$ by definition of union. If $y \in c$, $z \in \bar{c}$ and $z \prec y$, then, again by definition of union, there exists $c' \in \boldsymbol{c}_{x,\eta}$ such that $z \in c'$. If $z \notin c$—that is, $z \in c' \setminus c$—then by step IV c is an initial segment of c', a contradiction, since $y \in c$, $z \in c'$ and $z \prec y$, but $z \notin c$. This proves that $z \in c$, hence c is a proper initial segment of $\bar{c}$.

(d) Considering items (a) and (b), it remains to prove that, if $d \subset \bar{c}$ is a proper initial segment, then $\min(\bar{c} \setminus d) = \eta(d)$. We set $y := \min(\bar{c} \setminus d)$. By item (a) and proposition 8.17, $d = \mathrm{prec}_{\bar{c}}(y)$. There exists $c \in \boldsymbol{c}_{x,\eta}$ such that $y \in c$. It follows from item (c) that $d \subset c$, hence $d = \mathrm{prec}_c(y)$, therefore $y = \min(c \setminus d) = \eta(d)$ by hypothesis.

Step VI: The contradiction. By steps I and V there exists a strict upper bound of $\bar{c}$, hence $\eta(\bar{c}) \in \mathrm{subs}(\bar{c})$ is well-defined. We set $\bar{\bar{c}} := \bar{c} \cup \{\eta(\bar{c})\}$.

Exercise 9.14. Verify that $\bar{\bar{c}} \in \boldsymbol{c}_{x,\eta}$.

Since $\bar{c} \subsetneq \bar{\bar{c}}$, exercise 9.14 contradicts the maximality of $\bar{c}$. □

Exercise 9.15. By using the same idea leading to the proof of (ZL $\Rightarrow$ WO) in theorem 9.10, prove that (ZL $\Rightarrow$ AC4) without passing through WO.

Hint: Fixing $x \neq \emptyset$, consider the set of non-empty pre-functions ξ such that $\mathrm{dom}(\xi) \subset \mathcal{P}^*(x)$ and $\xi(a) \in a$ for every $a \in \mathrm{dom}(\xi)$. Such a set is not empty, since, for any $y \in x$, it contains the unique possible pre-function $\xi\colon \{\{y\}\} \to \{y\}$. Order the set of such pre-functions by inclusion and prove that it satisfies the hypothesis of Zorn's lemma, therefore it admits a maximal element $\bar{\xi}$. Conclude by showing that necessarily $\mathrm{dom}(\bar{\xi}) = \mathcal{P}^*(x)$, hence $(\mathcal{P}^*(x), x, \bar{\xi})$ is a choice function.

From now on the equivalent assumptions `(A8)`, `(A8')`, `(A8")` and `(A8"')` will hold for us, but we will always specify when they are necessary in the proof of a certain statement.

9.5 EXTENSION TO CLASSES

The choice axioms have been formulated about sets, therefore it is natural to inquire if they can be extended to classes. In principle it is possible, but the extended statements are strictly stronger than the ones about sets, hence they do not follow as a consequence (they should be imposed as stronger assumptions). Let us briefly analyse each one of them.

`The axiom of choice for classes.` The natural formulation of AC1 for classes is the following one: (AC1!) *For every class* c*, such that* $\emptyset \notin \mathsf{c}$*, there exists a function* $\xi\colon \mathsf{c} \to \bigcup \mathsf{c}$ *such that* $\xi(x) \in x$ *for every* $x \in \mathsf{c}$. Nevertheless, it is enough to formulate such statement for the universe u as follows (we set $\mathsf{u}^* := \mathsf{u} \setminus \{\emptyset\}$): (AC1!!) *There exists a function* $\xi\colon \mathsf{u}^* \to \mathsf{u}$ *such that* $\xi(x) \in x$ *for every set* $x \neq \emptyset$. In fact, (AC1!!) is a particular case of (AC1!), obtained by choosing $\mathsf{c} = \mathsf{u}^*$. Conversely, if (AC1!!) holds, we easily get a choice function for any class by restriction. Assumption (AC1!) or (AC1!!) is called *axiom of global choice.*

`The well-ordering principle for classes.` The natural formulation of WO for classes is the following one: (WO!) *Every class* c *can be endowed with a well-order.* Again, it is enough to formulate such statement for the universe u as follows: (WO!!) u *can be endowed with a well-order.*

`Zorn's lemma for classes.` The natural formulation of ZL for classes is the following one: (ZL!) *Let* $(\mathtt{c}, \preceq)$ *be an ordered class such that every chain admits an upper bound. Then there exists a maximal element in* c. Here a chain is a totally ordered sub*class*, since, if it were only a totally ordered sub*set*, then (ZL!) would be false.[5] In this case we cannot state (ZL!) only for u, since the latter does not satisfy the hypothesis[6] and in any case we could not deduce (ZL!) for any class starting from u. Finally, we can formulate HMP as follows: (HMP!) *Let* $(\mathtt{c}, \preceq)$ *be an ordered class. Then there exists a maximal chain in* c. The equivalence with (ZL!) can be proven as in the case of sets.

`Equivalence.` The three statements (AC1!), (ZL!) and (WO!) are equivalent, the proof being analogous to the one shown above for sets. Anyway, we do not need to assume that they hold and we prefer not to do it. Actually, it is possible to show that they would not add any new information within the framework of sets. Moreover, the formulation of such assumptions implicitly involves the use of hyper-classes, since, for example, (AC1!!) can be equivalently stated by affirming that the hyper-class of choice functions of $\mathtt{u}^*$ is not empty. The same consideration holds about (WO!). In the case of (ZL!), we have to consider *every* subclass that is a chain, hence we are dealing with the hyper-class of chains in c. For these reasons we simply ignore these statements.

---

[5]In order to prove the last affirmation, let us consider the universe u, ordered by set inclusion. Every chain c, that is a set, admits the upper bound $\bigcup c$, but we can easily show that there are no maximal elements in u. In fact, if $x \in \mathrm{u}$, then necessarily $x \subsetneq \mathrm{u}$ (since x is a set), hence there exists $y \in \mathrm{u} \setminus x$. It follows that $x \subsetneq x \cup \{y\}$, therefore x is not maximal.

[6]That's because a chain c, that is not a set, admits no upper bound. In fact, if u is an upper bound of c, then by definition $x \subset u$ for every $x \in c$. It follows from exercise 1-3 in section 4.5 (considering x and as a class in the statement of the latter) that $\bigcup c \subset u$, hence $\bigcup c$ is a set by assumption `(A4")`, thus c is a set by exercise 4-2 in section 6.5. For example, the class **Ord**, that will be defined in chapter 11, is a proper class and a chain, hence it admits no upper bound in the universe.

CHAPTER 10

Cardinality

In this chapter we show that in general there exists no bijection between two infinite sets. In this case, we say that they have *different cardinalities* and we explain how to compare them. Actually, we introduce this comparison between cardinalities *before* defining the notion of cardinality itself. Indeed, we will define what is the cardinality of a set in chapter 12, assuming that we already know when two sets have the same cardinality or not. There is no circularity in these definitions, since comparison will not rely on the definition of cardinality.

Notation 10.1. Since the distinction between meta-theoretical and set-theoretical natural numbers became quite irrelevant after the identification shown in section 7.3.1, from now on we will not use the bold notation (like '$\boldsymbol{n}$' or '$\mathbf{0}$') any more. We will indicate natural numbers with the usual notation (like 'n' or '0') usually thinking of the set-theoretical ones.

10.1 COMPARISON OF CARDINALITIES

Definition 10.2. Let x and y be two sets.

- We say that x and y *have the same cardinality* if there exists a bijection $f\colon x \xrightarrow{\approx} y$. We use the notation $|x| = |y|$.
- We say that *the cardinality of x is lower or equal to the cardinality of y* (equivalently, that the one of y is *bigger or equal* to the one of x) if there exists an injective function $f\colon x \hookrightarrow y$. We use the notation $|x| \leq |y|$.
- We say that *the cardinality of x is lower* (or *strictly lower*) *than the cardinality of y* (equivalently, that the one of y is *(strictly) bigger* than the one of x) if $|x| \leq |y|$ and $|x| \neq |y|$. We use the notation $|x| < |y|$. ◊

As we anticipated above, the symbol $|x|$ has no meaning up to now. We gave a meaning only to the corresponding equalities or inequalities. Moreover, we observe that the second item of definition 10.2 can be reformulated as follows: $|x| \leq |y|$ if and only if there exists $y' \subset y$ such that $|x| = |y'|$.

DOI: 10.1201/9781003565291-10

Remark 10.3. It follows from definitions 7.20-(i, iv) and 10.2 that

(i) a set x contains n elements if and only if $|x| = |n|$; and

(ii) a set x is denumerable if and only if $|x| = |\mathbb{N}|$.

Intuitively, a set is denumerable if it is infinite and we can "list" its elements, a list being a bijection with $\mathbb{N}$. This remark will be completed by exercise 10.19. ◇

Exercise 10.4. Prove that, for every sets x, y, and z:

- $|x| = |x|$.
- $|x| = |y|$ if and only if $|y| = |x|$.
- If $|x| = |y|$ and $|y| = |z|$, then $|x| = |z|$.

This means that the fact of having the same cardinality is an equivalence relation in the class **u** of all sets. *Remark:* The quotient class does not exist, as exercise 2 of section 6.5 shows. We can think of it as a hyperclass.

Exercise 10.5. Prove that:

(i) If $|x_1| = \cdots = |x_n|$, then $|x_1| = |x_n|$.

(ii) If $|x_1| \leq \cdots \leq |x_n|$, then $|x_1| \leq |x_n|$.

Exercise 10.6. Let x be a finite set with n elements. Prove that:

- If y is a finite set with m elements, then $|x| = |y|$ (respectively, $|x| \leq |y|$ or $|x| < |y|$) if and only if $n = m$ (respectively, $n \leq m$ or $n < m$).
- If y is an infinite set, then $|x| < |y|$.

Exercise 10.7. Prove that:

- If x is infinite and $|x| \leq |y|$, then y is infinite.
- If y is finite and $|x| \leq |y|$, then x is finite.

Exercise 10.8. Prove that:

- If $|x| = |y|$, then $|\mathcal{P}(x)| = |\mathcal{P}(y)|$.
- If $|x| \leq |y|$, then $|\mathcal{P}(x)| \leq |\mathcal{P}(y)|$.
- If $|x_1| = |y_1|, \ldots, |x_k| = |y_k|$, then $|x_1 \times \cdots \times x_k| = |y_1 \times \cdots \times y_k|$ and $|x_1 \sqcup \ldots \sqcup x_k| = |y_1 \sqcup \ldots \sqcup y_k|$.
- If $|x_1| \leq |y_1|, \ldots, |x_k| \leq |y_k|$, then $|x_1 \times \cdots \times x_k| \leq |y_1 \times \cdots \times y_k|$ and $|x_1 \sqcup \ldots \sqcup x_k| \leq |y_1 \sqcup \ldots \sqcup y_k|$.

Remark: We will see in exercise 2 of section 12.8 that, assuming AC, the fourth item holds with strict inequalities too.

10.1.1 Comparison through Surjectivity

Comparison of cardinalities through surjective functions is equivalent to the one through injective functions only assuming the axiom of choice, as the following proposition shows. The only exception is the empty set, since, given $y \neq \emptyset$, we have the injective empty function $\emptyset \hookrightarrow y$, but no surjective function $y \twoheadrightarrow \emptyset$.

Proposition 10.9 (AC$\Leftarrow$). *Let x and y be two sets such that $x \neq \emptyset$. The inequality $|x| \leq |y|$ holds if and only if there exists a surjective function $g\colon y \twoheadrightarrow x$.*

Proof. ($\Rightarrow$) Let $f\colon x \hookrightarrow y$ be an injective function. We fix $z_0 \in x$ and we define:

$$
\begin{aligned}
g\colon y &\twoheadrightarrow x \\
w &\mapsto \begin{cases} (f\|_{\mathrm{Im}(f)})^{-1}(w) & w \in \mathrm{Im}(f) \\ z_0 & w \in y \setminus \mathrm{Im}(f). \end{cases}
\end{aligned}
$$

The function g is surjective, since, for every $z \in x$, we have $z = g(f(z)) \in \mathrm{Im}(g)$. ($\Leftarrow$) This is the *partition principle*, that immediately follows from AC7 (see section 9.1.2). □

The direction ($\Leftarrow$) of proposition 10.9 is a reformulation of the partition principle. When we introduced the latter in section 9.1.2, we affirmed that we would justify its name in the present section. The following exercise accomplishes this task.

Exercise 10.10. Prove that (independently of the axiom of choice) the partition principle is equivalent to the following statement: if p is a partition of a set x, then $|p| \leq |x|$.

Without assuming the axiom of choice, we can state the following partial version of proposition 10.9.

Proposition 10.11. *Let x and y be two sets such that $x \neq \emptyset$ and y admits a well-order. The inequality $|x| \leq |y|$ holds if and only if there exists a surjective function $g\colon y \twoheadrightarrow x$.*

Proof. ($\Rightarrow$) The same proof of proposition 10.9 holds. ($\Leftarrow$) Let us endow y with a well-order. We get the function $f\colon x \to y$, $z \mapsto \min(\ddot{g}(\{z\}))$. It is straightforward to verify that $g \circ f = \mathrm{id}_x$, hence f is injective. □

Assuming AC, proposition 10.11 is an equivalent formulation of proposition 10.9, with an alternative proof.

10.1.2 Some Relevant Examples

In this section we assume that the reader knows Dedekind's construction of the field of real numbers, the notion of decimal representation of a real number, and the elementary real functions of one variable. These topics are usually covered by basic calculus courses. With these pre-requisites, we are going to show some relevant examples of comparison of cardinalities.

Example 10.12. $|\mathbb{N}| < |\mathbb{R}|$.

Proof. Let us fix any function $\varphi\colon \mathbb{N} \to \mathbb{R}$. We set $x_i := \varphi(i)$. We consider the decimal representation of each x_i:

$$\begin{aligned} x_0 &= \epsilon_0\, n_0\, .\, d_{0,0}\, d_{0,1}\, d_{0,2} \dots \\ x_1 &= \epsilon_1\, n_1\, .\, d_{1,0}\, d_{1,1}\, d_{1,2} \dots \\ x_2 &= \epsilon_2\, n_2\, .\, d_{2,0}\, d_{2,1}\, d_{2,2} \dots \\ &\vdots \end{aligned}$$

where:

- $n_i \in \mathbb{N}$
- $d_{i,j} \in \{0, \dots, 9\}$, in such a way that, for every $i \in \mathbb{N}$, there not exists $j_0 \in \mathbb{N}$ such that $d_{i,j} = 9\ \forall j \geq j_0$
- $\epsilon_i \in \{1, -1\}$ if $x_i \neq 0$, otherwise $\epsilon_i = 0$ [1]

Let us fix the following real number:

$$y := 0\,.\,e_0\, e_1\, e_2 \dots, \qquad e_i := \begin{cases} 0 & \text{if } d_{i,i} \neq 0 \\ 1 & \text{if } d_{i,i} = 0 \end{cases}$$

We show that $y \notin \mathrm{Im}(\varphi)$—that is, $y \neq x_i$ for every $i \in \mathbb{N}$. In fact, by construction, $e_i \neq d_{i,i}$ for every i, hence the i-th digit of y is different from the i-th digit of x_i for every i. It follows that the decimal representation of y cannot coincide with the decimal representation of x_i for any i. This proves that no function $\varphi\colon \mathbb{N} \to \mathbb{R}$ is surjective (in particular, bijective), thus $|\mathbb{N}| \neq |\mathbb{R}|$. Since the embedding $\mathbb{N} \hookrightarrow \mathbb{R}$ is injective, $|\mathbb{N}| \leq |\mathbb{R}|$. Therefore, $|\mathbb{N}| < |\mathbb{R}|$. □

The previous example, proved by Cantor, can be considered historically the starting point of modern set theory. In fact, for the first time it was discovered that an infinite set can be "bigger" than another one, hence Cantor realised that a sort of algebra of infinite sets could be developed. Until this fundamental discovery, mathematicians were used to compare finite sets, for which it is possible to count the number of elements, and infinite sets, that were

[1] We cannot eliminate ϵ_i by assuming $n_i \in \mathbb{Z}$, since, when $n_i = 0$ and $d_{i,j} \neq 0$ for some j, the sign ϵ_i has to be specified.

considered the "biggest" ones, without distinction between them. Actually, the mathematical community of that time resisted quite strongly against Cantor's ideas, that only after decades were recognised to be valid and revolutionary.

Example 10.13. $|\mathbb{N}| = |\mathbb{Z}| = |\mathbb{Q}|$.

Proof. We can list the elements of $\mathbb{Z}$ as follows:

$$0 \quad -1 \quad 1 \quad -2 \quad 2 \quad \cdots$$

Formally, we have the bijection $\varphi\colon \mathbb{N} \xrightarrow{\approx} \mathbb{Z}$, $2n \mapsto n$, $2n+1 \mapsto -n-1$. Now we construct a bijection between $\mathbb{N}$ and $\mathbb{Q}$. We start considering $\mathbb{Q}^{>0}$. Any positive rational number can be expressed uniquely as $\frac{n}{m}$, where $n, m \in \mathbb{N}^*$ and $\mathrm{GCD}(n,m) = 1$. We introduce the strict total order relation in $\mathbb{Q}^{>0}$ defined by $\frac{n}{m} \prec \frac{n'}{m'}$ if and only if one of the two following alternatives holds: (i) $n + m < n' + m'$; (ii) $n + m = n' + m'$ and $n < n'$. We get the following list of the positive rational numbers:

$$\tfrac{1}{1} \quad \tfrac{1}{2} \quad \tfrac{2}{1} \quad \tfrac{1}{3} \quad \tfrac{3}{1} \quad \tfrac{1}{4} \quad \tfrac{2}{3} \quad \tfrac{3}{2} \quad \tfrac{4}{1} \quad \cdots \tag{$\star$}$$

Now we get the following list of all rational numbers:

$$0 \quad \tfrac{1}{1} \quad -\tfrac{1}{1} \quad \tfrac{1}{2} \quad -\tfrac{1}{2} \quad \tfrac{2}{1} \quad -\tfrac{2}{1} \quad \tfrac{1}{3} \quad -\tfrac{1}{3} \quad \tfrac{3}{1} \quad -\tfrac{3}{1} \quad \cdots \tag{$\star\star$}$$

Let us formalise this bijection between $\mathbb{N}$ and $\mathbb{Q}$.

Exercise 10.14. Show that $(\mathbb{Q}^{>0}, \prec)$ is a well-ordered set such that any proper initial segment is finite.

Since $\mathbb{Q}$ is infinite because of the embedding $\mathbb{N} \hookrightarrow \mathbb{Q}$, it follows from corollary 8.22 that there exists a unique bijection $\psi'\colon \mathbb{N} \to \mathbb{Q}^{>0}$ that respects the two order relations. Such a bijection coincides with the list ($\star$). Finally, we get the bijection $\psi\colon \mathbb{N} \to \mathbb{Q}$, coinciding with the list ($\star\star$), defined by $\psi(0) := 0$, $\psi(2n-1) := \psi'(n-1)$ and $\psi(2n) := -\psi'(n-1)$ for every $n \in \mathbb{N}^*$. □

We can equivalently state example 10.13 by affirming that $\mathbb{Z}$ and $\mathbb{Q}$ are denumerable, while example 10.12 shows that $\mathbb{R}$ is infinite and not denumerable.

Example 10.15. *Every non-empty interval of $\mathbb{R}$, that does not coincide with a point, has the same cardinality of $\mathbb{R}$.*

Proof. It is easy to construct a bijection from any non-empty interval of $\mathbb{R}$, that does not coincide with a point, to an interval of the form $(-1,1)$, $[-1,1)$, $[-1,1]$, $(0,+\infty)$, $[0,+\infty)$ or $\mathbb{R}$, the bijection being an affine function—that is, a function of the form $x \mapsto ax + b$, with $a, b \in \mathbb{R}$ and $a \neq 0$. We have the bijections $f_1\colon \mathbb{R} \to (-1,1)$, $x \mapsto \frac{2}{\pi}\arctan(x)$, and $f_2\colon \mathbb{R} \to (0,+\infty)$, $x \mapsto e^x$. Let us consider the sequence $x_n = -\frac{1}{n}$ for $n \in \mathbb{N}^*$. We get the bijection $f_3\colon [-1,1) \to (-1,1)$ defined by $f_3(x_n) := x_{n+1}$ and $f_3(y) := y$ if $y \in [-1,1) \setminus \{x_n\}_{n\in\mathbb{N}^*}$. By applying the same construction with respect to the sequences $x_n := \frac{1}{n}$ and $x_n := 1 - \frac{1}{n}$, we get respectively the bijections $f_4\colon [-1,1] \to [-1,1)$ and $f_5\colon [0,+\infty) \to (0,+\infty)$. □

Example 10.16. $|\mathbb{N} \times \overset{k \text{ times}}{\cdots} \times \mathbb{N}| = |\mathbb{N} \sqcup \overset{k \text{ times}}{\cdots} \sqcup \mathbb{N}| = |\mathbb{N}|$ *for every* $k \in \mathbb{N}^*$.

Proof. We first prove that $|\mathbb{N} \times \mathbb{N}| = |\mathbb{N} \sqcup \mathbb{N}| = |\mathbb{N}|$. In the case of $\mathbb{N} \times \mathbb{N}$, we argue as we did in example 10.13 with respect to $\mathbb{Q}$. In particular, we introduce the strict total order relation in $\mathbb{N} \times \mathbb{N}$ defined by $(n, m) \prec (n', m')$ if and only if one of the two following alternatives holds: (i) $n + m < n' + m'$; (ii) $n + m = n' + m'$ and $n < n'$. We get the following list of all the pairs natural numbers:

$$(0,0) \quad (0,1) \quad (1,0) \quad (0,2) \quad (1,1) \quad (2,0) \quad \cdots \qquad (\#)$$

The reader can verify that '$\prec$' is a well-order in $\mathbb{N} \times \mathbb{N}$ such that every proper initial segment is finite, the proof being analogous to the resolution of exercise 10.14. It follows from corollary 8.22 that there exists a unique bijection $\varphi \colon \mathbb{N} \to \mathbb{N} \times \mathbb{N}$ respecting the order relations, that coincides with the list (#). In the case of $\mathbb{N} \sqcup \mathbb{N}$, we can list its elements as follows:

$$(0, \emptyset) \quad (0, \{\emptyset\}) \quad (1, \emptyset) \quad (1, \{\emptyset\}) \quad (2, \emptyset) \quad (2, \{\emptyset\}) \quad \cdots$$

Formally, we have the bijection $\psi \colon \mathbb{N} \to \mathbb{N} \sqcup \mathbb{N}$, $2n \mapsto (n, \emptyset)$, $2n+1 \mapsto (n, \{\emptyset\})$.

This proves the statement for $k = 2$. In general, assuming inductively that the result holds for k, we have $|\mathbb{N} \times \overset{(k+1) \text{ times}}{\cdots} \times \mathbb{N}| = |(\mathbb{N} \times \overset{k \text{ times}}{\cdots} \times \mathbb{N}) \times \mathbb{N}| = |\mathbb{N} \times \mathbb{N}| = |\mathbb{N}|$. In the last equality we applied the result for $k = 2$, while in the previous one we used the inductive hypothesis and exercise 10.8-iii (with $k = 2$, $x_1 = \mathbb{N} \times \overset{k \text{ times}}{\cdots} \times \mathbb{N}$ and $x_2 = y_1 = y_2 = \mathbb{N}$ in the statement of the latter). The same argument holds for disjoint union. □

Remark 10.17. It follows from exercise 11 of section 2.8 that $|\mathbb{N} \sqcup \overset{k \text{ times}}{\cdots} \sqcup \mathbb{N}| = |\mathbb{N} \times \{1, \dots, k\}|$. Hence, it follows from example 10.16 that $|\mathbb{N} \times \{1, \dots, k\}| = |\mathbb{N}|$. Actually, we could prove the latter statement as we did for $\mathbb{N} \times \mathbb{N}$ in example 10.16, through the well-order '$\prec$' defined in the same way, and deduce from exercise 11 of section 2.8 that $|\mathbb{N} \sqcup \overset{k \text{ times}}{\cdots} \sqcup \mathbb{N}| = |\mathbb{N}|$. If we consider $\bigsqcup_{n \in \mathbb{N}} \mathbb{N}$—that is, the disjoint union of a denumerable quantity of copies of $\mathbb{N}$—it follows from exercise 19 of section 4.5 that such set coincides with $\mathbb{N} \times \mathbb{N}$, hence $\left|\bigsqcup_{n \in \mathbb{N}} \mathbb{N}\right| = |\mathbb{N}|$ by example 10.16. ◇

10.2 COUNTABLE SETS

We recall that a set x is *denumerable* if $|x| = |\mathbb{N}|$ and it is *countable* or *at most denumerable* if it is finite or denumerable. We have seen that the existence of non-countable sets is the starting point of modern set theory. Therefore, denumerable sets are particular cases of infinite sets. In this section we show some of their basic properties.

Proposition 10.18. *If x is denumerable and $y \subset x$, then y is countable. Equivalently, if y is infinite, then it is denumerable.*

Proof. We start supposing that $x = \mathbb{N}$. If $y \subset \mathbb{N}$ is infinite, we restrict the order relation '$\leq$' from $\mathbb{N}$ to y and we get the well-order $(y, \leq)$. Any proper initial segment of y is of the form $\text{prec}_y(n)$ and $\text{prec}_y(n) \subset \text{prec}_{\mathbb{N}}(n) = n$, hence $\text{prec}_y(n)$ is finite. It follows from corollary 8.22 that y is denumerable.

In general, we fix a bijection $f\colon x \xrightarrow{\approx} \mathbb{N}$ and we set $y' := \dot{f}(y)$. By the first part of the proof, the set y' is countable. Since $f|_y||_y\colon y \xrightarrow{\approx} y'$ is a bijection, $|y| = |y'|$, hence y is countable. □

Exercise 10.19. Prove that:

- A set x is finite if and only if $|x| < |\mathbb{N}|$.
- A set x is countable if and only if $|x| \leq |\mathbb{N}|$.

Intuitively, a set is countable if we can "list" its elements, a list being a bijection with a natural number n or with $\mathbb{N}$. This completes remark 10.3.

Proposition 10.20. *A set x is countable if and only if there exist a countable set y and a surjective function $f\colon y \twoheadrightarrow x$.*

Proof. ($\Rightarrow$) Trivial by choosing $y = x$ and $f = \text{id}_x$. ($\Leftarrow$) If y is finite, then x is finite by exercise 7.22-iv. If y is denumerable—that is, there exists a bijection $g\colon \mathbb{N} \to y$—then we get the surjective function $f \circ g\colon \mathbb{N} \twoheadrightarrow x$, hence the result immediately follows form exercise 10.19-ii and proposition 10.11. □

Proposition 10.21. *If $x_1, \ldots, x_k$ are countable, then the following sets are countable:*

(i) $x_1 \times \cdots \times x_k$ (ii) $x_1 \sqcup \ldots \sqcup x_k$ (iii) $x_1 \cup \ldots \cup x_k$ (iv) $x_1 \cap \ldots \cap x_k$

Moreover:

- *If at least one set x_i is denumerable and every set x_i is not empty, then* (i) *is denumerable.*
- *If at least one set x_i is denumerable, then* (ii) *and* (iii) *are denumerable.*
- *If at least one set x_i is finite, then* (iv) *is finite.*

Proof. By hypothesis and exercise 10.19-ii, we have $|x_i| \leq |\mathbb{N}|$ for every $i \in \{1, \ldots, k\}$.

(i) It follows from exercise 10.8-iv and example 10.16 that $|x_1 \times \cdots \times x_k| \leq |\mathbb{N} \times \overset{k \text{ times}}{\cdots} \times \mathbb{N}| = |\mathbb{N}|$, hence $|x_1 \times \cdots \times x_k|$ is countable by exercise 10.19-ii. If (up to the order) x_1 is infinite and every x_i is not empty, then we fix $y_i \in x_i$ for every $i \in \{2, \ldots, k\}$ and we get the embedding $x_1 \hookrightarrow x_1 \times \cdots \times x_k$, $y \mapsto (y, y_2, \ldots, y_k)$. It follows that $x_1 \times \cdots \times x_k$ is infinite, thus it is denumerable.

(ii) It follows from exercise 10.8-iv and example 10.16 that $|x_1 \sqcup \ldots \sqcup x_k| \leq |\mathbb{N} \sqcup \overset{k \text{ times}}{\cdots} \sqcup \mathbb{N}| = |\mathbb{N}|$, hence $|x_1 \sqcup \ldots \sqcup x_k|$ is countable by exercise 10.19-ii. If (up to the order) x_1 is infinite, then $x_1 \sqcup \ldots \sqcup x_k$ is infinite too because of the embedding $x_1 \hookrightarrow x_1 \sqcup \ldots \sqcup x_k$, $y \mapsto (y, \emptyset)$, thus it is denumerable.

(iii) The union $x_1 \cup \ldots \cup x_k$ is countable because of proposition 10.20 and the surjective function $\eta\colon x_1 \sqcup \ldots \sqcup x_n \twoheadrightarrow x_1 \cup \ldots \cup x_n$ (see formula (5.2) in section 5.3).[2] If x_i is infinite, then $x_1 \cup \ldots \cup x_k$ is infinite too because of the inclusion $x_i \subset x_1 \cup \ldots \cup x_k$, thus it is denumerable.

(iv) We have the inclusion $x_1 \cap \ldots \cap x_k \subset x_1$, hence $|x_1 \cap \ldots \cap x_k| \leq |x_1| \leq |\mathbb{N}|$. It follows that $|x_1 \cap \ldots \cap x_k|$ is countable by exercise 10.19-ii. If (up to the order) x_1 is finite, then $x_1 \cap \ldots \cap x_k$ is finite too by exercise 7.22-ix. □

Proposition 10.22. *Let $x = \{x_i\}_{i \in I}$ be an indexed set such that I and every x_i are countable. Then:*

- *$\bigsqcup_i x_i$, $\bigcup_i x_i$ and $\bigcap_i x_i$ are countable (assuming $I \neq \emptyset$ in the case of intersection);*
- *if I is finite, then $\prod_i x_i$ is countable.*[3]

Moreover:

- *if at least one element x_i is denumerable, then $\bigsqcup_i x_i$ and $\bigcup_i x_i$ are denumerable;*
- *if there exists x_i finite, then $\bigcap_i x_i$ is finite;*
- *if I is finite, at least one element x_i is denumerable and every x_i is not empty, then $\prod_i x_i$ is denumerable.*

Proof. By hypothesis we have the embeddings $j\colon I \hookrightarrow \mathbb{N}$ and $\iota_i\colon x_i \hookrightarrow \mathbb{N}$. We get the embedding $J\colon \bigsqcup_i x_i \hookrightarrow \mathbb{N} \times \mathbb{N}$, $(y, i) \mapsto (\iota_i(y), j(i))$, hence $\left|\bigsqcup_i x_i\right| \leq |\mathbb{N} \times \mathbb{N}| = |\mathbb{N}|$. If x_{i_0} is denumerable, then $\bigsqcup_i x_i$ is infinite because of the embedding $x_{i_0} \hookrightarrow \bigsqcup_i x_i$, $y \mapsto (y, i_0)$, thus it is denumerable.

The union is countable because of proposition 10.20 and the surjective function $\bar{\eta}_x\colon \bigsqcup_i x_i \twoheadrightarrow \bigcup_i x_i$ (see formula (4.3) in section 4.2).[4] If x_{i_0} is denumerable, then $\bigcup_i x_i$ is infinite because of the inclusion $x_{i_0} \subset \bigcup_i x_i$, thus it is denumerable.

With respect to the intersection, we fix $i_0 \in I$ and we have $\bigcap_i x_i \subset x_{i_0}$, hence $\left|\bigcap_i x_i\right| \leq |x_{i_0}| \leq |\mathbb{N}|$. If we can choose x_{i_0} finite, then $\bigcap_i x_i$ is finite too by exercise 7.22-ix.

If I is finite, then there exist $n \in \mathbb{N}$ and a bijection $\phi\colon n \xrightarrow{\approx} I$. We can suppose $I \neq \emptyset$ without loss of generality, hence $n \in \mathbb{N}^*$. We get the embedding $K\colon \prod_i x_i \hookrightarrow \mathbb{N} \times \overset{n \text{ times}}{\cdots} \times \mathbb{N}$, $f \mapsto \big(\iota_{\phi(0)}(f(\phi(0))), \ldots, \iota_{\phi(n-1)}(f(\phi(n-1)))\big)$. Hence, $\left|\prod_i x_i\right| \leq |\mathbb{N} \times \overset{n \text{ times}}{\cdots} \times \mathbb{N}| = |\mathbb{N}|$. If x_{i_0} is denumerable and $x_i \neq \emptyset$ for every $i \in I$, then we fix $y_i \in x_i$ for every $i \in I \setminus \{i_0\}$ and we get the

[2] We can also construct an explicit embedding $x_1 \cup \ldots \cup x_k \hookrightarrow x_1 \sqcup \ldots \sqcup x_k$ (see exercise 12-i at the end of the chapter).

[3] Cfr. exercise 10 at the end of the chapter.

[4] We can also construct an explicit embedding $\bigcup_i x_i \hookrightarrow \bigsqcup_i x_i$ (see exercise 12-ii at the end of the chapter).

embedding $x_{i_0} \hookrightarrow \prod_i x_i$, $y_{i_0} \mapsto (i \mapsto y_i)$. It follows that $\prod_i x_i$ is infinite, thus denumerable. □

Corollary 10.23. *If x and every element of x are countable, then:*

- *$\bigsqcup x$, $\bigcup x$ and $\bigcap x$ are countable (assuming $x \neq \emptyset$ in the case of intersection);*
- *if x is finite, then $\prod x$ is countable.*

Moreover:

- *if x contains a denumerable element, then $\bigsqcup x$ and $\bigcup x$ are denumerable;*
- *if x contains a finite element, then $\bigcap x$ is finite;*
- *if x is finite, it contains a denumerable element and it does not contain the empty set, then $\prod x$ is denumerable.*

Proof. It is a particular case of proposition 10.22, thinking of x as indexed over itself. □

In particular, the previous proposition and corollary show that a denumerable union of (at most) denumerable sets is (at most) denumerable. This is a basic but very relevant fact in set theory.

Proposition 10.24. *If x is non-empty and countable, then* $\mathrm{FSeq}(x)$ *is denumerable.*[5]

Proof. We recall that $\mathrm{FSeq}_n(x) := \mathrm{Func}(s(n), x)$, thus $\mathrm{FSeq}(x) = \bigcup_{n \in \mathbb{N}} \mathrm{FSeq}_n(x)$. Hence, if we prove that $\mathrm{FSeq}_n(x)$ is countable for every $n \in \mathbb{N}$, then we get the result from proposition 10.22. We have the natural bijection $\varphi_n \colon \mathrm{FSeq}_n(x) \xrightarrow{\approx} x \times \overset{s(n)\text{ times}}{\cdots} \times x$, $f \mapsto (f(0), \dots, f(n))$. It follows that $|\mathrm{FSeq}_n(x)| = |x \times \overset{s(n)\text{ times}}{\cdots} \times x| \leq |\mathbb{N} \times \overset{s(n)\text{ times}}{\cdots} \times \mathbb{N}| = |\mathbb{N}|$, hence $\mathrm{FSeq}_n(x)$ is countable, as required. If we fix any $y \in x$, there exists the constant sequence of length $s(n)$, with image y, for every $n \in \mathbb{N}$, therefore $\mathrm{FSeq}(x)$ is infinite, thus denumerable. □

Proposition 10.25 (AC⇒). *A set x is infinite if and only if it admits a denumerable subset. Equivalently, a set x is infinite if and only if $|\mathbb{N}| \leq |x|$.*

Proof. (⇐) It is a particular case of exercise 7.22-x. (⇒) *Method I:* By AC, we fix a choice function $\xi \colon \mathcal{P}^*(x) \to x$. Moreover, we fix any $y_0 \in x$. We recursively set $y_{n+1} := \xi(x \setminus \{y_0, \dots, y_n\})$. By definition of choice function, $y_{n+1} \notin \{y_0, \dots, y_n\}$, hence the function $n \mapsto y_n$ is injective. It follows that the subset $\{y_n\}_{n \in \mathbb{N}}$ of x is denumerable. *Method II:* By WO, we fix a well-order in x. Comparing x and $y := \mathbb{N}$, one of the alternatives of theorem 8.21

[5]Cfr. ex. 4 at the end of the chapter.

holds. Alternative (ii) cannot hold, since x is infinite, while any proper initial segment of $\mathbb{N}$ is finite. Therefore, alternative (i) or (iii) holds, and in both cases we get an injective function $f\colon \mathbb{N} \hookrightarrow x$, whose image is a denumerable subset of x. □

Exercise 10.26. Prove the direction ($\Rightarrow$) of proposition 10.25 by applying ZL. *Hint:* Consider the set of injective functions $f\colon a \hookrightarrow x$, where $a \subset \mathbb{N}$ is a (not necessarily proper) initial segment. Endow this set with a natural order relation and verify that the hypothesis of ZL holds. Show that the domain of a maximal element is $\mathbb{N}$.

10.3 FUNDAMENTAL RESULTS

In the previous section we focused on countable sets. Now we show some fundamental results about comparison of cardinalities in general. Various other relevant results will be stated in chapter 12, using the theory of cardinal numbers.

10.3.1 Cantor-Bernstein Theorem

Working with finite sets, if $|x| \leq |y|$ and $|y| \leq |x|$, then $|x| = |y|$ (obviously the converse holds too). This follows from the antisymmetric property of the relation '$\leq$' between natural numbers, applied to the number of elements of x and y. It is natural to inquire if this property keeps on holding allowing for infinite sets too. The following theorem shows that the answer is positive, independently of the axiom of choice.

Theorem 10.27 (Cantor-Bernstein). *Given any two set x and y, the equality $|x| = |y|$ holds if and only if $|x| \leq |y|$ and $|y| \leq |x|$.*

Proof. ($\Rightarrow$) Obvious. ($\Leftarrow$) Let $f\colon x \hookrightarrow y$ and $g\colon y \hookrightarrow x$ be injective functions. We use the following notation:

- $(gf)^n := (g \circ f) \circ \overset{n \text{ times}}{\cdots} \circ (g \circ f)\colon x \hookrightarrow x$
- $(gf)^{n\cdot}\colon \mathcal{P}(x) \hookrightarrow \mathcal{P}(x)$ is the induced function between power sets
- we define $(fg)^n\colon y \hookrightarrow y$ and $(fg)^{n\cdot}\colon \mathcal{P}(y) \hookrightarrow \mathcal{P}(y)$ similarly

We set $a_0 := x$ and $b_0 := y$. Moreover, we recursively set $a_{n+1} := \dot{g}(b_n)$ and $b_{n+1} := \dot{f}(a_n)$. We get two chains $x = a_0 \supset a_1 \supset a_2 \supset \cdots$ and $y = b_0 \supset b_1 \supset b_2 \supset \cdots$ such that:

- $a_{2n} = (gf)^{n\cdot}(x)$ and $b_{2n} = (fg)^{n\cdot}(y)$
- $a_{2n+1} = \dot{g}((fg)^{n\cdot}(y)) = (gf)^{n\cdot}(\dot{g}(y))$ and $b_{2n+1} = \dot{f}((gf)^{n\cdot}(x)) = (fg)^{n\cdot}(\dot{f}(x))$

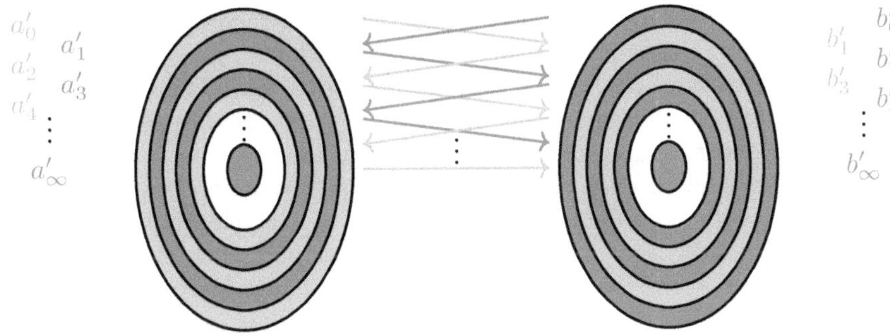

Figure 10.1 Cantor-Bernstein theorem.

We set $a'_n := a_n \setminus a_{n+1}$ and $b'_n := b_n \setminus b_{n+1}$. Moreover, we set $a'_\infty := \bigcap_{n\in\mathbb{N}} a_n$ and $b'_\infty := \bigcap_{n\in\mathbb{N}} b_n$. We have the following unions of *pairwise disjoint* sets:

$$x = \Big(\bigcup_{n\in\mathbb{N}} a'_n\Big) \cup a'_\infty \qquad\qquad y = \Big(\bigcup_{n\in\mathbb{N}} b'_n\Big) \cup b'_\infty .$$

Figure 10.1 represents these unions.

Since f and g are injective, the restrictions $f|_{a_n}\|_{b_{n+1}} : a_n \xrightarrow{\approx} b_{n+1}$ and $g|_{b_n}\|_{a_{n+1}} : b_n \xrightarrow{\approx} a_{n+1}$ are bijective, therefore also $f|_{a'_n}\|_{b'_{n+1}} : a'_n \xrightarrow{\approx} b'_{n+1}$ and $g|_{b'_n}\|_{a'_{n+1}} : b'_n \xrightarrow{\approx} a'_{n+1}$ are well-defined and bijective. Moreover, we claim that $f(a'_\infty) = b'_\infty$, hence $f|_{a'_\infty}\|_{b'_\infty} : a'_\infty \xrightarrow{\approx} b'_\infty$ is bijective. In fact, $x \in a'_\infty$ if and only if $x \in a_n$ for every $n \in \mathbb{N}$, if and only if $f(x) \in b_{n+1}$ for every $n \in \mathbb{N}$, if and only if $f(x) \in b'_\infty$. Thus, we get the following bijection:

$$\begin{aligned} h\colon x &\xrightarrow{\approx} y \\ z &\mapsto \begin{cases} f(z) & z \in a'_{2n} \\ g^{-1}(z) & z \in a'_{2n+1} \\ f(z) & z \in a'_\infty \end{cases} \end{aligned}$$

It follows that $|x| = |y|$, as required.[6] □

Corollary 10.28. *It is impossible that $|x| \le |y|$ and $|y| < |x|$ (equivalently, $|x| < |y|$ and $|y| \le |x|$). In particular, it is impossible that $|x| < |y|$ and $|y| < |x|$.*

Proof. If $|x| \le |y|$ and $|y| < |x|$, then trivially $|x| \le |y|$ and $|y| \le |x|$, hence $|x| = |y|$ by theorem 10.27. Nevertheless, the hypothesis $|y| < |x|$ implies by definition that $|x| \ne |y|$. □

---

[6] Analogously, we could also observe that $g(b'_\infty) = a'_\infty$, hence $g\colon b'_\infty \xrightarrow{\approx} a'_\infty$ is bijective, and define a similar bijection $k\colon y \xrightarrow{\approx} x$.

We can interpret Cantor-Bernstein theorem meta-theoretically as follows. Exercise 10.4 shows that having the same cardinality induces an equivalence relation in the class of all sets u, that we denote by '$\simeq$'. The corresponding quotient $\mathrm{u}/\simeq$ is not defined as a class, but it is defined as a hyperclass. Given a set x, we denote by $[x]$ the class formed by all the sets with the same cardinality of x, such a class being an element of the hyperclass $\mathrm{u}/\simeq$. In the latter, we have the well-defined relation $[x] \preceq [y]$ if and only if $|x| \leq |y|$. Cantor-Bernstein theorem affirms that this relation is antisymmetric. Since it is reflexive and transitive because of exercise 10.4, it is an order relation. Here we are not using the language of hyperclasses, but in this case it helps to visualise the meaning of the theorem.

Actually, Frege originally defined the cardinality $|x|$ of the set x as the class $[x]$ itself. Nevertheless, he was not aware of the distinction between classes and sets yet, hence he thought of $[x]$ as a set. With our more modern language, this definition of cardinality would be quite problematic. In fact, considering for example a finite set, it is reasonable to think that its cardinality is its number of elements, as we will see in chapter 12. With Frege's definition, the latter affirmation would be false, unless we define a natural number as a proper class and $\mathbb{N}$ as a hyperclass, that is quite cumbersome. For this reason, the approach of chapter 12, due to Von Neumann, is definitely more appropriate within the universe of sets.

Exercise 10.29. Prove that, if $|x_1| \leq \cdots \leq |x_n|$ and at least one of the inequalities is strict, then $|x_1| < |x_n|$. *Remark:* It follows that, if $|x_1| \leq \cdots \leq |x_n|$ and $|x_1| = |x_n|$, then $|x_1| = \cdots = |x_n|$.

10.3.2 Cantor's Theorem on Power Set

It is natural to inquire if there exists a set of "maximum size"—that is, a set x such that every set y verifies $|y| \leq |x|$. The following famous theorem by Cantor shows that the answer is negative, independently of the axiom of choice. It immediately follows that the universe u is not a set, otherwise it would be the "biggest" set.

Theorem 10.30 (Cantor's theorem). *For every set x, we have $|x| < |\mathcal{P}(x)|$.*

Proof. The result is trivial if $x = \emptyset$, hence we can suppose $x \neq \emptyset$. Let us fix any function $\psi\colon x \to \mathcal{P}(x)$. We set $a := \{y \in x : y \notin \psi(y)\}$. By construction, $a \in \mathcal{P}(x)$. Let us show that $a \notin \mathrm{Im}(\psi)$. By contradiction, let us suppose that there exists $y_0 \in x$ such that $\psi(y_0) = a$. Then, $y_0 \in a$ if and only if $y_0 \notin \psi(y_0)$ (by definition of a), if and only if $y_0 \notin a$ (since $a = \psi(y_0)$ by hypothesis), a contradiction. This shows that no function $\psi\colon x \to \mathcal{P}(x)$ is surjective (in particular, bijective), hence $|x| \neq |\mathcal{P}(x)|$. The canonical embedding $x \hookrightarrow \mathcal{P}(x)$, $y \mapsto \{y\}$, shows that $|x| \leq |\mathcal{P}(x)|$. Therefore, $|x| < |\mathcal{P}(x)|$. □

Corollary 10.31. *The class of all sets is proper.*

Proof. Let us suppose that $\mathrm{u} := \{x : x = x\}$ is a set. Then $\mathcal{P}(\mathrm{u}) \subset \mathrm{u}$, since each element of $\mathcal{P}(\mathrm{u})$ is a set. This implies that $|\mathcal{P}(\mathrm{u})| \leq |\mathrm{u}|$, contradicting proposition 10.30 because of corollary 10.28. □

With the meta-theoretical language introduced after corollary 10.28, Cantor's theorem implies that the relation '$\preceq$' in the hyperclass $\mathrm{u}/\simeq$ does not admit a maximum.

10.3.3 Comparing Sets

Another natural question is the following one: given any two sets x and y, is it possible to compare their cardinalities? The following proposition shows that the answer is trivially positive assuming the axiom of choice.

Proposition 10.32 (AC). *Given any two sets x and y, at least one of the two inequalities $|x| \leq |y|$ and $|y| \leq |x|$ holds.*

Proof. It immediately follows from WO and theorem 8.21. □

We already remarked that theorems 10.27 and 10.30 do not require the axiom of choice, hence the statement of proposition 10.32 is the first (and only) fundamental property in this section that relies on it. Nevertheless, many fundamental properties we will see in chapter 12 will depend on the axiom of choice (in some cases they will be equivalent to it).

Considering again the meta-theoretical language introduced after corollary 10.28, proposition 10.32 states that the relation '$\preceq$' in the hyperclass $\mathrm{u}/\simeq$ is a *total* order. We will see in chapter 12 that it is a well-order.

10.4 EXERCISES

1. Prove that $|\mathbb{R}| = |\mathcal{P}(\mathbb{N})|$.

2. Prove that $|\mathbb{R} \setminus \mathbb{Q}| = |\mathbb{R}|$.

3. Show that, if $|x_1| < |y_1|$ and $|x_2| = |y_2|$, it is possible that $|x_1 \times x_2| = |y_1 \times y_2|$ and $|x_1 \sqcup x_2| = |y_1 \sqcup y_2|$.

4. Prove that, if x contains at least two elements, then $\mathrm{Seq}(x)$ is not countable.

5. Prove that, if x is finite and not empty and y is denumerable, then $\mathrm{Func}(x, y)$ is denumerable.

Remark: It follows that, if x is finite and y is countable, then $\mathrm{Func}(x, y)$ is countable.

6. Prove that, assuming the axiom of choice, a set x is infinite if and only if there exists a proper subset $y \subsetneq x$ such that $|y| = |x|$.

7. Prove that, assuming the axiom of choice:

- If a set x is infinite and $y \subset x$ is finite, then $|x \setminus y| = |x|$.
- If a set x is not countable and $y \subset x$ is countable, then $|x \setminus y| = |x|$.

8. Prove that $|\mathbb{R} \times \mathbb{R}| = |\mathbb{R}|$ and deduce that $|\mathbb{R} \times \mathbb{N}| = |\mathbb{R}|$. *Hint:* Considering the corresponding decimal representations, construct a unique real number from two given ones.

9. Let x be a set of intervals of $\mathbb{R}$, such that every element of x is different from a point and any two elements of x are disjoint (in particular, x can be a set of pair-wise disjoint *open* intervals of $\mathbb{R}$). Assuming the axiom of choice, prove that x is countable.

10. Let $x = \{x_i\}_{i \in I}$ be an indexed set such that I and every x_i are countable. Show that we *cannot* deduce that $\prod_i x_i$ is countable, unless I is finite.

Remark: In particular, it follows that, if x is a countable set and every element of x is countable, then we *cannot* deduce that $\prod x$ is countable, unless x is finite.

11. Let x be denumerable and $y \subset \mathcal{P}(x)$ be the set of *finite* subsets of x. Prove that y is denumerable, even without assuming the axiom of choice.

12. (i) Given a finite sequence of sets $x_1, \ldots, x_k$, prove that $|x_1 \cup \ldots \cup x_k| \leq |x_1 \sqcup \ldots \sqcup x_k|$ (even without assuming the axiom of choice) by constructing an embedding $x_1 \cup \ldots \cup x_k \hookrightarrow x_1 \sqcup \ldots \sqcup x_k$.

(ii) Given an indexed set $x = \{x_i\}_{i \in I}$, such that I is countable, prove that $\left|\bigcup_i x_i\right| \leq \left|\bigsqcup_i x_i\right|$ (even without assuming the axiom of choice) by constructing an embedding $\bigcup_i x_i \hookrightarrow \bigsqcup_i x_i$.

13. Prove that $|\mathrm{Func}(\mathbb{N}, \mathbb{N})| = |\mathcal{P}(\mathbb{N})|$. Equivalently, $|\mathrm{Func}(\mathbb{N}, \mathbb{N})| = |\mathbb{R}|$.

Remark: We recall that $\mathrm{Seq}(\mathbb{N}) = \mathrm{Func}(\mathbb{N}, \mathbb{N})$, hence $|\mathrm{Seq}(\mathbb{N})| = |\mathbb{R}|$.

14. (i) Prove that $|\mathrm{Func}(\mathbb{N}, \mathbb{R})| = |\mathbb{R}|$. (ii) Let $C^0(\mathbb{R}; \mathbb{R})$ be the set of *continuous* functions from $\mathbb{R}$ to $\mathbb{R}$. Prove that $|C^0(\mathbb{R}; \mathbb{R})| = |\mathbb{R}|$.

Remark: We recall that $\mathrm{Seq}(\mathbb{R}) = \mathrm{Func}(\mathbb{N}, \mathbb{R})$, hence $|\mathrm{Seq}(\mathbb{R})| = |\mathbb{R}|$.

15. Is it possible to find a set of parts of $\mathbb{N}$—that is, a set $x \subset \mathcal{P}(\mathbb{N})$—such that:

- x is not countable
- for every $y, z \in x$, such that $y \neq z$, the intersection $y \cap z$ is finite?

CHAPTER 11

Ordinals

In chapter 10, we have seen that two sets can have different cardinalities, but we did not establish what is the cardinality of a set. We will introduce this concept in the next chapter by defining the notion of *cardinal number*, as a particular case of the notion of *ordinal number*, that we introduce here. In the framework of finite sets, these two notions coincide, but in the infinite framework they become different and non-equivalent concepts.

Remark 11.1. In this chapter we will never need to assume the axiom of choice. ◇

11.1 TRANSITIVE SETS

Ordinal numbers will be defined as particular cases of *transitive sets*, therefore we begin by considering this notion.

Definition 11.2. A set x is *transitive* if, for every $y \in x$, we have $y \subset x$. ◇

This means that a set is transitive if and only if it contains all the elements of its elements. The natural numbers (within set theory) are examples of transitive sets:

- $0 := \emptyset$ is trivially transitive.
- $1 := \{\emptyset\}$ is transitive, since $\emptyset \subset 1$.
- $2 := \{\emptyset, \{\emptyset\}\}$ is transitive, since $\emptyset \subset 2$ and $\{\emptyset\} \subset 2$.
- ...

The general proof can be realised by induction, as the following lemma shows.

Lemma 11.3. *Every natural number $n \in \mathbb{N}$ is a transitive set. Moreover, $\mathbb{N}$ is transitive.*

DOI: 10.1201/9781003565291-11

Proof. If $n = 0$, then it is trivially transitive. For every $n \geq 1$, we have $n = \{0, \ldots, n-1\}$, thus we have to prove that $m \subset n$ for every $m < n$. If $m = 0$, then trivially $m \subset n$. Otherwise, $m = \{0, \ldots, m-1\} \subset \{0, \ldots, n-1\} = n$. This proves the first part of the statement. About the second part, if $n \in \mathbb{N}$, then $n = 0$ or $n \geq 1$. In the first case, trivially $0 \subset \mathbb{N}$. In the second case, we have $n = \{0, \ldots, n-1\} \subset \mathbb{N}$. □

It is easy to construct non-transitive sets. For example, $x := \{\{\emptyset\}\}$ is not transitive, since $\{\emptyset\} \not\subset x$. Actually, most of ordinary sets, like $\mathbb{Z}$, $\mathbb{Q}$, $\mathbb{R}$, $\mathbb{C}$, $\{1, 9, 23\}$, and any other similar sets, are not transitive. Almost all of the groups, rings, topological spaces, etc. we deal with in ordinary mathematics are definitely non-transitive.

Remark 11.4. By definition, a set x is transitive if and only if $(y \in x) \Rightarrow (y \subset x)$. In general the converse does not hold, even if y is a proper subset of x. For example, $\{1\} \subset 2$ but $\{1\} \notin 2$. Similarly, $\{0, 2\} \subset 3$, but $\{0, 2\} \notin 3$. ◇

We can rephrase the definition of transitive set as follows.

Lemma 11.5. *A set x is transitive if and only if $x \subset \mathcal{P}(x)$.*

Proof. By definition, a set x is transitive if and only if $\forall y\big((y \in x) \Rightarrow (y \subset x)\big)$. This is equivalent to $\forall y\big((y \in x) \Rightarrow (y \in \mathcal{P}(x))\big)$—that is, $x \subset \mathcal{P}(x)$. □

11.1.1 Equivalent Characterisations

We saw that a set is transitive if and only if it contains all the elements of its elements. This property can be iterated. In fact, let us suppose that x is transitive and $y_n \in \cdots \in y_1 \in x$. From $y_1 \in x$, we deduce that $y_1 \subset x$, hence $y_2 \in x$. It follows that $y_2 \subset x$, hence $y_3 \in x$. Keeping on in this way, we conclude that $y_n \in x$, therefore x contains all the elements of the elements ... of its elements, at any order. More formally, $\left(\bigcup^n x\right) \subset x$ for every $n \in \mathbb{N}$, as the next proposition shows.

Proposition 11.6. *Given a set x, the following conditions are equivalent:*

(i) x is transitive.

(ii) For every y and z, if $z \in y \in x$, then $z \in x$.

(iii) $\left(\bigcup x\right) \subset x$.

(iv) $\left(\bigcup^n x\right) \subset x$ for every $n \in \mathbb{N}$.

Proof. (i) $\Rightarrow$ (ii) If $z \in y \in x$, then in particular $y \in x$, hence definition 11.2 implies that $y \subset x$, thus $z \in x$. (i) $\Leftarrow$ (ii) If $y \in x$, then, for every $z \in y$, by hypothesis $z \in x$. This means that $y \subset x$. (ii) $\Rightarrow$ (iii) If $z \in \bigcup x$, then by definition there exists $y \in x$ such that $z \in y$. By hypothesis, this implies that $z \in x$, hence $\left(\bigcup x\right) \subset x$. (ii) $\Leftarrow$ (iii) If $z \in y \in x$, then by definition $z \in \bigcup x$.

By hypothesis, this implies that $z \in x$. (iii) $\Rightarrow$ (iv). Induction on n. If $n = 0$, then the result is obvious, and, if $n = 1$, then it holds by hypothesis. Let us suppose that the statement holds for $n \geq 1$. Then

$$\left(\bigcup^{n+1} x\right) = \left(\bigcup^{n} \bigcup x\right) \subset \left(\bigcup^{n} x\right) \subset x.$$

The equality on the left follows from exercise 4.10, the first inclusion follows from hypothesis iii and lemma 4.12-2, and the second inclusion follows from the inductive hypothesis. (iii) $\Leftarrow$ (iv). Obvious. □

11.1.2 Proper and Hereditary Transitivity

We observe that, if $x = \{x\}$, then x is transitive. In fact, its only element is x itself and trivially $x \subset x$. We already pointed out that every set belonging to itself will be ruled out from our universe. Nevertheless, since we still have to exclude such sets, for the moment we strengthen definition 11.2 as follows.

Definition 11.7. A set x is *properly transitive* if it is transitive and $x \notin x$. ♢

The following definition will be the key to define ordinals.

Definition 11.8. A set x is *(properly) hereditarily transitive* if it is (properly) transitive and, for every $n \in \mathbb{N}$, every element of $\bigcup^n x$ is (properly) transitive too. ♢

We use the abbreviations "pt", "pht", "(p)t" and "(p)ht" to denote respectively the expressions "properly transitive", "properly hereditarily transitive", "(properly) transitive" and "(properly) hereditarily transitive".

Exercise 11.9. Prove that natural numbers and $\mathbb{N}$ are pht.

Example 11.10. The set $\{\emptyset, \{\emptyset\}, \{\{\emptyset\}\}\}$ is pt, but not hereditarily, since the element $\{\{\emptyset\}\}$ is not transitive. If we add $\emptyset$ to the latter, then we obtain the set 3, which is pht.

Proposition 11.11. *We have that*

(i) a set x is (p)ht if and only if x and every element of x are (p)t; and

(ii) if a set x is (p)ht, then, for every $n \in \mathbb{N}$, every element of $\bigcup^n x$ is (p)ht (in particular, every element of x is (p)ht).

Proof. (i) ($\Rightarrow$) Obvious from definition 11.8, considering the particular case $n = 0$. ($\Leftarrow$) By proposition 11.6, items (i) and (iv), every element of $\bigcup^n x$, for any n, belongs to x, hence it is (p)t by hypothesis.

(ii) Let us suppose that there exist $n \in \mathbb{N}$ and $y \in \bigcup^n x$ such that y is not (p)ht. By definition 11.8, y is (p)t, hence not hereditarily. This means that there exist $m \in \mathbb{N}$ and $z \in \bigcup^m y$ such that z is not (p)t. By lemma 4.12-3, we deduce $z \in \bigcup^{m+n+1} x$, that is a contradiction, since every element of $\bigcup^{m+n+1} x$ must be a (p)t set by definition 11.8. □

We remark that the converse of proposition 11.11-ii does not hold. For example, if $x = \{\{\emptyset\}\}$, then every element of $\bigcup^n x$ is pht for every $n \in \mathbb{N}$, but x is not transitive. The converse would trivially hold by adding the hypothesis that x is (p)t.

The following proposition is a simple but fundamental characterisation of pht sets among pt sets.

Proposition 11.12. *Let x be a pt set. The set x is pht if and only if the relation '$\in$' is a strict order relation in x.*

Proof. ($\Rightarrow$) By hypothesis every element of x is in particular proper—that is, $y \notin y$ for every $y \in x$—hence '$\in$' is an anti-reflexive relation. Let us suppose that $w \in z \in y$, with $w, z, y \in x$. Since $y \in x$, by hypothesis y is transitive, hence we get $w \in y$ by proposition 11.6-ii. This shows that '$\in$' is a transitive relation. ($\Leftarrow$) Since '$\in$' is anti-reflexive by hypothesis, every element of x is proper. By proposition 11.11-i, it is enough to prove that every $y \in x$ is transitive. Let us suppose that $w \in z \in y$, with $y \in x$. It easily follows from proposition 11.6-ii that $w, z \in x$ too. Since $w, z, y \in x$ and '$\in$' is transitive in x by hypothesis, we conclude that $w \in y$. This shows that y is transitive by proposition 11.6-ii. □

Example 11.13. In example 11.10 the relation '$\in$' is not transitive in x, since $\emptyset \in \{\emptyset\} \in \{\{\emptyset\}\}$, but $\emptyset \notin \{\{\emptyset\}\}$.

In the statement of proposition 11.12, it is essential to suppose that x is pt. Indeed, it can happen that '$\in$' is a strict order relation in a set x, but that x is not transitive. For example, if $x = \{3, 4, 5\}$, then '$\in$' coincides with '$<$', but x is not transitive. It is not necessary that every element of x is transitive either. For example, if $x = \{3, 4, \{3, 4\}\}$, then '$\in$' is a strict well-order in x, but x and $\{3, 4\}$ are not transitive.

11.2 DEFINITION AND FIRST PROPERTIES

Proposition 11.12 provides an interesting and useful characterisation of pht sets. Now we can ask the following question: if x is pht, is the (strict) order '$\in$' a well-order? We will show in theorem 13.36 that the answer is positive in the universe of well-founded sets, that is the reduced universe we are going to construct since chapter 2. Nevertheless, we will construct such universe using the notion of ordinal number, therefore the following definition is necessary up to now.

Definition 11.14. An *ordinal number* is a pht set in which the strict order relation '$\in$' is a well-order. ◇

Actually, we could give the following equivalent definition, that is the common one in literature.

Definition 11.15. An *ordinal number* is a transitive set in which '$\in$' is a strict well-order relation. ◇

Let us show the equivalence between these two definitions. Definition 11.14 trivially implies definition 11.15, since the latter is weaker (we only require that the set is transitive, not necessarily properly and hereditarily). Conversely, definition 11.15 excludes that $x \in x$, otherwise x would be an element of itself making '$\in$' not anti-reflexive, therefore x is properly transitive. Proposition 11.12 implies that x is pht, completing the equivalence. Even if definition 11.15 is simpler, we preferred to give definition 11.14 first, since we think that the intermediate step of pht sets (that will turn out to be an equivalent characterisation in the universe of well-founded sets) makes the definition of ordinal number more accessible at a first glance.

Example 11.16. Natural numbers are ordinals. In fact, we saw in example 11.9 that they are pht. In principle this is enough, but up to now it is necessary to check that '$\in$' is a well-order. This immediately follows from exercise 7.11-i (stating that '$\in$' coincides with '$<$', that is a total order in a finite set).

Example 11.17. The set $\mathbb{N}$ is an ordinal. In fact, we saw in example 11.9 that it is pht. Exercises 7.11-i (stating that '$\in$' coincides with '$<$') and 1.16 (stating that '$<$' is a well-order in $\mathbb{N}$) imply that $\mathbb{N}$ satisfies definition 11.15.

11.2.1 Structure of the Class of Ordinals

From now on we will denote ordinals by lower-case Greek letters and we will treat them as ordered sets, leaving implicit that the strict order relation is '$\in$'. Let us analyse more deeply their structure.

Lemma 11.18. *If α is an ordinal and $\alpha \neq \emptyset$, then $\emptyset \in \alpha$ and $\emptyset = \min(\alpha)$.*

Proof. Since α is a well-order by definition and it is not empty, it admits a minimum β. If $\beta \neq \emptyset$, then there exists $\gamma \in \beta$. Since $\gamma \in \beta \in \alpha$ and α is transitive, we have $\gamma \in \alpha$. Since '$\in$' is precisely the strict order relation in α, it follows that γ is strictly less than β in α, hence β is not the minimum, a contradiction. Therefore, $\beta = \emptyset \in \alpha$. □

The following proposition shows that ordinals are embedded in each other, forming an increasing class of sets, in such a way that every ordinal is an initial segment of any bigger one.

Proposition 11.19. *The following properties of ordinals hold.*

(i) If α is an ordinal and $\beta \in \alpha$, then β is an ordinal and $\beta = \mathrm{prec}_\alpha(\beta)$.

(ii) If α is an ordinal, then the proper initial segments of α are the elements of α (that are ordinals by item i).

(iii) If α and β are ordinals and they are isomorphic as ordered sets, then $\alpha = \beta$.

(iv) If α and β are ordinals, then one of the following alternatives holds:

I. $\alpha = \beta$

II. $\alpha \in \beta$ *(hence* $\alpha \subsetneq \beta$ *and it is a proper initial segment)*

III. $\beta \in \alpha$ *(hence* $\beta \subsetneq \alpha$ *and it is a proper initial segment)*

Proof. (i) First of all, β is pht by proposition 11.11-ii with $n = 0$. Since α is an ordinal, by definition it is transitive, hence $\beta \in \alpha$ implies that $\beta \subset \alpha$. The relation '$\in$' in β is obviously the restriction of the same relation in α (since in any case it is the restriction of the relation '$\in$' defined in all of the universe). Since $(\alpha, \in)$ is a well-order, its restriction to β is a well-order too, therefore β is an ordinal. Moreover, since '$\in$' is the strict order relation in α, by definition of 'prec' (stated in proposition 8.14-i we have $\mathrm{prec}_{\alpha}(\beta) = \{\gamma \in \alpha : \gamma \in \beta\} = \beta$.

(ii) If $\beta \in \alpha$, then it is a proper initial segment by item i. Conversely, if a is a proper initial segment of α, then we set $\beta := \min(\alpha \setminus a)$, that is an ordinal by item i, and by proposition 8.17 we have $a = \mathrm{prec}_{\alpha}(\beta) = \{\gamma \in \alpha : \gamma \in \beta\} = \beta$.

(iii) We can suppose that $\alpha, \beta \neq \emptyset$, otherwise the result is trivial. By hypothesis there exists an ordered-set isomorphism $\varphi \colon \alpha \to \beta$. It follows from lemma 11.18 that $\emptyset$ is the minimum of α and β, hence, since an ordered-set isomorphism sends the minimum to the minimum, we have $\varphi(\emptyset) = \emptyset$. Therefore, the set $x := \{\delta \in \alpha : \varphi(\delta) = \delta\}$ is not empty. By construction $x \subset \alpha \cap \beta$. Let us prove that $x = \alpha = \beta$. We observe that $x = \alpha$ if and only if $x = \beta$, since φ is bijective. By contradiction, let us suppose that $x \neq \alpha, \beta$. It follows that there exists $\gamma := \min(\alpha \setminus x)$. We have:

$$\delta \in \varphi(\gamma) \overset{(\star)}{\Longleftrightarrow} \varphi^{-1}(\delta) \in \gamma \overset{(\star\star)}{\Longleftrightarrow} \delta \in \gamma,$$

therefore $\varphi(\gamma) = \gamma$, a contradiction, since $\gamma \notin x$. The equivalence $(\star)$ follows from the fact that φ is an ordered-set isomorphism (hence φ^{-1} too), the strict order relation being '$\in$'. The equivalence $(\star\star)$ follows from the fact that $\varphi^{-1}(\delta) \in \gamma$ means that $\varphi^{-1}(\delta)$ is less than γ, therefore it belongs to x by definition of γ. The same consideration holds in the opposite direction.

(iv) It follows from proposition 8.21 that one the following alternatives holds: I. There exists an ordered-set isomorphism between α and β. In this case item iii of the present proposition implies that $\alpha = \beta$. II. There exists an ordered-set isomorphism between α and a proper initial segment of β. It follows from proposition 8.17 that there exists $\gamma \in \beta$ such that α is isomorphic to $\mathrm{prec}_{\beta}(\gamma)$. By item i of the present proposition, γ is necessarily an ordinal and α is isomorphic to γ. By item iii, this means that $\alpha = \gamma \in \beta$. It follows from item ii that α is a proper initial segment of β. III. Analogous to II, exchanging α and β. □

We denote by **Ord** the class of ordinals. It is actually a class, since being properly transitive and well-ordered by '$\in$' can be easily expressed through a wf. We will see soon that **Ord** is a proper class. The following proposition can be informally expressed as follows: if **Ord** were a set, then it would be an ordinal.

Proposition 11.20. *The following properties hold:*

(i) if $\alpha \in \mathbf{Ord}$, then $\alpha \subset \mathbf{Ord}$ (equivalently, if $\beta \in \alpha \in \mathbf{Ord}$, then $\beta \in \mathbf{Ord}$);

(ii) if $\alpha \in \mathbf{Ord}$, then $\alpha \notin \alpha$;

(iii) if $\alpha, \beta, \gamma \in \mathbf{Ord}$ and $\gamma \in \beta \in \alpha$, then $\gamma \in \alpha$;

(iv) if $a \subset \mathbf{Ord}$ is a non-empty class, then it admits a minimum μ with respect to '$\in$'.

The previous items can be summarised by affirming that **Ord** *is a transitive class and '$\in$' is a strict well-order relation in* **Ord**.

Proof. (i) It is a part of proposition 11.19-i. (ii) It is part of definition 11.14, since α is properly transitive. (iii) It follows from proposition 11.6-ii, since α is transitive by definition. (iv) Because of proposition 11.19-iv, the class **Ord** is totally ordered by '$\in$', hence the class a is too. We fix $\alpha \in a$. If α is the minimum of a, then $\mu = \alpha$. Otherwise, the set $b := \{\beta \in a : \beta \in \alpha\}$ is not empty and, if b admits a minimum, then $\mu = \min(b)$. Since $b \subset \alpha$ by construction and α is an ordinal (hence it is totally ordered by definition 11.14), b admits a minimum. □

Corollary 11.21. **Ord** *is a proper class and it does not admit a maximum.*

Proof. Let us suppose by contradiction that **Ord** is a set. Then, by proposition 11.20 and definition 11.15, **Ord** is an ordinal, hence $\mathbf{Ord} \in \mathbf{Ord}$. This is a contradiction, since an ordinal is *properly* transitive by definition 11.14. Moreover, if $\alpha \in \mathbf{Ord}$ were the maximum, then $\beta \in \alpha$ or $\beta = \alpha$ for every $\beta \in \mathbf{Ord}$, therefore $\beta \in \alpha \cup \{\alpha\}$ for every $\beta \in \mathbf{Ord}$—that is, $\mathbf{Ord} \subset \alpha \cup \{\alpha\}$. Since $\alpha \cup \{\alpha\}$ is a set, it would follow that **Ord** is a set too by assumption `(A4")`, a contradiction. □

The following proposition (after the lemma) enriches proposition 11.20-iv.

Lemma 11.22. *The (non-strict) order relation in* **Ord** *induced by '$\in$' is '$\subset$'.*

Proof. We have to prove that $\beta \subset \alpha$ if and only if $\beta = \alpha$ or $\beta \in \alpha$ for every $\alpha, \beta \in \mathbf{Ord}$. ($\Leftarrow$) If $\beta = \alpha$, then trivially $\beta \subset \alpha$. If $\beta \in \alpha$, then $\beta \subset \alpha$ by transitivity of α. ($\Rightarrow$) Let us suppose that $\beta \subsetneq \alpha$. Then β is not bigger than α, otherwise $\alpha \in \beta$, hence $\alpha \subset \beta$ by transitivity of β, a contradiction. Since we supposed that $\alpha \neq \beta$, it follows that β is lower than α—that is, $\beta \in \alpha$. □

Proposition 11.23. *Let $a \subset \mathbf{Ord}$ be a subclass.*

(i) If $a \neq \emptyset$, then it admits a minimum μ and $\mu = \bigcap a$.

(ii) If a is a set, then it admits a supremum ξ in **Ord** *(that can be a maximum or not) and $\xi = \bigcup a$.*

Proof. (i) By proposition 11.20-iv, there exists $\mu := \min(a)$. Since μ is the minimum, it follows from lemma 11.22 that $\mu \subset \alpha$ for every $\alpha \in a$, hence $\mu \subset \bigcap a$. Conversely, if $\alpha \in \bigcap a$, then $\alpha \in \beta$ for every $\beta \in a$. In particular, we can choose $\beta = \mu$, hence $\alpha \in \mu$. It follows that $\bigcap a \subset \mu$. (ii) We set $\xi := \bigcup a$ and we prove that $\xi = \sup(a)$. *Step I: ξ is an ordinal.* In fact, first of all ξ is a set, since a is a set by hypothesis; therefore, $\xi \subsetneq \mathbf{Ord}$. If $\beta \in \alpha \in \xi$, then there exists $\gamma \in a$ such that $\alpha \in \gamma$. It follows that $\beta \in \alpha \in \gamma$, hence $\beta \in \gamma$ by transitivity of γ. Since $\gamma \in a$, it follows that $\beta \in \xi$. This shows that ξ is transitive. Since $\xi \subset \mathbf{Ord}$, the relation '$\in$', restricted to ξ, is a well-order by proposition 11.20-iv, hence ξ is an ordinal by definition 11.15. *Step II: $\xi = \sup(a)$.* In fact, if $\alpha \in a$, then $\alpha \subset \left(\bigcup a\right) = \xi$ (see exercise 1-5 in section 4.5), hence α is less than or equal to ξ by lemma 11.22. It follows that ξ is an upper bound of a. If β is an upper bound of a, then, for every $\alpha \in a$, we have that $\alpha \subset \beta$ by lemma 11.22. It follows that $\left(\bigcup a\right) \subset \beta$—that is, $\xi \subset \beta$. By lemma 11.22, ξ is less than or equal to β, hence ξ is the minimum of the upper bounds of a. □

Corollary 11.24. *A subclass $a \subset$ **Ord** is a set if and only if it is bounded above.*

Proof. ($\Rightarrow$) It immediately follows from proposition 11.23-ii. ($\Leftarrow$) If $\alpha \in \mathbf{Ord}$ is an upper bound of a, then $s(\alpha)$ is a strict upper bound—that is, $\beta \in s(\alpha)$ for every $\beta \in a$. This means that $a \subset s(\alpha)$, hence it is a set by assumption (A4"). □

11.2.2 Ordinals and Well-Orders

Let us consider the class of well-orders. The existence of an isomorphism is an equivalence relation in such a class, therefore the corresponding equivalence (proper) classes are defined. Proposition 11.19-iii shows that, if such a class contains an ordinal, then the latter is unique. Now we are going to show that every equivalence class contains an ordinal—that is, ordinals can be thought of as canonical representatives of equivalence classes of well-orders. This means that studying ordinals is essentially equivalent to studying well-orders.

In the following lemma, we apply the notion of initial segment to a class instead of a set, the definition being the same. The statement of the lemma is analogous to proposition 11.19-(i, ii), replacing the ordinal α by the class **Ord**.

Lemma 11.25. *The following properties of the class* **Ord** *hold:*

*(i) If $\beta \in$ **Ord**, then $\beta = \mathrm{prec}_{\mathbf{Ord}}(\beta)$.*

(ii) The proper initial segments of **Ord** *are its elements (that is, the ordinal numbers).*

Proof. (i) Since '$\in$' is the strict order relation in **Ord**, we have $\mathrm{prec}_{\mathbf{Ord}}(\beta) = \{\gamma \in \mathbf{Ord} : \gamma \in \beta\} = \beta$, the latter equality being due to the fact that every element of β is an ordinal by proposition 11.19-i. (ii) If $\beta \in \mathbf{Ord}$, then it is a proper initial segment by item i. Conversely, let $a \subset \mathbf{Ord}$ be a proper initial segment. We set $\alpha := \min(\mathbf{Ord} \setminus a)$. By proposition 8.17, applied to the class $\mathbf{Ord}$ instead of a set without variations, we have $a = \mathrm{prec}_{\mathbf{Ord}}\alpha = \alpha$, the latter equality being due to item i. □

Proposition 11.26. *For every well-order $(x, \prec)$ there exists a unique ordinal α such that $(x, \prec)$ and α are isomorphic as well-ordered sets.*

Proof. Uniqueness immediately follows from proposition 11.19-iii, therefore we only have to prove existence. Let us consider the set $a := \{y \in x : \exists \alpha \in \mathbf{Ord} : \mathrm{prec}_x(y) \approx \alpha\}$, the isomorphism being between well-ordered sets. The set a is not empty, since $m := \min(x) \in a$, because of the empty isomorphism $\mathrm{prec}_x(m) \approx \emptyset$. We define $f\colon a \to \mathbf{Ord}$, $y \mapsto \alpha$, where $\mathrm{prec}_x(y) \approx \alpha$. The function is well-defined, since α is necessarily unique because of proposition 11.19-iii. By definition, $f(m) = \emptyset$.

Step I: We set $b := \mathrm{Im}(f)$. We prove that a is an initial segment of x and b is an initial segment of $\mathbf{Ord}$. With respect to a, if $z \prec y \in a$, then (by definition of a) there exists $\alpha \in \mathbf{Ord}$ such that $\mathrm{prec}_x(y) \approx \alpha$. We call $\varphi\colon \mathrm{prec}_x(y) \to \alpha$ the unique isomorphism. Since $z \in \mathrm{prec}_x(y)$, we set $\beta := \varphi(z)$. Since φ respects the order, $w \prec z$ if and only if $\varphi(w) \in \varphi(z)$ if and only if $\varphi(w) \in \beta$, therefore $\varphi|_{\mathrm{prec}_x(z)}\|_{\mathrm{prec}_\alpha(\beta)}\colon \mathrm{prec}_x(z) \to \mathrm{prec}_\alpha(\beta)$ is an isomorphism. Since $\mathrm{prec}_\alpha(\beta) = \beta$ by proposition 11.19-i, it follows that $\mathrm{prec}_x(z) \approx \beta$, hence $z \in a$ and $f(z) = \beta$. This show that a is an initial segment of x. With respect to b, first of all it is a set because of the replacement hypothesis. Moreover, if $\beta \in \alpha \in b$, then (by definition of b) there exists $y \in x$ such that $f(y) = \alpha$—that is, such that $\mathrm{prec}_x(y) \approx \alpha$. We call $\varphi\colon \mathrm{prec}_x(y) \to \alpha$ the unique isomorphism. Since $\beta \in \alpha = \mathrm{Im}(\varphi)$, we set $z := \varphi^{-1}(\beta)$. Since φ respects the order, $w \prec z$ if and only if $\varphi(w) \in \varphi(z)$ if and only if $\varphi(w) \in \beta$, therefore $\varphi|_{\mathrm{prec}_x(z)}\|_{\mathrm{prec}_\alpha(\beta)}\colon \mathrm{prec}_x(z) \to \mathrm{prec}_\alpha(\beta)$ is an isomorphism. Since $\mathrm{prec}_\alpha(\beta) = \beta$ by proposition 11.19-i, it follows that $\mathrm{prec}_x(z) \approx \beta$, hence $z \in a$ and $f(z) = \beta$, thus $\beta \in \mathrm{Im}(f) = b$. This show that b is an initial segment of $\mathbf{Ord}$, hence it is an ordinal because of lemma 11.25-ii.

Step II: We prove that $f\|_b\colon a \to b$ is bijective. It is surjective by construction, since $b := \mathrm{Im}(f)$. It is injective too, since f is. In fact, if $f(y) = f(z)$, then $\mathrm{prec}_x(y) \approx \mathrm{prec}_x(z) \approx \alpha$, where $\alpha \in \mathbf{Ord}$. If $y \neq z$, then, up to exchange them, we can suppose that $z \prec y$, hence $\mathrm{prec}_x(z) \subsetneq \mathrm{prec}_x(y)$ by transitivity. This is a contradiction by proposition 8.19.

Step III: Let us show that $a = x$, proving the thesis. By contradiction, let us suppose that $a \subsetneq x$. We proved is step I that a is an initial segment of x. We set $y := \min(x \setminus a)$. It follows from proposition 8.17 that $a = \mathrm{prec}_x(y)$, therefore $f\|_b\colon \mathrm{prec}_x(y) \to b$ is an isomorphism by step II. Since $b \in \mathbf{Ord}$ by step I and lemma 11.25-ii, by definition of a we have $y \in a$, a contradiction. □

Proposition 11.26 justifies the following definition.

Definition 11.27. Given a well-order $(x, \prec)$, we set $\mathrm{type}(x, \prec) := \alpha$, where α is the unique ordinal such that $(x, \prec) \approx \alpha$. ◇

11.3 SUCCESSOR AND NATURAL NUMBERS

We repeat the definition of *successor*, that has already been introduced in chapter 7, in the framework of ordinal numbers.

Definition 11.28. Let α be an ordinal. We set $s(\alpha) := \alpha \cup \{\alpha\}$. We call $s(\alpha)$ the *successor* of α. ♢

Proposition 11.29. *If α is an ordinal, then:*

(i) $s(\alpha)$ is an ordinal.

(ii) $s(\alpha) = s(\beta)$ if and only if $\alpha = \beta$.

(iii) $\alpha \in s(\alpha)$ and there exits no ordinal β such that $\alpha \in \beta \in s(\alpha)$.

(iv) For every ordinal β, we have that $\beta \in s(\alpha)$ if and only if $\beta \subset \alpha$ (cfr. lemma 11.22).

(v) If $\beta \in \alpha$, then $s(\beta) \subset \alpha$ and $s(\beta) \in s(\alpha)$.

Proof. (i) Let us show that $s(\alpha)$ is transitive. If $\gamma \in \beta \in s(\alpha)$, then we have two possibilities: $\beta \in \alpha$ or $\beta \in \{\alpha\}$. In the first case, we get $\gamma \in \beta \in \alpha$, hence $\gamma \in \alpha$ by transitivity of α. In the second case, we have $\beta = \alpha$, hence $\gamma \in \alpha$. In both cases, $\gamma \in \alpha$. Since $\alpha \subset s(\alpha)$, it follows that $\gamma \in s(\alpha)$, proving the transitivity of $s(\alpha)$. Moreover, $\alpha \subset \mathbf{Ord}$ by proposition 11.20-i, hence $s(\alpha) \subset \mathbf{Ord}$. It follows from proposition 11.20-iv that '$\in$', restricted to $s(\alpha)$, is a well-order.

(ii) ($\Leftarrow$) Obvious. ($\Rightarrow$) By hypothesis $\alpha \cup \{\alpha\} = \beta \cup \{\beta\}$, thus in particular $\{\alpha\} \subset \beta \cup \{\beta\}$—that is, $\alpha \in \beta \cup \{\beta\}$. If $\alpha \neq \beta$, then necessarily $\alpha \in \beta$. The same argument, exchanging α and β, shows that $\beta \in \alpha$, that is a contradiction, since '$\in$' is a strict order.

(iii) By definition of s, we have $\{\alpha\} \subset s(\alpha)$—that is, $\alpha \in s(\alpha)$. If $\alpha \in \beta \in s(\alpha)$, we have two possibilities: $\beta \in \alpha$ or $\beta \in \{\alpha\}$. The first case cannot hold, since '$\in$' is a strict order. The second one is equivalent to $\beta = \alpha$, that would imply $\alpha \in \alpha$ by hypothesis, again contradicting the fact that '$\in$' is a strict order.

(iv) ($\Rightarrow$) If $\beta \in s(\alpha) = \alpha \cup \{\alpha\}$, we have two possibilities: $\beta \in \alpha$ or $\beta = \alpha$. In the first case, the result follows from the transitivity of α. In the second case, trivially $\beta \subset \alpha$. ($\Leftarrow$) By lemma 11.22, if $\beta \subset \alpha$, then we have two possibilities: $\beta \in \alpha$ or $\beta = \alpha$. In both cases, $\beta \in \alpha \cup \{\alpha\}$.

(v) If $\beta \in \alpha$, then $\beta \subset \alpha$ by transitivity and $\{\beta\} \subset \alpha$, hence $\beta \cup \{\beta\} \subset \alpha$—that is, $s(\beta) \subset \alpha$. It follows from item iv that $s(\beta) \in s(\alpha)$. □

It follows from proposition 11.29-(i, ii) that the operation of successor among ordinals can be thought of as an injective function $s\colon \mathbf{Ord} \to \mathbf{Ord}$. Surely it is not surjective, since $\emptyset \notin \operatorname{Im}(s)$, but we are going to show that this is not the only reason.

Definition 11.30. An ordinal number is called *successor ordinal* (or simply *successor*) if it belongs to the image of $s\colon \mathbf{Ord} \to \mathbf{Ord}$. Otherwise, if it is not empty, it is called *limit ordinal.* ♢

Proposition 11.31. *An ordinal α is a successor if and only if it admits a maximum β. In this case, $\alpha = s(\beta)$.*

Proof. ($\Rightarrow$) If $\alpha = s(\beta)$, then by definition $\alpha = \beta \cup \{\beta\}$. Let us show that $\beta = \max(\alpha)$. In fact, if $\gamma \in \alpha$ and $\gamma \neq \beta$, then $\gamma \in \beta$ by definition of union. Since '$\in$' is the strict order relation in α, this proves the thesis. ($\Leftarrow$) Let α be an ordinal and $\beta = \max(\alpha)$. By proposition 11.19-i, we have $\beta = \mathrm{prec}_\alpha(\beta) = \alpha \setminus \{\beta\}$, the latter equality being due to the fact that β is the maximum of α. Hence, $\alpha = (\alpha \setminus \{\beta\}) \cup \{\beta\} = \beta \cup \{\beta\} = s(\beta)$. □

If α is an ordinal, then in particular it is a subset of **Ord** by proposition 11.20-i, hence it admits a supremum by proposition 11.23-ii.

Corollary 11.32. *A non-zero ordinal α is:*

(i) a successor ordinal if and only if $\sup(\alpha) \in \alpha$;

(ii) a limit ordinal if and only if $\sup(\alpha) = \alpha$.

In case (i), if $\alpha = s(\beta)$, then $\sup(\alpha) = \max(\alpha) = \beta$ *and* $\bigcup \alpha = \beta$. *In case (ii),* $\bigcup \alpha = \alpha$.

Proof. (i) ($\Rightarrow$) By proposition 11.31, α admits a maximum β, hence $\sup(\alpha) = \beta \in \alpha$. ($\Leftarrow$) If $\beta := \sup(\alpha) \in \alpha$, then $\beta = \max(\alpha)$, hence $\alpha = s(\beta)$ by proposition 11.31.

(ii) ($\Rightarrow$) By lemma 11.25-i, we have $\alpha = \mathrm{prec}_{\mathbf{Ord}}(\alpha)$. It follows from proposition 8.17 that $\alpha = \min(\mathbf{Ord} \setminus \alpha)$. By proposition 8.14-ii, the class of strict upper bounds of α is $\mathbf{Ord} \setminus \alpha$ and, since α does not admit a maximum by proposition 11.31, it coincides with the class of upper bounds. It follows that $\sup(\alpha) = \min(\mathbf{Ord} \setminus \alpha)$ by definition of sup, thus $\alpha = \sup(\alpha)$. ($\Leftarrow$) Since $\alpha \notin \alpha$, we have $\sup(\alpha) \notin \alpha$, hence α does not admit a maximum. By proposition 11.31, it is not a successor.

The last statement immediately follows from proposition 11.23-ii applied to α. □

Example 11.33. Every non-zero natural number is a successor ordinal by exercise 1.4-i. Of course 0 is not a successor, since the empty set does not admit a maximum.

Example 11.34. $\mathbb{N}$ is a limit ordinal, since it does not admit a maximum. In fact, for every $n \in \mathbb{N}$, we trivially have $n \in s(n)$.

11.3.1 Ordinals and Natural Numbers

The following proposition characterises natural numbers among ordinals.

Proposition 11.35. *An ordinal α is a natural number if and only if, for every ordinal $\beta \subset \alpha$, we have that $\beta = \emptyset$ or β is a successor.*

Proof. ($\Rightarrow$) If $\alpha = 0$, then the thesis trivially holds. By exercise 1.4-i, every non-zero natural number is a successor (cfr. example 11.33). If $n \in \mathbb{N}^*$ and $\beta \in \mathbf{Ord}$, then, by proposition 11.22, $\beta \subset n$ if and only if $\beta = n$ or $\beta \in n$. Since $n \subset \mathbb{N}$ (see exercise 7.10-i), the thesis holds in both cases. ($\Leftarrow$) Since α and $\mathbb{N}$ are both ordinals, by proposition 11.19-iv we have three possibilities: $\alpha \in \mathbb{N}$, $\alpha = \mathbb{N}$ or $\mathbb{N} \in \alpha$. In the second and third cases, we have $\mathbb{N} \subset \alpha$ (in the third one by transitivity), hence the hypothesis does not hold, since $\mathbb{N}$ is not a successor. The only possibility left is that $\alpha \in \mathbb{N}$—that is, α is a natural number. □

Proposition 11.36. *An ordinal α is infinite if and only if $\mathbb{N} \subset \alpha$.*

Proof. ($\Leftarrow$) It immediately follows from exercise 2-x of section 3.5. ($\Rightarrow$) Proposition 11.19-iv implies that one of the following alternatives holds: I. $\alpha \in \mathbb{N}$; II. $\alpha = \mathbb{N}$; III. $\mathbb{N} \in \alpha$. By transitivity of both $\mathbb{N}$ and α, we deduce respectively: I. $\alpha \subsetneq \mathbb{N}$; II. $\alpha = \mathbb{N}$; III. $\mathbb{N} \subsetneq \alpha$. Alternative I cannot hold, since α would be finite by exercise 2-ix of section 3.5, hence II or III holds. In both cases, $\mathbb{N} \subset \alpha$. □

We remark that all of the theory we exposed about ordinals (before proposition 11.35) holds without assuming the knowledge of natural numbers and does not rely on assumption `(A7)`. In fact, we referred to the set $\mathbb{N}$ and its elements only as examples. Therefore, we could ignore these examples and replace proposition 11.35 with the following definition.

Definition 11.37. A *natural number* is an ordinal n such that, for every ordinal $\beta \subset n$, we have that $\beta = \emptyset$ or β is a successor. ◇

Following this path, we define the following *class*:

$$\omega := \{n \in \mathbf{Ord} : \forall \beta \in \mathbf{Ord}(\beta \subset n \Rightarrow (\beta = 0 \vee \beta \in \mathrm{Im}(s)))\}. \tag{11.1}$$

The class ω is naturally endowed with the strict order relation '$\in$', hence we *define* $n < m$ if and only in if $n \in m$. We get a well-order, since $\omega \subset \mathbf{Ord}$ by definition. Now we need assumption `(A7)` to deduce that ω is a set, as we are going to show.

Lemma 11.38. *If $n \in \omega$ and β is an ordinal such that $n = s(\beta)$, then $\beta \in \omega$.*

Proof. We have $\beta \subset \beta \cup \{\beta\} = n$. Hence, if $\gamma \subset \beta$, then $\gamma \subset n$, therefore $\gamma = \emptyset$ or γ is a successor. □

Lemma 11.39. *If* $\mathtt{c} \in \mathtt{cs}$,[1] *then* $\omega \subset \mathtt{c}$.

[1]See section 7.2 for the definition of the class cs.

Proof. By contradiction, suppose that $\omega \not\subset \mathsf{c}$. It follows that $a := \omega \setminus \mathsf{c} \neq \emptyset$. Since $a \subset \omega$, it admits a minimum $n \in \omega$. It is not possible that $n = \emptyset$, since $\emptyset \in \mathsf{c}$ by definition of cs. Hence, by definition 11.37 and lemma 11.38, there exists $m \in \omega$ such that $n = s(m) = m \cup \{m\}$. It follows that $m \in n$—that is, $m < n$—and $m \notin a$, since n is the minimum of a. Hence, $m \in \mathsf{c}$, that implies $n = s(m) \in \mathsf{c}$ by definition of cs, a contradiction. □

Proposition 11.40. *Assuming* (A7), *we have* $\omega = \bigcap \mathsf{cs}$ *(that is,* $\omega = \mathbb{N}$*).*

Proof. By lemma 11.39, we have $\omega \subset \bigcap \mathsf{cs}$, hence ω is a set by remark 6.2 and assumption (A4"). Let us prove that $\omega \in \mathsf{cs}$, hence $\omega = \bigcap \mathsf{cs}$. In fact, by definition $\emptyset \in \omega$, hence we have to prove that ω is closed under successor. If $n \in \omega$, then $s(n)$ is an ordinal by proposition 11.29-i. Let us prove that $s(n)$ satisfies definition 11.37, hence $s(n) \in \omega$. If β is an ordinal such that $\beta \subset s(n)$, then, by lemma 11.22, we have two possibilities: I. $\beta \in s(n)$; II. $\beta = s(n)$. If II holds, then β is trivially a successor. If I holds, then $\beta \subset n$ by proposition 11.29-iv. Since $n \in \omega$ by hypothesis, it follows that $\beta = 0$ or β is a successor, hence in any case $s(n)$ satisfies definition 11.37. □

In the context of ordinal numbers, the notation 'ω' is more common than '$\mathbb{N}$'. The strict order relation in ω is defined to be '$\in$', and it coincides with the relation '$<$' in $\mathbb{N}$ by exercise 7.11.

Exercise 11.41. State examples 11.16, 11.17, 11.33, and 11.34 relying on definitions 11.37 and (11.1).

All of the results and constructions shown in Chapter 7, from section 7.2.1 on, hold without variations in this framework.[2] Moreover, the statement of proposition 11.36 can be used as the *definition* of infinite ordinal. Then, we say that a set x is *finite* if there exists a bijection between x and an element of ω, *infinite* if it is not finite, *denumerable* if there exists a bijection between x and ω and *countable* if it is finite or denumerable. Moreover, we will see in section 11.5 that we can naturally define the operations of sum, product and power on ordinals, in such a way that, when restricted to ω, they coincide with the ones defined in section 7.2.3. This will complete the construction of natural numbers through the language of ordinals.

11.4 TRANSFINITE INDUCTION AND RECURSION

In this section, we show that the principles of induction and recursion can be suitable extended from $\mathbb{N}$ to **Ord**.

[2]We only observe that proposition 7.4-ii follows directly from proposition 11.29-ii.

11.4.1 Transfinite Induction

The following proposition shows that the class **Ord** satisfies properties that extend the Peano's axioms of $\mathbb{N}$. In particular, the more interesting item is the third one, about the induction principle in the framework of ordinals.

Proposition 11.42. *The following properties of the class* **Ord** *hold:*

(i) $s(\alpha) \neq 0$ *for every* $\alpha \in \mathbf{Ord}$.

(ii) $s(\alpha) = s(\beta)$ *if and only if* $\alpha = \beta$.

(iii) TRANSFINITE INDUCTION ON ORDINALS - FORM I: *If* $a \subset \mathbf{Ord}$ *is a sub-class satisfying the following properties:*

- $0 \in a$
- $\alpha \in a \Rightarrow s(\alpha) \in a$ *for every* $\alpha \in \mathbf{Ord}$
- *if* α *is a limit ordinal and* $\alpha \subset a$, *then* $\alpha \in a$

then $a = \mathbf{Ord}$.

Proof. (i) $\alpha \in \alpha \cup \{\alpha\} = s(\alpha)$, hence $s(\alpha) \neq \emptyset$. (ii) Already proved in proposition 11.29, item ii. (iii) Let us suppose by contradiction that $a \neq \mathbf{Ord}$—that is, $\mathbf{Ord} \setminus a \neq \emptyset$. Since **Ord** is well-ordered, there exists $\alpha := \min(\mathbf{Ord} \setminus a)$. Surely $\alpha \neq 0$, since $0 \in a$. If $\alpha = s(\beta)$, then $\beta \notin \mathbf{Ord} \setminus a$, since $\beta \in \alpha$ and α is the minimum of $\mathbf{Ord} \setminus a$. It follows that $\beta \in a$, hence $\alpha = s(\beta) \in a$ by hypothesis, a contradiction. Thus, α is not a successor. If α is a limit ordinal and $\beta \in \alpha$, then $\beta \in a$, since α is the minimum of $\mathbf{Ord} \setminus a$. It follows that $\alpha \subset a$, hence $\alpha \in a$ by hypothesis, a contradiction. We proved that α is different from 0, is not a successor and is not a limit ordinal. This shows that α cannot exist, hence $\mathbf{Ord} \setminus a = \emptyset$—that is, $a = \mathbf{Ord}$. □

The third hypothesis in item iii is equivalent to assume that, if α is a limit ordinal and a contains all of the ordinals less than α, then it contains α too. For example, in the case $\alpha = \omega$, it states that, if a contains every natural number, then it contains ω too.

We can extend to ordinals the second form of induction as well. Actually, it turns out to be more natural in this context, since it unifies the hypotheses on limit ordinals and on successors.

Proposition 11.43 (Transfinite induction on ordinals - Form II). *If* $a \subset \mathbf{Ord}$ *is a sub-class satisfying the following properties:*

- $0 \in a$;
- *if* $\alpha \subset a$, *then* $\alpha \in a$, *for every* $\alpha \in \mathbf{Ord}$;

then $a = \mathbf{Ord}$.

Proof. Let us suppose by contradiction that $a \neq \mathbf{Ord}$—that is, $\mathbf{Ord} \setminus a \neq \emptyset$. Since $\mathbf{Ord}$ is well-ordered, it follows that there exists $\alpha := \min(\mathbf{Ord} \setminus a)$. Surely $\alpha \neq 0$, since $0 \in a$. If $\beta \in \alpha$, then $\beta \in a$, since α is the minimum of $\mathbf{Ord} \setminus a$ and '$\in$' is the strict order relation. Therefore, $\alpha \subset a$. By hypothesis this implies that $\alpha \in a$, a contradiction. □

As above, the second hypothesis is equivalent to assume that, if a contains all of the ordinals less than α, then it contains α too.

The first form of induction easily follows from the second one. The converse holds too, but it is not as relevant as in the framework of meta-theoretical natural numbers, since both forms of induction are results and not assumptions.

Exercise 11.44. Show that we can start transfinite induction (in both forms) from any fixed ordinal α_0 instead of 0.

Exercise 11.45. Prove *bounded transfinite induction*: if $\alpha \in \mathbf{Ord}$ and $a \subset \alpha$ is a subset satisfying the following properties:

- $0 \in a$;
- if $\beta \subset a$ and $\beta \in \alpha$, then $\beta \in a$;

then $a = \alpha$. *Remark:* We stated bounded induction in form II, but we can state it in form I as well. Moreover, exercise 11.44 holds in the bounded case as well.

11.4.2 Transfinite Recursion

Let us extend to ordinal numbers the two forms of recursion.

Definition 11.46. Given a class c, we define $\text{Func}^*(\mathbf{Ord}, \mathsf{c}) := \{f : \alpha \to \mathsf{c} : \alpha \in \mathbf{Ord}\}$. ◇

Exercise 11.47. Verify that $\text{Func}^*(\mathbf{Ord}, \mathsf{c})$ is a well-defined class for any class c.

We set $\mathbf{Ord}^* := \mathbf{Ord} \setminus \{0\}$.

Proposition 11.48 (Transfinite Recursion on Ordinals - Form I). *Given:*

- *a class* c;
- *a fixed element* $a \in \mathsf{c}$;
- *a function* $k : \mathsf{c} \to \mathsf{c}$;
- *a function* $h : \text{Func}^*(\mathbf{Ord}, \mathsf{c}) \to \mathsf{c}$;

there exists a unique function $f : \mathbf{Ord} \to \mathsf{c}$ *such that:*

I. $f(0) = a$;

II. $f(s(\alpha)) = k(f(\alpha))$ *for every* $\alpha \in \mathbf{Ord}$;

III. *if* α *is a limit ordinal, then* $f(\alpha) = h(f|_\alpha)$.

Proposition 11.49 (Transfinite Recursion on Ordinals - Form II). *Given:*

- *a class* c*;*
- *a fixed element* $a \in \mathsf{c}$*;*
- *a function* $h\colon \mathrm{Func}^*(\mathbf{Ord}, \mathsf{c}) \to \mathsf{c}$*;*

there exists a unique function $f\colon \mathbf{Ord} \to \mathsf{c}$ *such that:*

I. $f(0) = a$*;*

II. $f(\alpha) = h(f|_\alpha)$ *for every* $\alpha \in \mathbf{Ord}^*$*.*

Proof of 11.49. `Uniqueness`: Let us suppose that f and f' verify the statement and let us prove by induction, in the form 11.43, that $f(\alpha) = f'(\alpha)$ for every $\alpha \in \mathbf{Ord}$. If $\alpha = 0$, then $f(0) = f'(0) = a$ by hypothesis. If α is a non-zero ordinal and $f(\beta) = f'(\beta)$ for every $\beta \in \alpha$, then $f|_\alpha = f'|_\alpha$, therefore $f(\alpha) = h(f|_\alpha) = h(f'|_\alpha) = f'(\alpha)$.

`Existence`: Let us show by induction that, for every non-zero $\alpha \in \mathbf{Ord}$, there exists a unique function $f_{[\alpha]}\colon \alpha \to \mathsf{c}$ such that:

I_α. $f_{[\alpha]}(0) = a$;

II_α. for every non-zero $\beta \in \alpha$, we have $f_{[\alpha]}(\beta) = h(f_{[\alpha]}|_\beta)$.

For $\alpha = 1$, there exists the unique function $f_{[1]}\colon \{0\} \to \mathsf{c}$, defined by $f_{[1]}(0) := a$, trivially satisfying I_1 and II_1. We fix any ordinal α bigger than 1 and we assume inductively that, for every non-zero $\beta \in \alpha$, there exists unique $f_{[\beta]}$ satisfying I_β and II_β. We observe that, if $\gamma \in \beta \in \alpha$, then

$$f_{[\beta]}|_\gamma = f_{[\gamma]}. \tag{!}$$

In fact, since $\gamma \in \alpha$ by transitivity, $f_{[\gamma]}$ satisfies I_γ and II_γ by hypothesis. We set $f'_{[\gamma]} := f_{[\beta]}|_\gamma$. We have $f'_{[\gamma]}(0) = f_{[\beta]}(0) = a$ and, for every $\delta \in \gamma \setminus \{0\}$, we have $f'_{[\gamma]}(\delta) = f_{[\beta]}(\delta) = h(f_{[\beta]}|_\delta) = h(f'_{[\gamma]}|_\delta)$, hence $f'_{[\gamma]}$ satisfies I_γ and II_γ too. By the inductive hypothesis (in particular, by uniqueness), we get (!). It follows that the indexed set $\{f_{[\beta]}\}_{\beta\in\alpha}$ is an increasing set of functions, hence we get the well-defined function

$$g_{[\alpha]} := \bigcup_{\beta\in\alpha} f_{[\beta]}\colon \bigcup_{\beta\in\alpha} \beta \to \mathsf{c} \tag{$\star$}$$

by corollary 4.21. If α is a limit ordinal, then $\bigcup_{\beta\in\alpha} \beta = \alpha$ by corollary 11.32-ii, hence we set $f_{[\alpha]} := g_{[\alpha]}$. If α is a successor ordinal, then $\bigcup_{\beta\in\alpha} \beta$ is the predecessor of α by corollary 11.32-i, hence we set:

$$f_{[\alpha]}(\beta) := \begin{cases} g_{[\alpha]}(\beta) & s(\beta) \in \alpha \\ h(g_{[\alpha]}) & s(\beta) = \alpha. \end{cases} \tag{$*$}$$

We remark that $(*)$ holds if α is a limit ordinal as well. It follows from $(*)$ that:

$$f_{[\alpha]}|_\beta = g_{[\alpha]}|_\beta \quad \forall \beta \in \alpha \qquad\qquad f_{[\alpha]}|_\beta = g_{[\alpha]} \text{ if } s(\beta) = \alpha. \qquad (**)$$

In fact, if $\gamma \in \beta \in \alpha$, then $s(\gamma) \in s(\beta) \subset \alpha$, hence $s(\gamma) \in \alpha$. In this case, by $(*)$, we have $f_{[\alpha]}(\gamma) = g_{[\alpha]}(\gamma)$, proving the l.h.s. formula. Moreover, if $s(\beta) = \alpha$, then the domain of $g_{[\alpha]}$ is β (and $\beta \in \alpha$), hence we get the r.h.s. formula as a particular case of the l.h.s. one.

We have

$$f_{[\alpha]}(0) \overset{(*)}{=} g_{[\alpha]}(0) \overset{(\star)}{=} f_{[1]}(0) \overset{\mathrm{I}_1}{=} a,$$

hence I_α holds. If $s(\beta) \in \alpha$, then

$$f_{[\alpha]}(\beta) \overset{(*)}{=} g_{[\alpha]}(\beta) \overset{(\star)}{=} f_{[s(\beta)]}(\beta) \overset{\mathrm{II}_{s(\beta)}}{=} h(f_{[s(\beta)]}|_\beta) \overset{(\star)}{=} h(g_{[\alpha]}|_\beta) \overset{(**)}{=} h(f_{[\alpha]}|_\beta).$$

If $s(\beta) = \alpha$, then

$$f_{[\alpha]}(\beta) \overset{(*)}{=} h(g_{[\alpha]}) \overset{(**)}{=} h(f_{[\alpha]}|_\beta),$$

hence II_α holds too, proving existence. About uniqueness, let us suppose that $f'_{[\alpha]} \colon \alpha \to \mathsf{c}$ satisfies I_α and II_α too. It is immediate to verify that $f'_{[\alpha]}|_\beta \colon \beta \to \mathsf{c}$ satisfies I_β and II_β for every $\beta \in \alpha$, hence, by the inductive hypothesis, $f'_{[\alpha]}|_\beta = f_{[\beta]} = f_{[\alpha]}|_\beta$. If α is a limit ordinal, this proves that $f'_{[\alpha]} = f_{[\alpha]}$, since $\alpha = \bigcup_{\beta \in \alpha} \beta$ by corollary 11.32-ii. If there exists β such that $\alpha = s(\beta)$, then, by corollary 11.32-i, it remains to consider the value in β. We have $f'_{[\alpha]}(\beta) = h(f'_{[\alpha]}|_\beta) = h(f_{[\alpha]}|_\beta) = f_{[\alpha]}(\beta)$, proving uniqueness.

We have shown by induction that $f_{[\alpha]}$ exists unique for every $\alpha \in \mathbf{Ord}$. Now we set $f(\alpha) := f_{[s(\alpha)]}(\alpha)$ for every $\alpha \in \mathbf{Ord}$. It follows that $f(0) = f_{[1]}(0) = a$, proving I. Moreover, we observe that, if $\beta \subset \alpha$, then:

$$f|_\beta = f_{[\alpha]}|_\beta. \qquad (!!)$$

In fact, for every $\gamma \in \beta$ we have $f(\gamma) = f_{[s(\gamma)]}(\gamma) = f_{[\beta]}(\gamma)$, the last equality being due to (!) if $s(\gamma) \in \beta$ and being trivial if $s(\gamma) = \beta$. Therefore, for every non-zero ordinal α:

$$f(\alpha) = f_{[s(\alpha)]}(\alpha) \overset{(*)}{=} h(g_{[s(\alpha)]}) \overset{(**)}{=} h(f_{[s(\alpha)]}|_\alpha) \overset{(!!)}{=} h(f|_\alpha),$$

proving II. □

Proof of 11.48. In the thesis of 11.48, the function h is applied only to functions defined on limit ordinals. Therefore, there are no effects on the statement if we modify its value on successors. Thus, we start from the functions h and k, provided by the hypothesis of 11.48, and, given $g \colon s(\alpha) \to \mathsf{c}$, we set $h(g) := k(g(\alpha))$. Now we apply 11.49. We get a function $f \colon \mathbf{Ord} \to \mathsf{c}$ such that

- $f(0) = a$;
- $f(s(\alpha)) = h(f|_{s(\alpha)}) = k(f(\alpha))$; and

- if α is a successor ordinal, then $f(\alpha) = h(f|_\alpha)$, where the value of h remained the original one.

This means that f satisfies the thesis of 11.48, as required. □

As in the case of natural numbers, it may happen that the recursive definition of $f(s(\alpha))$ or $f(\alpha)$ depends on α too. It is also possible to define recursively a function $f(z, \alpha)$, where z ranges within a fixed class $\mathtt{d}$. The reader can find the details in appendix C.

Exercise 11.50. Show that we can start transfinite recursion (in both forms) from any fixed ordinal α_0 instead of 0 (cfr. exercise 11.44).

Exercise 11.51. State and prove *bounded* transfinite recursion (cfr. exercise 11.45).

11.4.3 Proof of AC ⇒ ZL by Transfinite Recursion

In section 9.2.1, we said that it is possible to prove the implication AC ⇒ ZL using the theory of ordinal numbers, in a way that is easier to visualise than the one shown in the proof of theorem 9.10. Here we provide this proof by using transfinite recursion. Basically, we show that, if we assume AC and negate ZL, then we can embed the proper class **Ord** in a set x.

Proposition 11.52. *AC ⇒ ZL.*

Proof. Assuming AC, let us suppose by contradiction that $(x, \preceq)$ is an ordered set that satisfies that hypothesis but not the thesis of ZL. By transfinite recursion in the form 11.49, we construct an embedding $\iota\colon \mathbf{Ord} \hookrightarrow x$. This is a contradiction, since the image of ι is a proper class by assumption (A6) and it is a subclass of x, contradicting assumption (A4").

Step I: If $c \subset x$ is a chain, then it admits a strict *upper bound.* In fact, since the hypothesis of ZL holds, there exists an upper bound $u \in x$ of c. This means that $y \preceq u$ for every $y \in c$. We have two possibilities: I. $u \notin c$; II. $u \in c$. In case I, u itself is a strict upper bound. In case II, since x has no maximal elements by hypothesis, in particular u is not maximal, hence there exists $v \in x$ such that $u \prec v$. It follows that $y \preceq u \prec v$ for every $y \in c$, thus v is a strict upper bound.

Step II: The function η. We fix a choice function $\xi\colon \mathcal{P}^*(x) \to x$. Let $\boldsymbol{c}_x$ be the set of chains of x. Given $c \in \boldsymbol{c}_x$, we call $\operatorname{subs}(c) \subset x$ the set of strict upper bounds of c. Step I affirms that $\operatorname{subs}(c) \neq \emptyset$. We define the function $\eta\colon \boldsymbol{c}_x \to x$, $c \mapsto \xi(\operatorname{subs}(c))$. Since ξ is a choice function, it follows that $\eta(c) \in \operatorname{subs}(c)$ for every chain c—that is, η associates to each chain of x a corresponding strict upper bound. We extend η to $\eta\colon \mathcal{P}(x) \to x$ by declaring that, if $a \in \mathcal{P}(x) \setminus \boldsymbol{c}_x$, then $\eta(a) := \eta(\emptyset)$.

Step III: Construction of the embedding. We set $\iota(0) := \eta(\emptyset)$. Assuming to know $\iota|_\alpha$, we set $\iota(\alpha) := \eta(\operatorname{Im}(\iota|_\alpha))$. This means that we apply proposition 11.49 with $\mathsf{c} = x$, $a = \eta(\emptyset)$ and $h(\varphi) := \eta(\operatorname{Im}\varphi)$, obtaining the function $f = \iota$.

Step IV: Proof of injectivity. Let us verify that ι is actually an embedding, completing the proof. We show by transfinite induction that, for every $\alpha \in$ **Ord**:

(i) if $\beta \in \alpha$, then $\mathrm{Im}(\iota|_\beta) \subsetneq \mathrm{Im}(\iota|_\alpha)$;

(ii) $\mathrm{Im}(\iota|_\alpha)$ is a chain in x.

If $\alpha = 0$, then $\mathrm{Im}(\iota|_\alpha) = \emptyset$, hence both results are trivial. *I. If (i) and (ii) hold for α, then they hold for $s(\alpha)$.* In order to prove (i) for $s(\alpha)$, we only have to prove that $\mathrm{Im}(\iota|_\alpha) \subsetneq \mathrm{Im}(\iota|_{s(\alpha)})$, since, for every other $\beta \in s(\alpha)$, we have $\beta \in \alpha \in s(\alpha)$, hence the result follows by the inductive hypothesis (i) and by transitivity. By the inductive hypothesis (ii), $\mathrm{Im}(\iota|_\alpha)$ is a chain in x, hence $\iota(\alpha) = \eta(\mathrm{Im}(\iota|_\alpha)) \in \mathrm{subs}(\mathrm{Im}(\iota|_\alpha))$. Since $\iota(\alpha)$ is a *strict* upper bound, it follows that $\mathrm{Im}(\iota|_\alpha) \subsetneq \mathrm{Im}(\iota|_\alpha) \cup \{\iota(\alpha)\} = \mathrm{Im}(\iota|_{s(\alpha)})$, proving (i). Moreover, it immediately follows that $\mathrm{Im}(\iota|_{s(\alpha)})$ is a chain too, proving (ii). *II. If α is a limit ordinal and (i) and (ii) hold for every $\beta \in \alpha$, then they hold for α too.* In fact, because of the inductive hypotheses (i) and (ii), the subset $\mathrm{Im}(\iota|_\alpha) = \bigcup\{\mathrm{Im}(\iota|_\beta) : \beta \in \alpha\}$ of x is the union of a chain of chains, hence it is a chain (see the proof of proposition 9.6, step II). This proves (ii). About (i), the weak inclusion is obvious, since $\beta \subset \alpha$ by transitivity. About the strict inclusion, since α is a limit ordinal, there exists γ such that $\beta \in \gamma \in \alpha$. It follows from the inductive hypothesis (i) that $\mathrm{Im}(\iota|_\beta) \subsetneq \mathrm{Im}(\iota|_\gamma) \subset \mathrm{Im}(\iota|_\alpha)$, proving (i).

Now we prove that ι respects the strict order (that immediately implies injectivity)—that is, if $\beta \in \alpha$, then $\iota(\beta) \prec \iota(\alpha)$ (the converse follows too). In fact, by item (ii) above, $\mathrm{Im}(\iota|_\alpha)$ is a chain in x, hence $\iota(\alpha) = \eta(\mathrm{Im}(\iota|_\alpha)) \in \mathrm{subs}(\mathrm{Im}(\iota|_\alpha))$. Since $\iota(\alpha)$ is a *strict* upper bound of $\mathrm{Im}(\iota|_\alpha)$ and $\iota(\beta) \in \mathrm{Im}(\iota|_\alpha)$, it follows that $\iota(\beta) \prec \iota(\alpha)$. □

11.5 OPERATIONS ON ORDINALS

The following definitions rely on definitions 8.2 and 8.3.

Definition 11.53. Given $\alpha, \beta \in$ **Ord**, we set:

- $\alpha + \beta := \mathrm{type}(\alpha \sqcup \beta)$
- $\alpha \cdot \beta := \mathrm{type}(\beta \times \alpha)$ ◇

We get the operations $+\,,\ \cdot :$ **Ord** $\times$ **Ord** $\to$ **Ord**, that satisfy the following properties.

Proposition 11.54. *For every $\alpha, \beta, \gamma \in$ **Ord**, we have:*

(i) $(\alpha + \beta) + \gamma = \alpha + (\beta + \gamma)$

(ii) $\alpha + 0 = 0 + \alpha = \alpha$

(iii) $\alpha + s(\beta) = s(\alpha + \beta)$*; in particular,* $\alpha + 1 = s(\alpha)$

(iv) $(\alpha\beta)\gamma = \alpha(\beta\gamma)$

(v) $\alpha \cdot 0 = 0 \cdot \alpha = 0$ *and* $\alpha \cdot 1 = 1 \cdot \alpha = \alpha$

(vi) $\alpha(\beta + \gamma) = \alpha\beta + \alpha\gamma$

(vii) $\alpha \cdot s(\beta) = \alpha\beta + \alpha$

Exercise 11.55. Prove proposition 11.54.

Items i and iv of proposition 11.54 show that sum and product of ordinals are associative. Moreover, items ii and v show that 0 and 1 are the corresponding identity elements and item vi shows that the sum is left-distributive with respect to the product. These properties extend to **Ord** the corresponding ones in $\mathbb{N}$. Such extension does not hold about commutativity and right distributivity, as we will see in exercise 11.61.

Proposition 11.56. *For every* $\alpha, \beta, \gamma \in$ **Ord**, *we have:*

(i) $\alpha + \beta = 0$ *if and only if* $\alpha = \beta = 0$.

(ii) If $\alpha \in \beta$, *then* $\gamma + \alpha \in \gamma + \beta$ *and* $\alpha + \gamma \subset \beta + \gamma$.

(iii) If $\alpha \subset \beta$, *then* $\gamma + \alpha \subset \gamma + \beta$ *and* $\alpha + \gamma \subset \beta + \gamma$.

(iv) If $\beta \neq 0$, *then* $\alpha + \beta$ *is a limit ordinal (respectively, a successor ordinal) if and only if* β *is a limit ordinal (respectively, a successor ordinal).*

(v) If $a \subset$ **Ord** *is any non-empty subset and* $\alpha \in$ **Ord**, *then, setting* $\alpha + a := \{\alpha + \beta : \beta \in a\}$, *we have* $\sup(\alpha + a) = \alpha + \sup(a)$.

(vi) If β *is a limit ordinal, then* $\alpha + \beta = \sup\{\alpha + \gamma : \gamma \in \beta\}$.

(vii) If $\alpha \subset \beta$, *then there exists unique* γ *such that* $\beta = \alpha + \gamma$ *(uniqueness implies that cancellation law holds on the right—that is,* $\alpha + \gamma = \alpha + \gamma'$ *implies* $\gamma = \gamma'$*).*

Exercise 11.57. Prove proposition 11.56. Show with an example that we cannot replace '$\subset$' by '$\in$' in item ii.

Exercise 11.58. Prove that every ordinal α can be expressed uniquely in the form $\alpha = \beta + n$, where β is a limit ordinal and $n \in \omega$.

Proposition 11.59. *For every* $\alpha, \beta, \gamma \in$ **Ord**, *we have:*

(i) $\alpha\beta = 0$ *if and only if* $\alpha = 0$ *or* $\beta = 0$*; moreover,* $\alpha\beta = 1$ *if and only if* $\alpha = \beta = 1$.

(ii) If $\alpha \in \beta$ *and* $\gamma \neq 0$, *then* $\gamma\alpha \in \gamma\beta$ *and* $\alpha\gamma \subset \beta\gamma$.

(iii) If $\alpha \subset \beta$, *then* $\gamma\alpha \subset \gamma\beta$ *and* $\alpha\gamma \subset \beta\gamma$.

(iv) *If* $\alpha, \beta \neq 0$*, then* $\alpha\beta$ *is a successor ordinal if and only if both* α *and* β *are successor ordinals.*

(v) *If* $a \subset \mathbf{Ord}$ *is any non-empty subset and* $\alpha \in \mathbf{Ord}$*, then, setting* $\alpha \bullet a := \{\alpha\beta : \beta \in a\}$*, we have* $\sup(\alpha \bullet a) = \alpha \cdot \sup(a)$*.*

(vi) *If* β *is a limit ordinal, then* $\alpha\beta = \sup\{\alpha\gamma : \gamma \in \beta\}$*.*

(vii) *If* $\alpha \neq 0$ *and* β *is any ordinal, then there exist unique* γ *and* δ *such that* $\beta = \alpha \cdot \gamma + \delta$ *and* $\delta \in \alpha$ *(uniqueness with* $\delta = 0$ *implies that cancellation law holds on the right—that is,* $\alpha\gamma = \alpha\gamma'$ *implies* $\gamma = \gamma'$*).*

Exercise 11.60. Prove proposition 11.59. Show with an example that we cannot replace '$\subset$' by '$\in$' in item ii.

Exercise 11.61. Prove that:

- $1 + \omega = \omega$, while $\omega + 1 \neq \omega$
- $2 \cdot \omega = \omega$, while $\omega \cdot 2 \neq \omega$
- $(1 + 1)\omega = \omega$, while $1\omega + 1\omega \neq \omega$
- $\omega + 1 \neq \sup\{\gamma + 1 : \gamma \in \omega\}$
- $\omega \cdot 2 \neq \sup\{\gamma \cdot 2 : \gamma \in \omega\}$

We can also define sum and product of ordinals by recursion. In this way, power can be defined naturally too.

Definition 11.62. Given $\alpha, \beta \in \mathbf{Ord}$, we set:

- $\alpha + 0 := \alpha$
- $\alpha + s(\beta) := s(\alpha + \beta)$
- if β is a limit ordinal, $\alpha + \beta := \sup\{\alpha + \gamma : \gamma \in \beta\}$ ◇

Definition 11.63. Given $\alpha, \beta \in \mathbf{Ord}$, we set:

- $\alpha \cdot 0 := 0$
- $\alpha \cdot s(\beta) := \alpha \cdot \beta + \alpha$
- if β is a limit ordinal, $\alpha \cdot \beta := \sup\{\alpha \cdot \gamma : \gamma \in \beta\}$ ◇

Definition 11.64. Given $\alpha, \beta \in \mathbf{Ord}$, we set:

- $\alpha^0 := 1$
- $\alpha^{s(\beta)} := \alpha^\beta \cdot \alpha$
- if β is a limit ordinal, $\alpha^\beta := \sup\{\alpha^\gamma : \gamma \in \beta\}$ ◇

Exercise 11.65. Prove that definitions 11.62 and 11.63 are equivalent to definition 11.53.

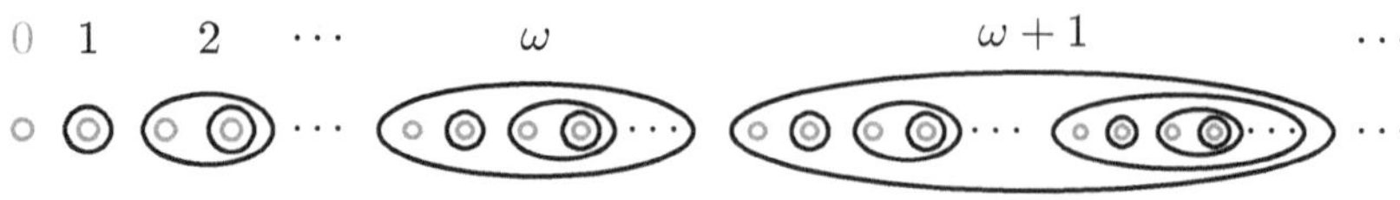

Figure 11.1 First ordinals.

Exercise 11.66. Prove that:

(i) $\alpha^1 = \alpha$; $\alpha^\beta = 0$ if and only if $\alpha = 0$ and $\beta \neq 0$; and $\alpha^\beta = 1$ if and only if $\alpha = 1$ or $\beta = 0$.

(ii) If $\alpha \in \beta$, then $\gamma^\alpha \in \gamma^\beta$ for every $\gamma \supset 2$ and $\alpha^\gamma \subset \beta^\gamma$ for every γ.

(iii) If $\alpha \subset \beta$, then $\gamma^\alpha \subset \gamma^\beta$ (except in the case $\gamma = \alpha = 0$ and $\beta \neq 0$) and $\alpha^\gamma \subset \beta^\gamma$.

(iv) If $\alpha \supset 2$ and $\beta \neq 0$, then α^β is a successor ordinal if and only if α is a successor ordinal and β is a natural number.

(v) If $a \subset \mathbf{Ord}$ is any non-empty subset and $\alpha \in \mathbf{Ord}^*$, then, setting $\alpha^a := \{\alpha^\beta : \beta \in a\}$, we have $\sup(\alpha^a) = \alpha^{\sup(a)}$.

Show with an example that we cannot replace '$\subset$' by '$\in$' in item ii.

Exercise 11.67. Prove that, for every $\alpha, \beta, \gamma \in \mathbf{Ord}$:

- $\alpha^1 = \alpha$
- $\alpha^{\beta+\gamma} = \alpha^\beta \cdot \alpha^\gamma$
- $(\alpha^\beta)^\gamma = \alpha^{\beta\gamma}$

Show that in general $(\alpha\beta)^\gamma \neq \alpha^\gamma\beta^\gamma$.

It is possible to provide a direct definition of power too, without relying on transfinite recursion, but it is much less natural than the ones of sum and product (see exercise 4 at the end of the chapter).

Now we can think more deeply about the structure of the first ordinals. We have natural numbers, then $\omega, \omega+1, \omega+2, \ldots, \omega+\omega = \omega 2, \omega 2+1, \omega 2+2, \ldots, \omega 3, \omega 3+1, \omega 3+2, \ldots$. Keeping on in this way, we get $\omega n + m$, where $n, m \in \omega$. Afterwards, we get $\omega\omega = \omega^2, \omega^2+1, \omega^2+2, \ldots, \omega^2+\omega, \omega^2+\omega+1, \ldots, \omega^2+\omega 2, \omega^2+\omega 2+1, \ldots, \omega^2+\omega 3, \ldots$. Keeping on in this way, we get $\omega^2+\omega n+m$, then $\omega^2 2+\omega n+m$, $\omega^2 3+\omega n+m$, and so on until ω^3. Then we have $\omega^3+\omega^2 n+\omega m+k$, and so on. Keeping on again, we get $\omega^n m_0 + \omega^{n-1} m_1 + \cdots + \omega m_{n-1} + m_n$. After all these ordinals, we have ω^ω, then $\omega^\omega k + \omega^n m_0 + \omega^{n-1} m_1 + \cdots + \omega m_{n-1} + m_n$. Then we have $\omega^{\omega 2}, \ldots, \omega^{\omega 3}, \ldots$ and $\omega^{(\omega^2)}$. Afterwards, we have all the intermediate ordinals until $\omega^{(\omega^3)}$, and so on until $\omega^{(\omega^\omega)}$, and we can start again, but we prefer to stop here. Figure 11.1 represents the first few ordinals.

11.5.1 Ordinal Operations on Natural Numbers

The next proposition shows that the operations of sum, product, and power among ordinals can be restricted to ω, and these restrictions coincide with the usual operations among natural numbers, defined in section 7.2.3.

Proposition 11.68. *If $n, m \in \omega$, then $n + m, nm, n^m \in \omega$.*

Proof. Induction on m, considering definitions 11.62, 11.63 and 11.64. *Sum:* For $m = 0$, we have $n + 0 = n$, and $n \in \omega$ by hypothesis. If $n + m \in \omega$, then $n + s(m) = s(n + m) \in \omega$, since $n + m \in \omega$ by the inductive hypothesis and ω is closed by successor because of proposition 11.40. *Product:* For $m = 0$, we have $n0 = 0$, and $0 \in \omega$, since $\omega \in \mathtt{cs}$ by proposition 11.40. If $nm \in \omega$, then $n \cdot s(m) = nm + n \in \omega$ because of the previous result about the sum. *Power:* For $m = 0$, we have $n^0 = 1$, and $1 = s(0) \in \omega$, since $\omega \in \mathtt{cs}$ by proposition 11.40. If $n^m \in \omega$, then $n^{s(m)} = n^m n \in \omega$ because of the previous result about the product. □

Corollary 11.69. *The operations of sum, product, and power among ordinals, restricted to ω, coincide with the usual operations among natural numbers, as defined in section 7.2.3.*

Proof. Assuming that $\alpha, \beta \in \omega$ in definitions 11.62, 11.63, and 11.64, we get well-defined operations in ω by proposition 11.68. Since ω contains no limit ordinals by definition (11.1), definitions 11.62, 11.63, and 11.64, restricted to ω, coincide with the definitions given in section 7.2.3. □

It follows from proposition 11.68 that we can *define* the basic operations among natural numbers as the restrictions of the corresponding operations among ordinals. As anticipated in section 11.3.1, this completes the construction of natural numbers as particular cases of ordinals, without relying on the construction shown in Chapter 7.

With this approach, the order relation is provided by '$\subset$' or, in the strict from, by '$\in$'. Moreover, we observe that proposition 7.4 is a particular case of the corresponding properties in the whole class **Ord** (item iii is bounded induction in ω, see exercise 11.45). Similarly, recursion among natural numbers is ordinal bounded recursion in ω (see exercise 11.51). Using induction and recursion, we can prove all the properties stated in chapter 7. Nevertheless, we will see in section 12.5 that some of these properties (in particular, associativity, commutativity, and distributivity of sum and product) can be proven easier in the context of cardinal operations.

11.6 EXERCISES

1. Let $a \subset \mathbf{Ord}$ be any subset. Prove that:

(i) The subset a admits a maximum if and only if $\left(\bigcup a\right) \in a$. In this case, $\max(a) = \bigcup a$.

(ii) The subset a does not admit a maximum if and only if $a \subset \bigcup a$.

(iii) $a \subset s\left(\bigcup a\right)$.

2. Prove that, if α and β are denumerable (or countable or finite), then $\alpha + \beta$, $\alpha\beta$ and α^β are denumerable (or countable or finite).

3. (i) Prove that, fixing any ordinal $\xi \supset 2$, every $\alpha \in \mathbf{Ord}^*$ can be uniquely represented in the following form:

$$\alpha = \xi^{\beta_k} \cdot \gamma_k + \cdots + \xi^{\beta_1} \cdot \gamma_1,$$

where $k \in \omega \setminus \{0\}$, $0 \in \gamma_1, \ldots, \gamma_k \in \xi$ and $\beta_k \ni \cdots \ni \beta_1$.

(ii) Prove that, if $\xi \supset 2$, $\beta_k \ni \cdots \ni \beta_1$, $\beta'_h \ni \cdots \ni \beta'_1$, $\beta_k = \beta'_h$ and $\gamma_k \in \gamma'_h$, then

$$\xi^{\beta_k} \cdot \gamma_k + \cdots + \xi^{\beta_1} \cdot \gamma_1 \in \xi^{\beta'_h} \cdot \gamma'_h + \cdots + \xi^{\beta'_1} \cdot \gamma'_1$$

for any $\gamma_1, \ldots, \gamma_k, \gamma'_1, \ldots, \gamma'_h \in \xi$.

Remark: Choosing $\xi = \omega$, we represent any non-zero ordinal α uniquely in the form

$$\alpha = \omega^{\beta_k} \cdot n_k + \cdots + \omega^{\beta_1} \cdot n_1,$$

where $k, n_1, \ldots, n_k \in \omega \setminus \{0\}$ and $\beta_k \ni \cdots \ni \beta_1$. This is called *Cantor normal form* of α.

Remark: In the case $\alpha = \xi^{\beta_1}$, uniqueness implies that, if $\xi^{\beta_1} = \xi^{\beta'_1}$, then $\beta_1 = \beta'_1$.

4. Given $\alpha, \beta \in \mathbf{Ord}$ and a function $f\colon \alpha \to \beta$, the *support* of f is the set $\mathrm{supp}(f) := \{\gamma \in \alpha : f(\gamma) \neq 0\}$. We denote by $\mathrm{Func}^f(\alpha, \beta)$ the set of functions from α to β whose support is *finite*. We endow $\mathrm{Func}^f(\alpha, \beta)$ with the following order. Given $f, g \in \mathrm{Func}^f(\alpha, \beta)$, we define $D(f, g) := \{\gamma \in \alpha : f(\gamma) \neq g(\gamma)\}$. Note that $f \neq g$ if and only if $D(f, g) \neq \emptyset$. Moreover, $D(f, g)$ is finite, since $D(f, g) \subset \mathrm{supp}(f) \cup \mathrm{supp}(g)$. Therefore, if $f \neq g$, then there exists $\xi := \max(D(f, g))$ (see exercise 3.33). In this case, we set $f \prec g$ if and only if $f(\xi) \in g(\xi)$.

- Prove that '$\prec$' is a strict well-order in $\mathrm{Func}^f(\alpha, \beta)$.
- Prove that $\alpha^\beta = \mathrm{type}(\mathrm{Func}^f(\beta, \alpha), \preceq)$.
- Prove that the ordered-set isomorphism $\varphi\colon \mathrm{Func}^f(\beta, \alpha) \xrightarrow{\simeq} \alpha^\beta$ is the following one. If $f \in \mathrm{Func}^f(\beta, \alpha)$ and $\mathrm{supp}(f) = \{\delta_1, \ldots, \delta_k\}$, with $\delta_1 \in \cdots \in \delta_k$, then $\varphi(f) = \alpha^{\delta_k} \cdot f(\delta_k) + \cdots + \alpha^{\delta_1} \cdot f(\delta_1)$.

5. (i) Let $f\colon \mathbf{Ord} \to \mathbf{Ord}$ be a function that is

- *strictly increasing*—that is, $\alpha \in \beta$ implies $f(\alpha) \in f(\beta)$; and
- *continuous*—that is, for every limit ordinal α, we have $f(\alpha) = \sup\{f(\beta) : \beta \in \alpha\}$.

Prove that the class of fixed points of f—that is, $\{\alpha \in \mathbf{Ord} : f(\alpha) = \alpha\}$—is not bounded above (in particular, it is infinite, hence of course not empty).

(ii) Prove that there exist infinitely many ordinals α such that $\omega^\alpha = \alpha$. Show that the minimum of such ordinals is denumerable.[3]

[3] This is exercise 30.14 in the lectures notes (in Italian) by Prof. Alessandro Berarducci, available at `https://people.dm.unipi.it/berardu/Didattica/2012-13ETI/Insiemi2012-13/insiemi2016.pdf`

CHAPTER 12

Cardinals

We introduce the notion of cardinal number, as a particular case of the concept of ordinal number, and we use it to define the cardinality of a set, coherently with the comparison already defined in chapter 10.

Remark 12.1. Contrary to the case of ordinals, many important results in this chapter rely on the axiom of choice. ♢

12.1 DEFINITION AND FIRST PROPERTIES

Two ordinal numbers are in particular sets, hence we can compare their cardinalities. This is always possible, even without assuming the axiom of choice, since they are well-ordered by definition (see theorem 8.21).

Definition 12.2. An ordinal number α is called *cardinal number* if, for every ordinal $\beta \in \alpha$, we have $|\beta| < |\alpha|$. ♢

Considering that $\beta \in \alpha$ if and only if β is a proper initial segment of α (see lemma 11.25), this means that an ordinal number is a cardinal if and only if it has not the same cardinality of any proper initial segment of itself.

Example 12.3. Every natural number is a cardinal because of exercise 7.21-vii. The ordinal ω is a cardinal too, because of exercise 7.21-viii.

Exercise 12.4. Prove that:

- $\omega + n$ is not a cardinal for every $n \in \omega \setminus \{0\}$.
- $\omega \cdot n$ is not a cardinal for every $n \in \omega \setminus \{0, 1\}$.
- ω^n is not a cardinal for every $n \in \omega \setminus \{0, 1\}$.
- ω^ω is not a cardinal.

We denote by **Card** the class of cardinals. By definition, **Card** $\subset$ **Ord**.

Lemma 12.5. *If* $\kappa, \lambda \in$ **Card**, *then:*

(i) $\kappa = \lambda$ *if and only if* $|\kappa| = |\lambda|$*;*

DOI: 10.1201/9781003565291-12

(ii) $\kappa \subset \lambda$ *if and only if* $|\kappa| \leq |\lambda|$;

(ii) $\kappa \in \lambda$ *if and only if* $|\kappa| < |\lambda|$.

Proof. (i) ($\Rightarrow$) Obvious. ($\Leftarrow$) If $\kappa \neq \lambda$, then, since they are both ordinals, we have $\kappa \in \lambda$ or $\lambda \in \kappa$. By definition 12.2 we get respectively $|\kappa| < |\lambda|$ or $|\lambda| < |\kappa|$. (ii) ($\Rightarrow$) Obvious. ($\Leftarrow$) Since κ and λ are ordinals, we have $\kappa \subset \lambda$ or $\lambda \in \kappa$. In the latter case, we have $|\lambda| < |\kappa|$ by definition 12.2, a contradiction. (iii) It immediately follows from i and ii. □

Proposition 12.6. *For every* $\alpha \in \mathbf{Ord}$, *there exists a unique* $\kappa \in \mathbf{Card}$ *such that* $|\alpha| = |\kappa|$. *We have* $\kappa = \min\{\beta \in \mathbf{Ord} : (\beta \subset \alpha) \wedge (|\beta| = |\alpha|)\}$ *(hence* $|\alpha| \subset \alpha$*).*

Proof. We set $a := \{\beta \in \mathbf{Ord} : (\beta \subset \alpha) \wedge (|\beta| = |\alpha|)\}$. The class a is not empty, since trivially $\alpha \in a$, therefore it admits a minimum κ. Since $|\kappa| = |\alpha|$ by definition of a, we only have to prove that κ is a cardinal. If by contradiction $\gamma \in \kappa$ and $|\gamma| = |\kappa|$, then $\gamma \in \kappa \subset \alpha$ and $|\gamma| = |\kappa| = |\alpha|$, hence $\gamma \subset \alpha$ and $|\gamma| = |\alpha|$. It follows that $\gamma \in a$, contradicting the minimality of κ. Uniqueness immediately follows from lemma 12.5-i. □

Corollary 12.7. *For every set* x *that admits a well-order, there exits a unique cardinal* κ *such that* $|x| = |\kappa|$.

Proof. We fix any well-order in x. By proposition 11.26, x is isomorphic as a well-ordered set to an ordinal α. In particular, $|x| = |\alpha|$. If κ is the unique cardinal such that $|\alpha| = |\kappa|$, then we get $|x| = |\kappa|$. Uniqueness immediately follows from lemma 12.5-i. □

The previous corollary justifies the following definition.

Definition 12.8. If x is a set that admits a well-order, the *cardinality* of x is the unique cardinal κ such that $|x| = |\kappa|$. We denote the cardinality of x by $|x|$—that is, we set $|x| := \kappa$. ◊

Exercise 12.9. Show that, if x and y admit a well-order, then $|x| = |y|$, $|x| \leq |y|$ or $|x| < |y|$, in the sense of definition 10.2, if and only if respectively $|x| = |y|$, $|x| \subset |y|$ or $|x| \in |y|$, where $|x|$ and $|y|$ are the cardinalities of x and y as defined in 12.8.

Proposition 12.10. *For every* $\alpha, \beta \in \mathbf{Ord}$*:*

(i) α *is a cardinal if and only if* $|\alpha| = \alpha$.

(ii) $|\alpha|$ *coincides with the cardinal* κ *in the statement of proposition 12.6.*

(iii) If $|\alpha| \subset \beta \subset \alpha$, *then* $|\beta| = |\alpha|$.

(iv) If $|\alpha| < |\beta|$, *then* $\alpha \in \beta$.

Proof. (i) ($\Rightarrow$) Straightforward by choosing $x = \alpha$ in definition 12.8. ($\Leftarrow$) By definition $|\alpha|$ is a cardinal and by hypothesis it coincides with α. (ii) By proposition 12.6 we have $|\alpha| = |\kappa|$ and by item i we have $|\kappa| = \kappa$. Considering exercise 12.9, the result immediately follows. (iii) Since $|\alpha| \subset \beta$, we have $||\alpha|| \leq |\beta|$—that is, $|\alpha| \leq |\beta|$ by item i—hence $|\alpha| \subset |\beta|$ by exercise 12.9. Since $\beta \subset \alpha$, we have $|\beta| \leq |\alpha|$—that is, $|\beta| \subset |\alpha|$ by exercise 12.9—hence equality follows. (iv) If $\beta \subset \alpha$, then $|\beta| \leq |\alpha|$, contradicting the hypothesis. By proposition 11.19-iv, the only possibility left is $\alpha \in \beta$. □

Up to now we never needed to assume the axiom of choice, since we assumed the existence of a well-order by hypothesis when it was necessary.

Remark 12.11. Assuming the axiom of choice, definition 12.8 can by applied to any set, hence *every set has a well-defined cardinality*. In particular, exercise 12.9 applies to any pair of sets, hence the comparison between cardinalities, in the sense of definition 10.2, is always equivalent to the comparison between the cardinalities of the two sets, as defined in 12.8.

If we do not assume AC, then, considering two sets x and y, such that at least one of the two does not admit a well-order, it is possible that $|x| = |y|$, $|x| \leq |y|$ or $|x| < |y|$, in the sense of definition 10.2, even if $|x|$ and $|y|$ are not defined. Hence, definition 10.2 turns out to be more general than the comparison between cardinalities as defined in 12.8.[1] ◇

12.2 SUCCESSORS, LIMITS AND THE FUNCTION $\aleph$ (OR ω)

We have seen in example 12.3 that every natural number is a cardinal. The corollary of the next lemma shows that natural numbers are the only cardinals that are successor ordinals.

Lemma 12.12. *For every infinite ordinal α, we have $|s(\alpha)| = |\alpha|$.*

Proof. It follows from lemma 11.36 that $\omega \subset \alpha \subset s(\alpha)$. We have the bijection $\varphi\colon s(\alpha) \xrightarrow{\approx} \alpha$ defined by $n \mapsto n+1$ for every $n \in \omega$, $\beta \mapsto \beta$ for every $\beta \in \alpha\setminus\omega$, $\alpha \mapsto 0$. It is easy to show that φ is actually a bijection, since its inverse is given by $n \mapsto n-1$ for every $n \in \omega\setminus\{0\}$, $\beta \mapsto \beta$ for every $\beta \in \alpha\setminus\omega$, $0 \mapsto \alpha$. □

Corollary 12.13. *Every infinite cardinal is a limit ordinal.*

Proof. If $\alpha = s(\beta)$ is an infinite ordinal, then β is infinite too, otherwise $\alpha = \beta \cup \{\beta\}$ would be finite. It follows from lemma 12.12 that $|\alpha| = |\beta|$, hence α is not a cardinal, since $\beta \in \alpha$. □

---

[1]For example, let x be a set that does not admit a well-order. If y is any set such that there exists a bijection between x and y, then $|x| = |y|$, even if both x and y do not admit a well-order. If we fix any $z \in x$ and we set $y := \{z\}$, then $|y| < |x|$, even if only y admits a well-order. If we set $y := x \sqcup \{0\}$, then $|x| \leq |y|$, even if both x and y do not admit a well-order.

12.2.1 Successor and Limit Cardinals

We can introduce the notion of successor among cardinals, that is different from the one among ordinals.

Definition 12.14. Given two cardinals κ and λ, we say that λ is the *successor* (as a cardinal) of κ, and we use the notation $\lambda = \kappa^+$, if $\kappa \in \lambda$ and there not exists any cardinal σ such that $\kappa \in \sigma \in \lambda$. ◇

Proposition 12.15. *For every cardinal κ there exists its successor κ^+.*

Proof. We consider the class $a := \{\alpha \in \mathbf{Ord} : |\alpha| = \kappa\}$. This class is actually a set. In fact, let us call $\mathrm{wo}(\kappa)$ the set of well-order relations in the set κ, that is a set by assumption (A4"), since $\mathrm{wo}(\kappa) \subset \mathcal{P}(\kappa \times \kappa)$. We have the natural function $t\colon \mathrm{wo}(\kappa) \to \mathbf{Ord}$, $\preceq \mapsto \mathrm{type}(\kappa, \preceq)$. Let us show that $a = \mathrm{Im}(t)$, hence it is a set by assumption (A6). In fact, if $\preceq \in \mathrm{wo}(\kappa)$, then $t(\preceq) \in a$, since by definition there exists an ordered-set isomorphism (in particular, a bijection) between $(\kappa, \preceq)$ and $\mathrm{type}(\kappa, \preceq)$. This proves that $\mathrm{Im}(t) \subset a$. Conversely, given $\alpha \in a$, by definition of a there exists a bijection $\varphi\colon \kappa \xrightarrow{\approx} \alpha$, hence we define a well-order in κ by $\gamma \prec \delta$ if and only if $\varphi(\gamma) \in \varphi(\delta)$. It is actually a well-order, since φ is a bijection, and by construction it makes φ an ordered-set isomorphism. Therefore, $\mathrm{type}(\kappa, \preceq) = \alpha$—that is, $t(\preceq) = \alpha$. This proves that $a \subset \mathrm{Im}(t)$, hence equality holds.

By proposition 11.23, the ordinal $\lambda := \sup(a)$ is well-defined and $\lambda = \bigcup a$. Let us prove that $\lambda = \kappa^+$. We prove that:

(i) $|\kappa| < |\lambda|$

(ii) $\lambda = |\lambda|$, hence λ is a cardinal

(iii) $\lambda = \kappa^+$

(i) Trivially $\kappa \in a$, hence $\kappa \subset \sup(a) = \lambda$, therefore $|\kappa| \leq |\lambda|$. If by contradiction $|\lambda| = |\kappa| = \kappa$, then $\lambda \in a$, hence $\lambda = \max(a)$. Nevertheless, $s(\lambda) \in a$ by lemma 12.12, a contradiction. (ii) We set $\lambda' := |\lambda|$. It follows from item i and proposition 12.10-iv that $a \subset \lambda'$, hence $\lambda = \sup(a) \subset \lambda'$. Conversely, $\lambda' \subset \lambda$ by proposition 12.6, hence $\lambda' = \lambda$. In particular, λ is a cardinal. (iii) If by contradiction there exists a cardinal σ such that $\kappa \in \sigma \in \lambda$, then $a \subset \sigma$ by proposition 12.10-iv, thus $\lambda = \sup(a) \subset \sigma$, a contradiction. □

Corollary 12.16. *There not exists a maximum cardinal. More generally, for every ordinal α, there exists a cardinal κ such that $\alpha \in \kappa$.*

Proof. For every cardinal κ, we have $\kappa \in \kappa^+$. For every ordinal α, we have $\alpha \in |\alpha|^+$ by proposition 12.10-iv. □

Remark 12.17. Assuming the axiom of choice, proposition 12.15 immediately follows from Cantor's theorem 10.30. In fact, if κ is a cardinal, then $\lambda := |\mathcal{P}(\kappa)| > |\kappa| = \kappa$, hence the class $u_\kappa := \{\sigma \in \mathbf{Card} : \kappa \in \sigma\}$ is not empty. Since **Card** is well-ordered (being a subclass of **Ord**), there exists $\kappa^+ := \min(u_\kappa)$, proving the thesis. This proof requires the axiom of choice, since otherwise $\mathcal{P}(\kappa)$ may not admit a well-order, hence $|\mathcal{P}(\kappa)|$ may not be well-defined. $\diamondsuit$

As in the framework of ordinals, the definition of successor naturally leads to the following definition.

Definition 12.18. A non-zero cardinal κ is called *weak limit cardinal* it is not a successor. It is called *limit cardinal* if it is a weak limit cardinal different from ω. $\diamondsuit$

Proposition 12.19. *Let $a \subset$ **Card** be a subclass.*

(i) If $a \neq \emptyset$, then it admits a minimum μ and $\mu = \bigcap a$.

*(ii) If a is a set, then it admits a supremum ξ in **Card** (that can be a maximum or not) and $\xi = \bigcup a$.*

*Therefore, comparing with proposition 11.23, the minimum and the supremum of a in **Ord** and in **Card** coincide.*

Proof. (i) Since $a \subset$ **Ord**, it admits a minimum μ in **Ord** by proposition 11.23-i, and $\mu = \bigcap a$. Since the minimum of a belongs to a and $a \subset$ **Card**, necessarily μ is a cardinal, hence it is the minimum of a as a subset of **Card**. (ii) Since $a \subset$ **Ord**, it admits a supremum ξ in **Ord** by proposition 11.23-ii, and $\xi = \bigcup a$. Let us prove that ξ is a cardinal. If $\xi \in a$, then it is a cardinal, since $a \subset$ **Card**. Otherwise, let us suppose that $\alpha \in \xi$. Since $\xi = \sup(a)$, α cannot be an upper bound of a, hence there exists $\kappa \in a$ such that $\alpha \in \kappa \in \xi$. Since κ is a cardinal (because it belongs to a), we have $|\alpha| \leq |\kappa| = \kappa \in \xi = |\xi|$, hence $|\alpha| < |\xi|$. This proves that ξ is a cardinal. Since $\xi = \sup(a)$ in **Ord**, no ordinal α, such that $\alpha \in \xi$, is an upper bound of a. This holds in particular if α is a cardinal, hence $\xi = \sup(a)$ in **Card** too. $\square$

Corollary 12.20. *A subclass $a \subset$ **Card** is a set if and only if it is bounded above.*

Proof. ($\Rightarrow$) It immediately follows from proposition 12.19-ii. ($\Leftarrow$) If $\kappa \in$ **Card** is an upper bound of a, then κ^+ is a strict upper bound—that is, $\lambda \in \kappa^+$ for every $\lambda \in a$. This means that $a \subset \kappa^+$, hence it is a set by assumption `(A4")`. $\square$

Corollary 12.21. *For every cardinal κ, there exists a limit cardinal λ such that $\kappa \in \lambda$. More generally, for every ordinal α, there exists a limit cardinal λ such that $\alpha \in \lambda$.*

Proof. If $\kappa \in \omega$, we replace κ by ω itself. We construct recursively the following sequence $\{\kappa_n\}_{n \in \mathbb{N}}$ of cardinals: $\kappa_0 := \kappa$ and $\kappa_{n+1} := \kappa_n^+$. We call a the image of this sequence, that is a set by assumption `(A6)`, and we define $\lambda := \sup(a)$.

Necessarily $\kappa \in \lambda$, since $\kappa \subset \sup(a) = \lambda$ and κ is not the maximum of a (because $\kappa \in \kappa^+ \in a$). It remains to prove that λ is a limit cardinal. By construction $\omega \subset \kappa \in \lambda$, hence $\omega \in \lambda$. If by contradiction $\lambda = \sigma^+$, then $\sigma \in \lambda$, hence σ is not an upper bound of a. It follows that there exists $n \in \mathbb{N}$ such that $\sigma \in \kappa_n \subset \lambda$. Necessarily $\kappa_n \in \lambda$, since κ_n is not the maximum of a (because $\kappa_n \in \kappa_{n+1} \in a$), therefore $\sigma \in \kappa_n \in \lambda$, a contradiction, since $\lambda = \sigma^+$.

Considering an ordinal α, by the first part of the statement there exists a limit cardinal λ such that $|\alpha|^+ \in \lambda$. It trivially follows that $\alpha \in \lambda$. □

An infinite cardinal κ is always a limit ordinal by corollary 12.13, hence it does not admit a maximum and $\kappa = \sup_{\mathbf{Ord}}(\kappa)$ by proposition 11.31 and corollary 11.32-ii. We can also consider the set $\kappa \cap \mathbf{Card}$. In this case, we can formulate a proposition and a corollary analogous to 11.31 and 11.32.

Proposition 12.22. *A cardinal κ is a successor if and only if $\kappa \cap \mathbf{Card}$ admits a maximum λ. In this case, $\kappa = \lambda^+$.*

Proof. ($\Rightarrow$) If $\kappa = \lambda^+$, then $\lambda \in \kappa \cap \mathbf{Card}$. If $\sigma \in \kappa \cap \mathbf{Card}$, then it is not possible that $\lambda \in \sigma$, since otherwise $\lambda \in \sigma \in \lambda^+$, a contradiction. Hence, $\sigma \subset \lambda$, therefore $\lambda = \max(\kappa \cap \mathbf{Card})$. ($\Leftarrow$) We set $\lambda := \max(\kappa \cap \mathbf{Card})$. In particular, $\lambda \in \kappa$. If by contradiction there exists $\sigma \in \mathbf{Card}$ such that $\lambda \in \sigma \in \kappa$, then necessarily $\sigma \in \kappa \cap \mathbf{Card}$, contradicting the maximality of λ. By definition, it follows that $\kappa = \lambda^+$. □

Corollary 12.23. *A non-zero cardinal κ is*

(i) a successor cardinal if and only if $\sup(\kappa \cap \mathbf{Card}) \in \kappa$*; and*

(ii) a weak limit cardinal if and only if $\sup(\kappa \cap \mathbf{Card}) = \kappa$.

By proposition 12.19, the cardinal $\sup(\kappa \cap \mathbf{Card})$ *can be thought of indifferently as the supremum in* **Card** *or in* **Ord**.

Proof. (i) ($\Rightarrow$) By proposition 12.22, $\kappa \cap \mathbf{Card}$ admits a maximum λ, hence $\sup(\kappa \cap \mathbf{Card}) = \lambda \in \kappa$. ($\Leftarrow$) If $\lambda := \sup(\kappa \cap \mathbf{Card}) \in \kappa$, then $\lambda = \max(\kappa \cap \mathbf{Card})$, hence $\kappa = \lambda^+$ by proposition 12.22.

(ii) ($\Rightarrow$) Since κ is a successor ordinal by lemma 12.13, we have $\sup_{\mathbf{Ord}}(\kappa) = \kappa$ by corollary 11.32-ii. If $\lambda \in \kappa \cap \mathbf{Card}$, then in particular $\lambda \in \kappa$, hence κ is an upper bound of $\kappa \cap \mathbf{Card}$ too. Let us suppose that σ is an upper bound of $\kappa \cap \mathbf{Card}$ in **Card**. If by contradiction $\sigma \in \kappa$, then $\sigma^+ \subset \kappa$. Since κ is not a successor cardinal, the equality cannot hold, hence $\sigma^+ \in \kappa$. It follows that $\sigma \in \sigma^+ \in \kappa$, a contradiction, since σ is an upper bound of $\kappa \cap \mathbf{Card}$. Therefore, $\kappa \subset \sigma$ for every upper bound of $\kappa \cap \mathbf{Card}$, hence $\kappa = \sup(\kappa \cap \mathbf{Card})$. ($\Leftarrow$) Since $\kappa \notin \kappa$, we have $\sup(\kappa \cap \mathbf{Card}) \notin \kappa \cap \mathbf{Card}$, hence $\kappa \cap \mathbf{Card}$ does not admit a maximum. By proposition 12.22, κ is not a successor cardinal. □

12.2.2 The Function $\aleph$ (or ω)

We denote by $\mathbf{Card}_\infty$ the class of infinite cardinals—that is, $\mathbf{Card}_\infty := \{\kappa \in \mathbf{Card} : \omega \subset \kappa\}$. We have the natural bijection

$$\begin{aligned} \aleph\colon \mathbf{Ord} &\xrightarrow{\approx} \mathbf{Card}_\infty \\ \alpha &\mapsto \aleph_\alpha, \end{aligned} \tag{12.1}$$

defined by transfinite recursion on ordinals as follows:

- $\aleph_0 := \omega$
- $\aleph_{s(\alpha)} := \aleph_\alpha^+$
- if α is a limit ordinal, then $\aleph_\alpha := \sup\{\aleph_\beta : \beta \in \alpha\}$

In the third item, the supremum can be taken indifferently in **Ord** or in **Card**.

Exercise 12.24. Formalise the definition of the function $\aleph$, by applying the statement of proposition 11.48.

We are going to prove that $\aleph$ is an ordered-class isomorphism (in particular, a bijection). First of all it is well-defined, since, when α is a limit ordinal, the class $a := \{\aleph_\beta : \beta \in \alpha\}$ is a set by assumption `(A6)`. It follows that $\sup(a)$ is a well-defined cardinal by proposition 12.19-ii.

Proposition 12.25. *We have that:*

(i) The function $\aleph$ respects the order—that is, $\alpha \in \beta$ if and only if $\aleph_\alpha \in \aleph_\beta$.

*(ii) $\alpha \subset \aleph_\alpha$ for every $\alpha \in$ **Ord**.*

(iii) The function $\aleph$ is a bijection, hence an ordered-class isomorphism.

Proof. (i) ($\Rightarrow$) We fix α and we apply transfinite induction on β, starting from $s(\alpha)$ (see exercise 11.44). If $\beta = s(\alpha)$, then $\aleph_\alpha \in \aleph_\alpha^+ = \aleph_{s(\alpha)} = \aleph_\beta$, hence $\aleph_\alpha \in \aleph_\beta$. If the thesis holds for β, such that $s(\alpha) \subset \beta$, then $\aleph_\alpha \in \aleph_\beta \in \aleph_\beta^+ = \aleph_{s(\beta)}$, hence $\aleph_\alpha \in \aleph_{s(\beta)}$, as required. Now we suppose that β is a limit ordinal, such that $\alpha \in \beta$, and the thesis holds for every γ such that $\alpha \in \gamma \in \beta$. We fix any such γ_0. By the inductive hypothesis, $\aleph_\alpha \in \aleph_{\gamma_0}$. Moreover, since by definition $\aleph_\beta = \sup\{\aleph_\gamma : \gamma \in \beta\}$, we have $\aleph_{\gamma_0} \subset \aleph_\beta$, hence $\aleph_\alpha \in \aleph_\beta$, as required. ($\Leftarrow$) If by contradiction $\beta = \alpha$ or $\beta \in \alpha$, then we get respectively $\aleph_\beta = \aleph_\alpha$ or $\aleph_\beta \in \aleph_\alpha$, against the hypothesis.

(ii) Induction of α. If $\alpha = 0$, then trivially $0 \subset \omega = \aleph_0$. If $\alpha \subset \aleph_\alpha$, then, since $\aleph_\alpha \in \aleph_{s(\alpha)}$ by item i, we have $\alpha \in \aleph_{s(\alpha)}$. Since $\aleph_{s(\alpha)}$ is a limit ordinal by corollary 12.13, necessarily $s(\alpha) \in \aleph_{s(\alpha)}$, thus $s(\alpha) \subset \aleph_{s(\alpha)}$. If α is a limit ordinal and $\beta \subset \aleph_\beta$ for every $\beta \in \alpha$, then $\beta \subset \aleph_\beta \in \aleph_\alpha$ for every $\beta \in \alpha$ by item i. It follows that $\beta \in \aleph_\alpha$ for every $\beta \in \alpha$, therefore $\alpha \subset \aleph_\alpha$.

(iii) Injectivity immediately follows from item i. Let us suppose by contradiction that $\mathrm{Im}(\aleph) \subsetneq \mathbf{Card}_\infty$—that is, $a := \mathbf{Card}_\infty \setminus \mathrm{Im}(\aleph) \neq \emptyset$. Since

$a \subset \mathbf{Ord}$, the class a admits a minimum κ, that is an infinite cardinal by construction. Obviously $\kappa \neq \omega$, since $\omega = \aleph_0$. If κ is a successor ordinal—that is, $\kappa = \lambda^+$—then $\lambda \in \mathrm{Im}(\aleph)$ by minimality of κ. It follows that there exists $\alpha \in \mathbf{Ord}$ such that $\lambda = \aleph_\alpha$, hence by definition $\kappa = \lambda^+ = \aleph_\alpha^+ = \aleph_{s(\alpha)} \in \mathrm{Im}(\aleph)$, a contradiction. Since κ is not ω and is not a successor, it can only be a limit cardinal. We set $\alpha := \{\beta \in \mathbf{Ord} : \aleph_\beta \in \kappa\}$. We prove that α is a limit ordinal. If $\gamma \in \beta \in \alpha$, then $\aleph_\gamma \in \aleph_\beta \in \kappa$ by item i and the definition of α, hence $\aleph_\gamma \in \kappa$, therefore $\gamma \in \alpha$. This proves that α is an initial segment of $\mathbf{Ord}$. It is proper, since $\kappa \notin \alpha$ by item ii, hence α is an ordinal by lemma 11.25-ii. It has no maximum, since, if $\beta \in \alpha$—that is, $\aleph_\beta \in \kappa$—then $\aleph_{s(\beta)} \in \kappa$ because κ is a limit cardinal, hence $s(\beta) \in \alpha$. This proves that α is a limit ordinal by proposition 11.31, hence by definition $\aleph_\alpha = \sup\{\aleph_\beta : \beta \in \alpha\} = \sup\{\aleph_\beta : \aleph_\beta \in \kappa\}$. By minimality of κ, every element $\lambda \in \kappa$ belongs to the image of $\aleph$, hence $\{\aleph_\beta : \aleph_\beta \in \kappa\} = \{\lambda \in \mathbf{Card} : \lambda \in \kappa\} = \kappa \cap \mathbf{Card}$, therefore $\aleph_\alpha = \sup(\kappa \cap \mathbf{Card}) = \kappa$, the last equality being due to corollary 12.23-ii. Hence, $\kappa \in \mathrm{Im}(\aleph)$, a contradiction. □

Remark 12.26. The fact that $\aleph$ is an ordered-class isomorphism shows that **Card** is isomorphic to $\omega \sqcup \mathbf{Ord}$ as a well-ordered class. Hence, except for one copy more of ω, the class of cardinals has the same structure of the class of ordinals. ◊

Corollary 12.27. *$\aleph_\alpha$ is a limit cardinal if and only if α is a limit ordinal.*

Proof. Since $\aleph$ is a bijection, $\aleph_\alpha = \omega$ if and only if $\alpha = 0$ and there exists $\aleph_\beta \in \mathbf{Card}_\infty$ such that $\aleph_\alpha = \aleph_\beta^+ = \aleph_{s(\beta)}$ if and only if there exists $\beta \in \mathbf{Ord}$ such that $\alpha = s(\beta)$. □

Since a cardinal is in particular an ordinal, when we think of $\aleph_\alpha$ as an ordinal, we denote it by ω_α. We get the map

$$\begin{aligned} \omega\colon \mathbf{Ord} &\hookrightarrow \mathbf{Ord} \\ \alpha &\mapsto \omega_\alpha, \end{aligned} \tag{12.2}$$

that coincides with $\aleph$ (in particular, $\omega_0 = \omega$). Hence, for example, $\aleph_1 = \omega_1$ is at the same time the first uncountable cardinal and the first uncountable ordinal. Introducing two symbols for the same function is useful to stress in which framework we are working, and it will be more relevant from section 12.4 on, since we will see that the operations among cardinals are different from the ones among ordinals.

Remark 12.28. In the statement of proposition 12.25-ii, the inclusion is not strict in general. In fact, it follows from exercise 5-i in section 11.6 (by choosing $f = \aleph$) that there exist infinitely many ordinals α such that $\aleph_\alpha = \alpha$. ◊

12.3 TRANSFINITE INDUCTION AND RECURSION

Let us show that the principles of induction and recursion hold in **Card** too.

12.3.1 Transfinite Induction

The following proposition shows that the class **Card** satisfies properties that extend the Peano's axioms of $\mathbb{N}$ and restrict the analogous ones of **Ord**. In particular, the induction principle can be proved by adapting in the straightforward way the proof of proposition 11.42-iii. Nevertheless, we prefer to deduce it from proposition 11.42-iii itself, in two equivalent ways.

Proposition 12.29. *The following properties of the class* **Card** *hold:*

(i) $\kappa^+ \neq 0$ *for every* $\kappa \in \mathbf{Card}$.

(ii) $\kappa^+ = \lambda^+$ *if and only if* $\kappa = \lambda$.

(iii) TRANSFINITE INDUCTION ON CARDINALS: *If* $a \subset \mathbf{Card}$ *is a subclass satisfying the following properties:*

- $0 \in a$
- $\kappa \in a \Rightarrow \kappa^+ \in a$ *for every* $\kappa \in \mathbf{Card}$
- *if* κ *is a weak limit cardinal and* $\kappa \cap \mathbf{Card} \subset a$, *then* $\kappa \in a$

then $a = \mathbf{Card}$.

Proof. (i) We have $\kappa \in \kappa^+$, hence $\kappa^+ \neq \emptyset$. (ii) ($\Leftarrow$) Obvious. ($\Rightarrow$) If $\kappa \neq \lambda$, up to exchange them we can suppose $\kappa \in \lambda$. It follows that $\kappa^+ \subset \lambda \in \lambda^+$, hence $\kappa^+ \in \lambda^+$, thus $\kappa^+ \neq \lambda^+$. (iii) *Method I:* By induction in ω (that is, in $\mathbb{N}$), we have $\omega \subset a$, hence $\omega \in a$. Therefore, it is enough to consider $a \cap \mathbf{Card}_\infty$. We set $a' := \{\alpha \in \mathbf{Ord} : \aleph_\alpha \in a\}$. Since the function $\aleph$ by definition respects successors and limits, a' satisfies the hypotheses of 11.42-iii, hence the thesis holds too—that is, $a' = \mathbf{Ord}$. Therefore, $a \cap \mathbf{Card}_\infty = \mathrm{Im}(\aleph) = \mathbf{Card}_\infty$, hence $a = \mathbf{Card}$. *Method II:* We set $a' := \{\alpha \in \mathbf{Ord} : |\alpha| \in a\}$. It follows from proposition 12.10-iii that $a = a' \cap \mathbf{Card}$. The reader can prove that a' satisfies the hypotheses of 11.42-iii, hence the thesis holds too—that is, $a' = \mathbf{Ord}$. Therefore, $a = a' \cap \mathbf{Card} = \mathbf{Ord} \cap \mathbf{Card} = \mathbf{Card}$. □

We can formulate induction in the second form too. In this case we prefer to adapt the proof of proposition 11.43, since it is quite simple.

Proposition 12.30 (Transfinite induction on cardinals). *If* $a \subset \mathbf{Card}$ *is a subclass satisfying the following properties:*

- $0 \in a$
- *if* $\kappa \cap \mathbf{Card} \subset a$, *then* $\kappa \in a$

then $a = \mathbf{Card}$.

Proof. Let us suppose by contradiction that $a \neq \mathbf{Card}$—that is, $\mathbf{Card} \setminus a \neq \emptyset$. Since **Card** is well-ordered, it follows that there exists $\kappa := \min(\mathbf{Card} \setminus a)$. Surely $\kappa \neq 0$, since $0 \in a$. Since κ is the minimum of $\mathbf{Card} \setminus a$ and '$\in$' is the strict order relation, $\lambda \in a$ for every $\lambda \in \kappa$, hence $\kappa \cap \mathbf{Card} \subset a$. By hypothesis, this implies that $\kappa \in a$, a contradiction. □

12.3.2 Transfinite Recursion

As in the setting of ordinals, we introduce the following definition.

Definition 12.31. Given a class c, we define $\mathrm{Func}^*(\mathbf{Card}, \mathsf{c}) := \{f \colon \kappa \cap \mathbf{Card} \to \mathsf{c} : \kappa \in \mathbf{Card}\}$ and $\mathrm{Func}^*(\mathbf{Card}_\infty, \mathsf{c}) := \{f \colon \kappa \cap \mathbf{Card} \to \mathsf{c} : \kappa \in \mathbf{Card}_\infty\}$. ♢

We set $\mathbf{Card}^* := \mathbf{Card} \setminus \{0\}$. We state the two versions of recursion.

Proposition 12.32 (Transfinite Recursion on Cardinals - Form I). *Given*

- *a class* c*;*
- *a fixed element* $a \in \mathsf{c}$*;*
- *a function* $k \colon \mathsf{c} \to \mathsf{c}$*; and*
- *a function* $h \colon \mathrm{Func}^*(\mathbf{Card}, \mathsf{c}) \to \mathsf{c}$

there exists a unique function $f \colon \mathbf{Card} \to \mathsf{c}$ *such that*

I. $f(0) = a$*;*

II. $f(\kappa^+) = k(\kappa)$ *for every* $\kappa \in \mathbf{Card}$*; and*

III. if κ *is weak limit cardinal, then* $f(\kappa) = h(f|_{\kappa \cap \mathbf{Card}})$.

Proposition 12.33 (Transfinite Recursion on Cardinals - Form II). *Given*

- *a class* c
- *a fixed element* $a \in \mathsf{c}$
- *a function* $h \colon \mathrm{Func}^*(\mathbf{Card}, \mathsf{c}) \to \mathsf{c}$

there exists a unique function $f \colon \mathbf{Card} \to \mathsf{c}$ *such that*

I. $f(0) = a$*; and*

II. $f(\kappa) = h(f|_{\kappa \cap \mathbf{Card}})$ *for every* $\kappa \in \mathbf{Card}^*$.

Proof of 12.33. Given a non-zero ordinal α, we observe that, if it is a cardinal, then trivially $\alpha \cap \mathbf{Card} = |\alpha| \cap \mathbf{Card}$, otherwise $\alpha \cap \mathbf{Card} = |\alpha|^+ \cap \mathbf{Card}$. In any case, $\alpha \cap \mathbf{Card}$ is the intersection between a cardinal and $\mathbf{Card}$. Given a function $\varphi \colon \alpha \to \mathsf{c}$, so that $\varphi \in \mathrm{Func}^*(\mathbf{Ord}, \mathsf{c})$, we get the function $\varphi^\# := \varphi|_{\alpha \cap \mathbf{Card}}$, so that $\varphi^\# \in \mathrm{Func}^*(\mathbf{Card}, \mathsf{c})$. Considering the function h in the statement, we define the function $h' \colon \mathrm{Func}^*(\mathbf{Ord}, \mathsf{d}) \to \mathsf{c}$, $\varphi \mapsto h(\varphi^\#)$. We apply proposition 11.49 to a and h', getting a function $f' \colon \mathbf{Ord} \to \mathsf{c}$. We set $f := f'|_{\mathbf{Card}}$. We have $f(0) = f'(0) = a$ and $f(\kappa) = f'(\kappa) = h'(f'|_\kappa) = h((f'|_\kappa)^\#) = h((f'|_\kappa)|_{\kappa \cap \mathbf{Card}}) = h(f'|_{\kappa \cap \mathbf{Card}}) = h(f|_{\kappa \cap \mathbf{Card}})$ for every $\kappa \in \mathbf{Card}^*$. □

Proof of 12.32. It is straightforward to adapt the proof of 11.48, deducing the statement from proposition 12.33. □

We could also prove 12.33 by adapting the proof of 11.49, but we preferred to deduce it from the latter. All of the other forms of induction and recursion (bounded and/or starting from a cardinal κ_0 instead of 0, including the dependence on another class $\mathtt{d}$, etc.) hold in the framework of cardinals too.

12.4 OPERATIONS ON CARDINALS

We define the fundamental operations on cardinals as follows.

Definition 12.34. Given $\kappa, \lambda \in \mathbf{Card}$ we set:

- $\kappa \mathbin{\boldsymbol{+}} \lambda := |\kappa \sqcup \lambda|$;
- $\kappa \mathbin{\boldsymbol{\cdot}} \lambda := |\kappa \times \lambda|$;
- (AC) $\kappa^\lambda := |\mathrm{Func}(\lambda, \kappa)|$. ◇

The definition of κ^λ requires the axiom of choice, since otherwise $\mathrm{Func}(\lambda, \kappa)$ might not admit a well-order. Hence, every statement or proof that involves cardinal exponentiation relies on AC, except for the case κ^n, with $n \in \aleph_0$, that can be easily defined by recursion as $\kappa^0 := 1$ and $\kappa^{n+1} := \kappa^n \mathbin{\boldsymbol{\cdot}} \kappa$ (cfr. exercise 12.39-vi below).

Assuming AC, exercise 3 of section 7.4 implies that $|\mathcal{P}(x)| = 2^{|x|}$. In particular, for every cardinal κ, we have $2^\kappa = |\mathcal{P}(\kappa)|$.

Exercise 12.35. Show that $\aleph_0 \mathbin{\boldsymbol{+}} \aleph_0$ is different from $\omega + \omega$ and $\aleph_0 \mathbin{\boldsymbol{\cdot}} \aleph_0$ is different from $\omega \cdot \omega$ (we recall that $\aleph_0 = \omega$, but we think of the former as a cardinal and of the latter as an ordinal). Verify that $\omega + \omega$ and $\omega \cdot \omega$ are not cardinals.

Exercise 12.36. (AC) Show that, considering ordinal power, we have $2^\omega = \omega$, while, considering cardinal power, we have $2^{\aleph_0} = |\mathcal{P}(\aleph_0)| > \aleph_0$. Hence, $2^{\aleph_0} \neq |2^\omega| = \aleph_0$.

Exercise 12.37. (AC) Show that $\aleph_0^{\aleph_0} = 2^{\aleph_0} \neq \omega^\omega$, considering cardinal power in the first two cases and ordinal power in the third one. Verify that ω^ω is not a cardinal.

Exercises 12.35–12.37 show that the operations on cardinals are *not* the restriction of the corresponding ones on ordinals. Actually, we saw that, if we apply sum, product or power of ordinals to two cardinals, the result is not a cardinal in general. For this reason, we used the bold symbols '$\boldsymbol{+}$' and '$\boldsymbol{\cdot}$' to denote cardinal sum and product, and the ordinary symbols '$+$' and '$\cdot$' to denote ordinal sum and product. Unfortunately, we do not have this possibility with power. Whenever it is not clear from the context, we will specify whether we are dealing with ordinal or cardinal power. The same consideration holds when we denote the product by juxtaposition (like $\alpha\beta$ or $\kappa\lambda$).

Remark 12.38. It immediately follows from definitions 11.53 and 12.34 that $\kappa+\lambda = |\kappa+\lambda|$ and $\kappa\cdot\lambda = |\kappa\cdot\lambda|$. Nevertheless, exercises 12.36 and 12.37 show that the analogous statement about power is false. ◇

Exercise 12.39. Prove that:

(i) Sum and product of cardinals are associative and commutative.

(ii) Product of cardinals is distributive with respect to sum.

(iii) $\kappa+0=\kappa$, $\kappa\cdot 0=0$, $\kappa\cdot 1=\kappa$, $\kappa^0=1$, $\kappa^1=\kappa$ and $1^\kappa=1$ for every $\kappa \in \mathbf{Card}$; moreover, $0^0=1$ and, if $\kappa\neq 0$, then $0^\kappa=0$.

(iv) (AC) $\kappa^{\lambda+\sigma}=\kappa^\lambda\cdot\kappa^\sigma$ and $\kappa^{\lambda\cdot\sigma}=(\kappa^\lambda)^\sigma$ for every $\kappa,\lambda,\sigma\in\mathbf{Card}$.

(v) (AC) $(\kappa\lambda)^\sigma=\kappa^\sigma\lambda^\sigma$ for every $\kappa,\lambda,\sigma\in\mathbf{Card}$.

(vi) $n\kappa=\kappa+\overset{n\text{ times}}{\cdots}+\kappa=|\kappa\sqcup\overset{n\text{ times}}{\cdots}\sqcup\kappa|$ and $\kappa^n=\kappa\cdot\overset{n\text{ times}}{\cdots}\cdot\kappa=|\kappa\times\overset{n\text{ times}}{\cdots}\times\kappa|$ for every $\kappa\in\mathbf{Card}$ and $n\in\aleph_0\setminus\{0\}$.

Exercise 12.40. Prove that, if $\kappa\subset\kappa'$ and $\lambda\subset\lambda'$, then $\kappa+\lambda\subset\kappa'+\lambda'$, $\kappa\cdot\lambda\subset\kappa'\cdot\lambda'$ and, assuming AC and $\kappa\neq 0$, $\kappa^\lambda\subset\kappa'^{\lambda'}$.

Exercise 12.41. Show that, if $|x|=|y|$ and $|z|=|w|$, then, assuming AC, $|\mathrm{Func}(x,y)|=|\mathrm{Func}(z,w)|$. Conclude that $|\mathrm{Func}(x,y)|=|y|^{|x|}$ (cfr. exercise 10.8-iii about sum and product).

Exercise 12.39 shows that the basic properties of the operations between natural numbers extend to cardinals, contrary to the case of ordinals. Nevertheless, we are going to see that sum and product of cardinals are not particularly interesting when at least one of the two terms is infinite (thus they are interesting only among natural numbers). On the contrary, exponentiation of cardinals is extremely interesting.

Proposition 12.42. *If κ is an infinite cardinal, then $\kappa\cdot\kappa=\kappa$.*

Proof. We have to construct a bijection between κ and $\kappa\times\kappa$. If $\kappa=\aleph_0$, then the result follows from example 10.16. Now we apply induction in the form 12.30. Let us suppose that $\kappa\ni\aleph_0$ and the statement holds for every $\lambda\in\kappa\cap\mathbf{Card}$. We construct the well-order relation in $\kappa\times\kappa$ defined by $(\alpha,\beta)\preceq(\alpha',\beta')$ if and only if one of the following alternatives holds:

- $\max\{\alpha,\beta\}\in\max\{\alpha',\beta'\}$
- $\max\{\alpha,\beta\}=\max\{\alpha',\beta'\}$ and $\alpha\in\alpha'$
- $\max\{\alpha,\beta\}=\max\{\alpha',\beta'\}$, $\alpha=\alpha'$ and $\beta\subset\beta'$

If we imagine $\kappa\times\kappa$ as a lattice in a plane of coordinates (x,y), we call H_α the discrete half-boundary of a square formed by a horizontal segment of y-coordinate α and a vertical segment of x-coordinate α. Formally, $H_\alpha:=\{(\beta,\alpha):\beta\subset\alpha\}\cup\{(\alpha,\beta):\beta\subset\alpha\}$. We are affirming that, if $\alpha\in\beta$, then the elements of H_α are lower than the elements of H_β. Within H_α, we first have the elements of the horizontal segment without the right-extreme, ordered by the x-coordinate, then the elements of the vertical segment, ordered by the y-coordinate. The reader can verify that it is actually a well-order (see exercise 12.43 below). Let us consider $\text{type}(\kappa\times\kappa, \preceq)$. We fix any $(\alpha, \beta) \in \kappa\times\kappa$ and we set $\xi := \max\{\alpha, \beta\}$. The proper initial segment $\text{prec}_{(\kappa\times\kappa,\preceq)}(\alpha, \beta)$ is formed by the half-squares H_γ, with $\gamma \in \xi$, and by some elements of H_ξ. Hence, the cardinality of $\text{prec}_{(\kappa\times\kappa,\preceq)}(\alpha, \beta)$ is less than or equal to the one of $\bigcup_{\gamma\subset\xi} H_\gamma = s(\xi) \times s(\xi)$. Since $(\alpha, \beta) \in \kappa \times \kappa$—that is, $\alpha, \beta \in \kappa$—necessarily $\xi = \max\{\alpha, \beta\} \in \kappa$, hence $s(\xi) \subset \kappa$. Since κ is an infinite cardinal, it is a limit ordinal by corollary 12.13, thus $s(\xi) \in \kappa$, therefore $|s(\xi)| \in \kappa$. The inductive hypotheses implies that $|s(\xi) \times s(\xi)| = |s(\xi)| \cdot |s(\xi)| = |s(\xi)| \in \kappa$. We proved that the cardinality of every proper initial segment of $\text{type}(\kappa\times\kappa, \preceq)$ is less than κ, hence it is not possible that $\kappa \in \text{type}(\kappa\times\kappa, \preceq)$, therefore $\text{type}(\kappa\times\kappa, \preceq) \subset \kappa$. It follows that $\kappa \cdot \kappa = |\kappa \times \kappa| = |\text{type}(\kappa \times \kappa, \preceq)| \subset \text{type}(\kappa \times \kappa, \preceq) \subset \kappa$, hence $\kappa \cdot \kappa \subset \kappa$. Since we have the embedding $\iota\colon \kappa \hookrightarrow \kappa \times \kappa$, $\alpha \mapsto (\alpha, 0)$, necessarily $\kappa \cdot \kappa = \kappa$. □

Exercise 12.43. Prove that the relation '$\preceq$' in $\kappa \times \kappa$, introduced in the proof of proposition 12.42, is actually a well-order.

Corollary 12.44. *For every $\kappa, \lambda \in$ **Card***, if at least one of the two is infinite, then we have:*

$$\kappa + \lambda = \kappa \cdot \lambda = \max\{\kappa, \lambda\}.$$

Proof. Up to exchange κ and λ, we suppose that $\kappa \subset \lambda$, hence λ is infinite (otherwise both would be finite, against the hypothesis). We have $\lambda = 1 \cdot \lambda \subset \kappa \cdot \lambda \subset \lambda \cdot \lambda = \lambda$, therefore all of the terms are equal. In particular, $\kappa \cdot \lambda = \lambda$. Hence, $\lambda = 0 + \lambda \subset \kappa + \lambda \subset \lambda + \lambda = \lambda \cdot 2 = \lambda$, therefore all of the terms are equal. In particular, $\kappa + \lambda = \lambda$. □

Corollary 12.45 (AC). *For every $\kappa \in$ **Card** and $\lambda \in$ **Card**$_\infty$, such that $2 \subset \kappa \subset \lambda$, we have*

$$2^\lambda = \kappa^\lambda = \lambda^\lambda.$$

Proof. It immediately follows from exercise 12.40 that $2^\lambda \subset \kappa^\lambda \subset \lambda^\lambda$, hence we have to prove that $\lambda^\lambda \subset 2^\lambda$. We have $\lambda^\lambda = |\text{Func}(\lambda, \lambda)| \leq |\mathcal{P}(\lambda \times \lambda)| = 2^{\lambda\cdot\lambda} = 2^\lambda$. □

Corollary 12.46. *For every infinite cardinal κ and $n \in \aleph_0 \setminus \{0\}$, we have:*

(i) $\kappa^n = \kappa$.

(ii) $|\text{FSeq}_n(\kappa)| = \kappa$.[2]

(iii) $|\text{FSeq}(\kappa)| = \kappa$.

[2] In item ii the case $n = 0$ is admitted too, since (finite) sequences start from the index 0 by definition.

(iv) (AC) *If* $\mathcal{P}^{(n)}(\kappa)$ *denotes the set of subsets of* κ *with* n *elements, then* $|\mathcal{P}^{(n)}(\kappa)| = \kappa$.

(v) (AC) *If* $\mathcal{P}^f(\kappa)$ *denotes the set of* finite *subsets of* κ, *then* $|\mathcal{P}^f(\kappa)| = \kappa$.

Proof. (i) If trivially follows from proposition 12.42 and exercise 12.39-vi by induction on n.

(ii) For every $n \in \mathbb{N}$ we have the canonical bijection $\psi_n\colon \mathrm{FSeq}_n(\kappa) \xrightarrow{\approx} \kappa \times \overset{s(n)\text{ times}}{\cdots} \times \kappa$, $(\alpha_0, \dots, \alpha_n) \mapsto (\alpha_0, \dots \alpha_n)$, hence the result follows from item i and exercise 12.39-vi.

(iii) By item i, for every $n \in \mathbb{N}^*$ we have a bijection $\varphi_n\colon \kappa^n \xrightarrow{\approx} \kappa$. In order not to rely on AC in the choice $n \mapsto \varphi_n$, we construct φ_n recursively as follows. We start from $\varphi_1 := \mathrm{id}_\kappa$ and any fixed φ_2. Assuming to have constructed φ_n for $n \geq 2$, we define $\varphi_{n+1}\colon \kappa^{n+1} \xrightarrow{\approx} \kappa$, $(\alpha_1, \dots, \alpha_{n+1}) \mapsto \varphi_2(\varphi_n(\alpha_1, \dots, \alpha_n), \alpha_{n+1})$. Since $\mathrm{FSeq}(\kappa) = \bigcup_{n\in\mathbb{N}} \mathrm{FSeq}_n(\kappa)$ and the sets $\mathrm{FSeq}_n(\kappa)$ are pair-wise disjoint, we get $\varphi\colon \mathrm{FSeq}(\kappa) \xrightarrow{\approx} \aleph_0 \times \kappa$, $(\alpha_0, \dots \alpha_n) \mapsto \big(n, \varphi_{s(n)}(\alpha_0, \dots, \alpha_n)\big)$. It follows that $|\mathrm{FSeq}(\kappa)| = \aleph_0 \cdot \kappa = \kappa$, the last equality being due to corollary 12.44.

(iv) For every $n \in \mathbb{N}^*$, we fix a subset of κ that contains $n-1$ elements, that we denote by $a = \{\alpha_1, \dots, \alpha_{n-1}\}$. This is possible by exercise 7.22-viii. We get the embedding $\iota\colon \kappa \setminus a \hookrightarrow \mathcal{P}^{(n)}(\kappa)$, $\alpha \mapsto a \cup \{\alpha\}$. It follows that $|\kappa| = |\kappa \setminus a| \leq |\mathcal{P}^{(n)}(\kappa)|$, the first equality being due to exercise 7-i of section 10.4, that requires AC. Conversely, by AC we can fix a total order in every element of $\mathcal{P}^{(n)}(\kappa)$. More precisely, we consider the function that associates with each element of $\mathcal{P}^{(n)}(\kappa)$ the set of its total orders and we apply AC4. We get the embedding $j\colon \mathcal{P}^{(n)}(\kappa) \hookrightarrow \mathrm{FSeq}_{n-1}(\kappa)$, $\{\alpha_1, \dots, \alpha_n\} \mapsto (\alpha_1, \dots, \alpha_n)$, where $\alpha_1 \prec \cdots \prec \alpha_n$ is the total order fixed in the corresponding element of the domain. It follows that $|\mathcal{P}^{(n)}(\kappa)| \leq |\mathrm{FSeq}_{n-1}(\kappa)| = |\kappa|$, hence equality holds.

(v) For any fixed $n \in \mathbb{N}^*$, we have $\mathcal{P}^{(n)}(\kappa) \subset \mathcal{P}^f(\kappa)$, hence from item iv we get $|\kappa| = |\mathcal{P}^{(n)}(\kappa)| \leq |\mathcal{P}^f(\kappa)|$. Conversely, by AC and item iv we fix a bijection $\varphi_n\colon \mathcal{P}^{(n)}(\kappa) \xrightarrow{\approx} \kappa$ for every $n \in \mathbb{N}^*$. Since $\mathcal{P}^f(\kappa) = \bigcup_{n\in\mathbb{N}} \mathcal{P}^{(n)}(\kappa)$ and the sets $\mathcal{P}^{(n)}(\kappa)$ are pair-wise disjoint, we get the embedding $\iota\colon \mathcal{P}^f(\kappa) \hookrightarrow \aleph_0 \times \kappa$, $a \in \mathcal{P}^{(n)}(\kappa) \mapsto (n, \varphi_n(a))$ if $n \neq 0$ and $\emptyset \in \mathcal{P}^{(0)}(\kappa) \mapsto (0, 0)$. It follows that $|\mathcal{P}^f(\kappa)| \leq |\aleph_0 \times \kappa| = \aleph_0 \cdot \kappa = \kappa$, hence equality follows. □

We have seen that $|\mathrm{FSeq}(\kappa)| = \kappa$ for every infinite cardinal κ. The particular case $\kappa = \aleph_0$ has been proven in proposition 10.24 too. If we consider infinite sequences, then we do not have a similar result. By definition $\mathrm{Seq}(\kappa) = \mathrm{Func}(\mathbb{N}, \kappa)$, hence, assuming AC, $|\mathrm{Seq}(\kappa)| = \kappa^{\aleph_0}$. This cardinal can be equal or bigger than κ, as the following examples show.

Example 12.47 (AC). We have $\aleph_0^{\aleph_0} = 2^{\aleph_0} > \aleph_0$ by corollary 12.45, hence $|\mathrm{Seq}(\aleph_0)| > \aleph_0$. This does not happen only with $\aleph_0$. For example, the next exercise and (equivalently) example 12.68 below show that $\aleph_\omega^{\aleph_0} > \aleph_\omega$, hence $|\mathrm{Seq}(\aleph_\omega)| > \aleph_\omega$.

On the contrary, $(2^{\aleph_0})^{\aleph_0} = 2^{\aleph_0 \cdot \aleph_0} = 2^{\aleph_0}$, hence $|\mathrm{Seq}(2^{\aleph_0})| = 2^{\aleph_0}$. Actually, the same argument shows that, for every $\kappa \in \mathbf{Card}$ and $\lambda \in \mathbf{Card}_\infty$, we have $(\kappa^\lambda)^{\aleph_0} = \kappa^{\lambda \cdot \aleph_0} = \kappa^\lambda$, hence $|\mathrm{Seq}(\kappa^\lambda)| = \kappa^\lambda$.

Exercise 12.48 (AC). By a suitable adaptation of Cantor's diagonal argument, used in example 10.12, show that $\aleph_\omega^{\aleph_0} > \aleph_\omega$.

Another important consequence of proposition 12.42 is the following one.

Proposition 12.49 (AC). *Let $x = \{x_i\}_{i \in I}$ be an indexed set and κ be an infinite cardinal such that $|x_i| \leq \kappa$ for every $i \in I$ and $|I| \leq \kappa$. Then $\left|\bigcup_i x_i\right| \leq \left|\bigsqcup_i x_i\right| \leq \kappa$.*

Proof. The inequality $\left|\bigcup_i x_i\right| \leq \left|\bigsqcup_i x_i\right|$ immediately follows from the surjective function $\bar{\eta}_x$, defined in (4.3) (section 4.2), and from proposition 10.9, the latter requiring AC. In order to prove that $\left|\bigsqcup_i x_i\right| \leq \kappa$, by AC we choose an embedding $\iota_i \colon x_i \hookrightarrow \kappa$ for every $i \in I$. Moreover, we fix an embedding $j \colon I \hookrightarrow \kappa$. We get the embedding $\iota \colon \bigsqcup_i x_i \hookrightarrow \kappa \times \kappa$, $(y, i) \mapsto (\varphi_i(y), j(i))$. It follows that $\left|\bigsqcup_i x_i\right| \leq \kappa \cdot \kappa = \kappa$. □

Corollary 12.50 (AC). *Let x be a set and κ be an infinite cardinal such that $|y| \leq \kappa$ for every $y \in x$ and $|x| \leq \kappa$. Then $\left|\bigcup x\right| \leq \left|\bigsqcup x\right| \leq \kappa$.*

Proof. It is enough to think of x as indexed over itself—that is, $x = \{y\}_{y \in x}$—and apply proposition 12.49. □

Assuming AC, proposition 12.42 can be equivalently stated as follows: for every infinite set x, we have $|x| = |x \times x|$. It follows by induction that

$$|x| = |x \times \overset{n \text{ times}}{\cdots} \times x|$$

for every $n \in \mathbb{N}^*$, that is equivalent to corollary 12.46-i. If we do not assume AC, we can state proposition 12.42 by affirming that, if x admits a well-order, then $|x| = |x \times x|$. It turns out that the general statement $\forall x(|x| = |x \times x|)$ is *equivalent* to the axiom of choice (that is, it is a fifth equivalent formulation, beyond AC, ZL, HMP, and WO). We do not show the proof of the equivalence here.

12.5 OPERATIONS ON NATURAL NUMBERS

It immediately follows from exercises 1-(i, v) and 2 in section 7.4 that the operations of sum, product, and power among cardinals, restricted to $\aleph_0$, coincide with the usual ones among natural numbers, as defined in section 7.2.3. Hence, it follows from corollary 11.69 that they coincide with ordinal sum, product and power too.

Actually, we can prove the coincidence between ordinal operations in ω and cardinal operations in $\aleph_0$ directly from the corresponding definitions, without relying on the theory of natural numbers, as the following lemma and proposition show. Part of the statements has already been proved above, but here we avoid any reference to chapters 1 and 7.

Lemma 12.51. *For every $n \in \omega$, we have:*

(i) $|n| < |s(n)|$;

(ii) for every ordinal α, we have $(|\alpha| = |n|) \Leftrightarrow (\alpha = n)$;

(iii) n is a cardinal;

(iv) $n^+ = s(n)$.

Proof. (i) Since $n \subset s(n)$, we have $|n| \leq |s(n)|$. We prove that $|n| \neq |s(n)|$ by ordinal induction on n. If $n = 0$, then we have no bijection between $\emptyset$ and $\{\emptyset\}$, since there is no function from the latter to the former. Assuming the result for n, let us suppose by contradiction that $|s(n)| = |s(s(n))|$. This means that we have a bijection $\varphi\colon s(n) \xrightarrow{\approx} s(s(n))$, where $s(n) = n \cup \{n\}$ and $s(s(n)) = s(n) \cup \{s(n)\}$. We set $k := \varphi(n)$ and we compose φ with the transposition $\tau_{k,s(n)}$ in $s(s(n))$. We get the bijection $\psi := \tau_{k,s(n)} \circ \varphi$ such that $\psi(n) = s(n)$. Since ψ is bijective, we get the bijective restriction $\psi|_n||_{s(n)}\colon n \xrightarrow{\approx} s(n)$, contradicting the inductive hypothesis.

(ii) ($\Leftarrow$) Obvious because of the identity function. ($\Rightarrow$) Suppose by contradiction that $\alpha \neq n$. We have two possibilities: I. $\alpha \in n$; II. $n \in \alpha$. In case I, we have $\alpha \in n \in \omega$, hence $\alpha \in \omega$. We set $m := \alpha$. Since $m \in n$, we have $m \in s(m) \subset n$. By hypothesis there exits a bijection $\varphi\colon n \xrightarrow{\approx} m$, hence, from the inclusion $s(m) \subset n$, we get the injective function $\varphi|_{s(m)}\colon s(m) \hookrightarrow m$, contradicting item i. In case II, we have $n \in s(n) \subset \alpha$. By hypothesis there exits a bijection $\varphi\colon \alpha \xrightarrow{\approx} n$, hence, from the inclusion $s(n) \subset \alpha$, we get the injective function $\varphi|_{s(n)}\colon s(n) \hookrightarrow n$, contradicting item i. In any case we get a contradiction.

(iii) If $m \in n$, then $m \neq n$, hence $|m| \neq |n|$ by item ii.

(iv) Both n and $s(n)$ are cardinals by item iii. Since there are no ordinals between n and $s(n)$, in particular there are no cardinals between them, hence $s(n) = n^+$. □

Proposition 12.52. *For every $n, m \in \aleph_0$, we have that $n \mathbin{\boldsymbol{+}} m = n + m$, $n \mathbin{\boldsymbol{\cdot}} m = n \cdot m$ and the cardinal exponentiation n^m coincide with the ordinal exponentiation n^m. In particular, $n \mathbin{\boldsymbol{+}} m, n \mathbin{\boldsymbol{\cdot}} m, n^m \in \aleph_0$.*

Proof. By proposition 11.68, the three operations on ordinals restrict to ω. About sum and product, we have seen in remark 12.38 that $n \mathbin{\boldsymbol{+}} m = |n + m|$ and $n \mathbin{\boldsymbol{\cdot}} m = |n \cdot m|$, that is equivalent to $|n \mathbin{\boldsymbol{+}} m| = |n + m|$ and $|n \mathbin{\boldsymbol{\cdot}} m| = |n \cdot m|$. Therefore, lemma 12.51-ii, respectively with $\alpha = n \mathbin{\boldsymbol{+}} m$ and $\alpha = n \mathbin{\boldsymbol{\cdot}} m$, implies that $n \mathbin{\boldsymbol{+}} m = n + m$ and $n \mathbin{\boldsymbol{\cdot}} m = n \cdot m$, as required. About power, we prove the result by induction on m. If $m = 0$, on the ordinal side we have $n^0 = 1$ by definition, and, on the cardinal side, we have $n^0 = |\mathrm{Func}(0, n)| = 1$. Assuming the result for n^m, by definition we have $n^{s(m)} = n^m n$ on the ordinal side and $n^{s(m)} = |\mathrm{Func}(s(m), n)|$ on the cardinal side. We have the bijection $\Phi\colon \mathrm{Func}(s(m), n) \xrightarrow{\approx} \mathrm{Func}(m, n) \times n$, $f \mapsto (f|_m, f(m))$. It is actually a bijection, since its inverse is given by $(g, k) \mapsto g'$, where $g'(h) := g(h)$ if $h \in m$ and $g'(m) := k$. It follows that, using cardinal exponentiation, $n^{s(m)} = |\mathrm{Func}(s(m), n)| = |\mathrm{Func}(m, n) \times n| = n^m n$. Since n^m coincides with ordinal exponentiation by the inductive hypothesis, and considering that we already proved the identity for product, we get the result. □

Proposition 12.52 shows that it is possible to *define* the basic operations among natural numbers as the restrictions of the corresponding ones among cardinals. In this case, associativity, commutativity, and distributivity are just particular cases of the same properties of cardinal operations (see exercise 12.39), whose proofs are quite faster and easy to visualise than the ones by induction. This completes section 11.5.1, showing that we can think of natural numbers as ordinals or cardinals without distinction, using ordinal operations or cardinal operations depending on the convenience. In this setting, exercises 7.21 and 7.22 (or exercises 4 of section 1.5 and 2 of section 3.5, and, as particular cases of the former, 1.25 and 1.26) become straightforward through the notion of cardinality. Similarly, exercise 1.24 and exercises 1 and 2 of section 1.5 can be solved quite naturally using the language of cardinality. In other cases, like exercises 1.21, 1.23, 1.29, and 5–9 in section 1.5, the proof by induction keeps on being the most natural one. This is perfectly coherent with the present approach, since induction and recursion for natural numbers are particular cases of the same concepts for ordinal or cardinal numbers.

12.6 COFINALITY

The notion of *cofinality* is particularly relevant in the theory of ordinals and cardinals. Let us show the definition.

Lemma 12.53. *Given $\alpha, \beta \in \mathbf{Ord}^*$ and $f\colon \alpha \to \beta$, the following conditions are equivalent:*

(i) $\mathrm{Im}(f)$ *is not strictly bounded above in β*

(ii) for every $\delta \in \beta$ there exists $\gamma \in \alpha$ such that $\delta \subset f(\gamma)$

If β is a limit ordinal, then the two previous conditions are equivalent to the following ones:

(iii) $\mathrm{Im}(f)$ *is not bounded above in β*

(iv) for every $\delta \in \beta$ there exists $\gamma \in \alpha$ such that $\delta \in f(\gamma)$

If β is a successor ordinal, then the first two conditions are equivalent to the following one:

(v) $\max(\beta) \in \mathrm{Im}(f)$

Proof. (i) $\Rightarrow$ (ii) We prove the contrapositive. If there exists $\delta \in \beta$ such that $f(\gamma) \in \delta$ for every $\gamma \in \alpha$, then, by definition, δ is a strict upper bound of $\mathrm{Im}(f)$. (ii) $\Rightarrow$ (i) Again we prove the contrapositive. If $\delta \in \beta$ is a strict upper bound of $\mathrm{Im}(f)$, then, by definition, $f(\gamma) \in \delta$ for every $\gamma \in \alpha$. Hence, there not exists $\gamma \in \alpha$ such that $\delta \subset f(\gamma)$.

(i) $\Leftrightarrow$ (iii) Since β is a limit ordinal, it does not admit a maximum by proposition 11.31. In this case, a subset $a \subset \beta$ is bounded above if and only if it is strictly bounded above (since, if δ is an upper bound of a, then there exists δ' such that $\delta \in \delta'$, hence δ' is a strict upper bound of a). (iii) $\Leftrightarrow$ (iv) Similar to (i) $\Leftrightarrow$ (ii).

(i) $\Rightarrow$ (v) If $\max(\beta) \notin \mathrm{Im}(f)$, then $\max(\beta)$ is a strict upper bound of $\mathrm{Im}(f)$. (v) $\Rightarrow$ (i) If δ is a strict upper bound of $\mathrm{Im}(f)$, then $\max(\beta) \in \delta$, a contradiction. □

Definition 12.54. Given $\alpha, \beta \in \mathbf{Ord}^*$, a function $f \colon \alpha \to \beta$ is *cofinal* if $\mathrm{Im}(f)$ is not strictly bounded above in β—that is, if the equivalent conditions of lemma 12.53 hold. ◇

For every fixed $\beta \in \mathbf{Ord}^*$, we set

$$\mathrm{CofOrd}(\beta) := \{\alpha \in \mathbf{Ord}^* : \exists f \colon \alpha \to \beta \text{ cofinal}\}.$$

Since the identity $\mathrm{id}_\beta \colon \beta \to \beta$ is trivially cofinal, the set $\mathrm{CofOrd}(\beta)$ is not empty. Since $\mathbf{Ord}^*$ is well-ordered, $\mathrm{CofOrd}(\beta)$ admits a minimum.

Definition 12.55. The *cofinality* of an ordinal β is the minimum of the set $\mathrm{CofOrd}(\beta)$. We denote it by $\mathrm{cof}(\beta)$. ◇

Remark 12.56. For every $\alpha \supset \mathrm{cof}(\beta)$, there exists a cofinal map $f \colon \alpha \to \beta$. In fact, we fix a cofinal map $g \colon \mathrm{cof}(\beta) \to \beta$ and we set $f(\gamma) := g(\gamma)$ if $\gamma \in \mathrm{cof}(\beta)$ and $f(\gamma) := 0$ otherwise. Therefore, $\mathrm{CofOrd}(\beta) = \{\alpha \in \mathbf{Ord} : \alpha \supset \mathrm{cof}(\beta)\} = \mathbf{Ord} \setminus \mathrm{cof}(\beta)$. ◇

12.6.1 Basic Properties

We state some basic properties of cofinality.

Proposition 12.57. *For every* $\beta \in \mathbf{Ord}^*$*:*

(i) $\mathrm{cof}(\beta) \subset \beta$

(ii) β *is a successor ordinal if and only if* $\mathrm{cof}(\beta) = 1$

(iii) β *is a limit ordinal if and only if* $\mathrm{cof}(\beta)$ *is a limit ordinal*

(iv) $\mathrm{cof}(\beta) = 1$ *or* $\mathrm{cof}(\beta)$ *is a limit ordinal*

(v) $\mathrm{cof}(\omega) = \omega$

Proof. (i) Since $\mathrm{id}_\beta \colon \beta \to \beta$ is cofinal, we have $\beta \in \mathrm{CofOrd}(\beta)$, hence $\mathrm{cof}(\beta) = \min \mathrm{CofOrd}(\beta) \subset \beta$. (ii) ($\Rightarrow$) The function $f \colon 1 \to \beta$, $0 \mapsto \max(\beta)$, is cofinal. ($\Leftarrow$) Let $f \colon 1 \to \beta$ be a cofinal function. If $f(0)$ is not the maximum of β, then trivially the image of f is strictly bounded above in β, a contradiction. It follows that β admits a maximum, hence it is a successor ordinal by proposition 11.31. (iii) ($\Leftarrow$) If β is not a limit ordinal, then $\mathrm{cof}(\beta) = 1$ by item ii, thus $\mathrm{cof}(\beta)$ is not a limit ordinal. ($\Rightarrow$) We fix a cofinal function $f \colon \mathrm{cof}(\beta) \to \beta$. Let us suppose by contradiction that $\mathrm{cof}(\beta) = s(\gamma)$, where $\gamma = \max(\mathrm{cof}(\beta))$ by proposition 11.31. Since β is a limit ordinal, by lemma 12.53-iii the image of f is not bounded above. Hence, $\mathrm{Im}(f) \setminus \{f(\gamma)\}$ is not bounded above either, otherwise the maximum between $f(\gamma)$ and an upper bound of $\mathrm{Im}(f) \setminus \{f(\gamma)\}$ would be an upper bound of $\mathrm{Im}(f)$. It follows that $f|_\gamma \colon \gamma \to \beta$ is cofinal, thus $\mathrm{cof}(\beta) \subset \gamma \in s(\gamma) = \mathrm{cof}(\beta)$, a contradiction. (iv) It immediately follows from ii and iii. (v) It immediately follows from i and iii. □

Proposition 12.58. *For every $\beta \in$ **Ord***, there exists a cofinal map $f \colon \mathrm{cof}(\beta) \to \beta$ that is strictly increasing—that is, such that, if $\gamma \in \gamma' \in \mathrm{cof}(\beta)$, then $f(\gamma) \in f(\gamma') \in \beta$.*

Proof. If β is a successor ordinal, then the unique cofinal map $f \colon 1 \to \beta$, $0 \mapsto \max(\beta)$, is trivially strictly increasing, hence we can suppose that β is a limit ordinal. Thus, $\mathrm{cof}(\beta)$ is a limit ordinal too by proposition 12.57-iii. Let us fix any cofinal map $g \colon \mathrm{cof}(\beta) \to \beta$. We define $f \colon \mathrm{cof}(\beta) \to \beta$ by bounded recursion as follows:

- $f(0) := g(0)$
- $f(s(\gamma)) := \max\{g(s(\gamma)), s(f(\gamma))\}$ for every $\gamma \in \mathrm{cof}(\beta)$
- if $\gamma \in \mathrm{cof}(\beta)$ is a limit ordinal, then $f(\gamma) := \max\big\{g(\gamma), \sup\{f(\xi) : \xi \in \gamma\}\big\}$

Let us verify that f is well-defined, strictly increasing and cofinal.

Step I: f is well-defined. Trivially $f(0)$ is well-defined. In the case of $f(s(\gamma))$, since $\mathrm{cof}(\beta)$ is a limit ordinal, we have $s(\gamma) \in \mathrm{cof}(\beta)$, as required by the definition. Moreover, since β is a limit ordinal and $f(\gamma) \in \beta$ by definition of f, we deduce that $s(f(\gamma)) \in \beta$, hence the definition is coherent. In the case of limit ordinals, we have to verify that, if $\gamma \in \mathrm{cof}(\beta)$, then $\delta := \sup\{f(\xi) : \xi \in \gamma\} \in \beta$. Let us suppose by contradiction that $\delta \notin \beta$. Then, since $f(\xi) \in \beta$ for every $\xi \in \gamma$ (by definition of f), we have $\delta \subset \beta$, hence $\delta = \beta$. It follows that $f|_\gamma \colon \gamma \to \beta$ is cofinal, hence $\mathrm{cof}(\beta) \subset \gamma \in \mathrm{cof}(\beta)$, a contradiction.

Step II: f is strictly increasing. For every $\gamma \in \mathrm{cof}(\beta)$, we have

$$f(\gamma) \in s(f(\gamma)) \subset \max\big\{g(s(\gamma)), s(f(\gamma))\big\} = f(s(\gamma)), \qquad (*)$$

hence $f(\gamma) \in f(s(\gamma))$. Let us suppose that $\gamma \in \gamma' \in \mathrm{cof}(\beta)$. By proposition 11.56-vii, there exists a unique $\xi \in \mathbf{Ord}^*$ such that $\gamma' = \gamma + \xi$. Let us prove that $f(\gamma) \in f(\gamma+\xi)$ by induction on ξ. If $\xi = 1$, then $f(\gamma+1) = f(s(\gamma))$, hence the result follows from $(*)$. If $\xi = s(\eta)$, with $\eta \in \mathbf{Ord}^*$, then by the inductive hypothesis $f(\gamma) \in f(\gamma + \eta)$ and by $(*)$ we have $f(\gamma + \eta) \in f(s(\gamma + \eta)) = f(\gamma + s(\eta)) = f(\gamma + \xi)$, hence the result follows by transitivity. If ξ is a limit ordinal, then $\gamma + \xi$ is a limit ordinal too by proposition 11.56-iv. We choose any $\eta \in \xi \setminus \{0\}$. By the inductive hypothesis, $f(\gamma) \in f(\gamma+\eta)$. Moreover, since $\gamma + \eta \in \gamma + \xi$ by proposition 11.56-ii, we have $f(\gamma + \eta) \subset \sup\{f(\theta) : \theta \in \gamma + \xi\} \subset \max\{g(\gamma + \xi), \sup\{f(\theta) : \theta \in \gamma + \xi\}\} = f(\gamma + \xi)$, hence the result follows by transitivity.

Step III: f is cofinal. It follows from the definition of f that $g(\gamma) \subset f(\gamma)$ for every $\gamma \in \mathrm{cof}(\beta)$. In fact, $g(0) = f(0)$ by definition, $g(s(\gamma)) \subset \max\{g(s(\gamma)), s(f(\gamma))\} = f(s(\gamma))$ and, if γ is a limit ordinal, then $g(\gamma) \subset \max\{g(\gamma), \sup\{f(\xi) : \xi \in \gamma\}\} = f(\gamma)$. Given $\delta \in \beta$, by hypothesis and lemma 12.53-ii there exists $\gamma \in \mathrm{cof}(\beta)$ such that $\delta \subset g(\gamma)$, hence $\delta \subset g(\gamma) \subset f(\gamma)$, that implies $\delta \subset f(\gamma)$. This proves that f is cofinal. □

Proposition 12.59. *Given $\alpha, \beta \in \mathbf{Ord}^*$, if $f : \alpha \to \beta$ is cofinal and strictly increasing, then* $\mathrm{cof}(\alpha) = \mathrm{cof}(\beta)$.

Proof. Let us fix a cofinal map $g : \mathrm{cof}(\alpha) \to \alpha$. We prove that $f \circ g : \mathrm{cof}(\alpha) \to \beta$ is cofinal, hence $\mathrm{cof}(\beta) \subset \mathrm{cof}(\alpha)$. In fact, given $\delta \in \beta$, since f is cofinal, there exists $\gamma \in \alpha$ such that $\delta \subset f(\gamma)$. Moreover, since g is cofinal, there exists $\xi \in \mathrm{cof}(\alpha)$ such that $\gamma \subset g(\xi)$. Since f is increasing, we have $f(\gamma) \subset f \circ g(\xi)$, hence $\delta \subset f \circ g(\xi)$. This proves that, for every $\delta \in \beta$, there exists $\xi \in \mathrm{cof}(\alpha)$ such that $\delta \subset f \circ g(\xi)$, therefore $f \circ g$ is cofinal.

Now we fix a cofinal map $h : \mathrm{cof}(\beta) \to \beta$. Since f is cofinal, for every $\xi \in \mathrm{cof}(\beta)$ there exists $\gamma \in \alpha$ such that $h(\xi) \subset f(\gamma)$, hence the set $a_\xi := \{\gamma \in \alpha : h(\xi) \subset f(\gamma)\}$ is not empty. We define the function $k : \mathrm{cof}(\beta) \to \alpha$, $\xi \mapsto \min(a_\xi)$. Let us show that k is cofinal, therefore $\mathrm{cof}(\alpha) \subset \mathrm{cof}(\beta)$, completing the proof. In fact, since h is cofinal, given $\gamma \in \alpha$, there exists $\xi \in \mathrm{cof}(\beta)$ such that $f(\gamma) \subset h(\xi)$. Since by definition $k(\xi) \in a_\xi$, we have $h(\xi) \subset f(k(\xi))$, hence $f(\gamma) \subset f(k(\xi))$. Since f is strictly increasing, it follows that $\gamma \subset k(\xi)$, otherwise we would have $k(\xi) \in \gamma$, hence $f(k(\xi)) \in f(\gamma))$. This shows that, for every $\gamma \in \alpha$, there exists $\xi \in \mathrm{cof}(\beta)$ such that $\gamma \subset k(\xi)$, hence k is cofinal. □

Corollary 12.60. *For every $\beta \in \mathbf{Ord}^*$, we have* $\mathrm{cof}(\mathrm{cof}(\beta)) = \mathrm{cof}(\beta)$.

Proof. It immediately follows from propositions 12.58 and 12.59. □

Corollary 12.61. *For every limit ordinal α, we have* $\mathrm{cof}(\omega_\alpha) = \mathrm{cof}(\alpha)$.

Proof. We have the strictly increasing cofinal map $f : \alpha \to \omega_\alpha$, $\beta \mapsto \omega_\beta$, hence the result follows from proposition 12.59. □

12.6.2 Regular Ordinals

By lemma 12.57-i, the cofinality of β is at most β itself.

Definition 12.62. An infinite ordinal β is called *regular* if $\mathrm{cof}(\beta) = \beta$. $\diamondsuit$

We observe that

- by lemma 12.57-ii, requiring that β is an infinite ordinal only excludes the case $\beta = 1$;
- again by lemma 12.57-ii, a regular ordinal is necessarily a limit ordinal; and
- by corollary 12.60 and lemma 12.57-iii, for every limit ordinal β, the ordinal $\mathrm{cof}(\beta)$ is regular.

Example 12.63. The ordinal ω is regular by proposition 12.57-v, hence it is the minimum regular ordinal. On the contrary, ω_ω is not regular, since $\mathrm{cof}(\omega_\omega) = \omega$ by corollary 12.61 (alternatively, the cofinal map $f\colon \omega \to \omega_\omega$, $n \mapsto \omega_n$, shows that $\mathrm{cof}(\omega_\omega) \subset \omega$, and it immediately follows from lemma 12.57-iii that $\omega \subset \mathrm{cof}(\omega_\omega)$, hence equality holds).

Proposition 12.64. *We have that*

(i) every regular ordinal is a cardinal; and

(ii) (AC) *every successor infinite cardinal is a regular ordinal.*

Proof. (i) Let β be a regular ordinal and let us suppose by contradiction that there exists $\alpha \in \beta$ such that $|\alpha| = |\beta|$—that is, such that there exists a bijection $\varphi\colon \alpha \xrightarrow{\approx} \beta$. A bijection is trivially cofinal, therefore $\mathrm{cof}(\beta) \subset \alpha \in \beta$, contradicting the regularity of β.

(ii) Let κ be an infinite cardinal and let us suppose by contradiction that there exist $\alpha \in \kappa^+$ and a cofinal function $f\colon \alpha \to \kappa^+$. Since κ^+ is a limit ordinal by corollary 12.13, we deduce from corollary 11.32-ii that $\kappa^+ = \sup(\kappa^+) = \bigcup(\kappa^+)$. Let us prove that $\kappa^+ = \bigcup \mathrm{Im}(f)$. In fact, $\mathrm{Im}(f) \subset \kappa^+$, hence $\bigcup \mathrm{Im}(f) \subset \bigcup(\kappa^+)$ by lemma 4.12-ii. Conversely, if $\xi \in \bigcup(\kappa^+)$, then there exists $\eta \in \kappa^+$ such that $\xi \in \eta$. Since f is cofinal, there exists $\gamma \in \alpha$ such that $\eta \subset f(\gamma)$, hence $\xi \in f(\gamma) \in \mathrm{Im}(f)$, therefore $\xi \in \bigcup \mathrm{Im}(f)$. It follows that $\kappa^+ = \bigcup \mathrm{Im}(f)$. Since $f|^{\mathrm{Im}(f)}\colon \alpha \to \mathrm{Im}(f)$ is surjective, we have $|\mathrm{Im}(f)| \leq |\alpha| \leq |\kappa|$, the first inequality requiring AC and the second one being due to the fact that $\alpha \in \kappa^+$ and κ^+ is a cardinal. Moreover, for every $\gamma \in \alpha$, we have $f(\gamma) \in \kappa^+$, hence $|f(\gamma)| \leq |\kappa|$, again because κ^+ is a cardinal. It follows from corollary 12.50 (requiring AC) that $\left|\bigcup \mathrm{Im}(f)\right| \leq |\kappa|$, a contradiction, since $\kappa^+ = \bigcup \mathrm{Im}(f)$. $\square$

We call $\mathbf{Card}_\infty$ the class of infinite cardinals, $\mathbf{Card}_\infty^+$ the class of infinite successor cardinals and **RegOrd** the class of regular ordinals. By propositions 12.64 and 12.57-v we have

$$\mathbf{Card}_\infty^+ \cup \{\aleph_0\} \subset \mathbf{RegOrd} \subset \mathbf{Card}_\infty.$$

Are the inclusions proper? The second one is surely, since we have seen in example 12.63 that $\mathrm{cof}(\aleph_\omega) = \aleph_0$, hence $\aleph_\omega$ is not regular. About the first one, asking if it is proper is equivalent to asking if there exists a regular limit cardinal. This is a quite delicate matter, that we will briefly discuss in section 12.7. Here we just make a couple of remarks.

Remark 12.65. If $\aleph_\alpha$ is a regular limit cardinal, then $\aleph_\alpha = \alpha$ (hence α is a cardinal). In fact, α is a limit ordinal by corollary 12.27, and by hypothesis $\mathrm{cof}(\aleph_\alpha) = \aleph_\alpha$. By corollary 12.61 we have $\mathrm{cof}(\aleph_\alpha) = \mathrm{cof}(\alpha)$, thus $\aleph_\alpha = \mathrm{cof}(\alpha)$. Since $\mathrm{cof}(\alpha) \subset \alpha \subset \aleph_\alpha$ and $\aleph_\alpha = \mathrm{cof}(\alpha)$, necessarily $\mathrm{cof}(\alpha) = \alpha = \aleph_\alpha$. Nevertheless, this condition is not sufficient (see exercise 4 at the end of the chapter). Informally speaking, if a regular limit cardinal exists, then it is a "huge" cardinal. ◇

A regular limit cardinal is also called *weakly inaccessible* cardinal.

12.6.3 König's Lemma

We conclude the discussion about cofinality with the following meaningful result.

Proposition 12.66 (König's Lemma). (AC)[3] *If κ and λ are infinite cardinals and $\lambda \geq \mathrm{cof}(\kappa)$, then $\kappa^\lambda > \kappa$.*

Proof. Let us fix any confinal function $f\colon \lambda \to \kappa$, that exists by remark 12.56, and any function $g\colon \kappa \to \mathrm{Func}(\lambda, \kappa)$. We show that g cannot be surjective, proving the thesis. We define a function $h\colon \lambda \to \kappa$ as follows. Let us fix $\alpha \in \lambda$, so that $f(\alpha) \in \kappa$. For every $\beta \in f(\alpha)$, we have the function $g(\beta)\colon \lambda \to \kappa$, hence $(g(\beta))(\alpha) \in \kappa$. We set $a_\alpha := \{(g(\beta))(\alpha) : \beta \in f(\alpha)\} \subset \kappa$. We have the surjective function $\varphi\colon f(\alpha) \to a_\alpha$, $\beta \mapsto (g(\beta))(\alpha)$, hence $|a_\alpha| \leq |f(\alpha)| < |\kappa|$, the last inequality being due to the fact that $f(\alpha) \in \kappa$ and κ is a cardinal. Hence, $a_\alpha \subsetneq \kappa$, thus $\kappa \setminus a_\alpha \neq \emptyset$. We define $h\colon \lambda \to \kappa$, $\alpha \mapsto \min(\kappa \setminus a_\alpha)$.

Let us prove that $h \notin \mathrm{Im}(g)$, so that g is not surjective. In fact, let us suppose by contradiction that there exists $\beta \in \kappa$ such that $g(\beta) = h$. Since f is cofinal and κ is a limit ordinal, by lemma 12.53-iv there exists $\alpha \in \lambda$ such that $\beta \in f(\alpha)$, hence $(g(\beta))(\alpha) \in a_\alpha$. Nevertheless, $(g(\beta))(\alpha) = h(\alpha) \in \kappa \setminus a_\alpha$ by definition of h, a contradiction. □

Corollary 12.67. (AC) *If λ is an infinite cardinal, then* $\mathrm{cof}(2^\lambda) > \lambda$.

[3] We recall that, without AC, κ^λ is not defined in general. Moreover, in the proof, we will deal with inequalities between cardinalities through surjective functions.

Proof. We set $\kappa := 2^\lambda$. If by contradiction $\lambda \geq \mathrm{cof}(\kappa)$, then by proposition 12.66 we have $\kappa^\lambda > \kappa$. Nevertheless, $\kappa^\lambda = (2^\lambda)^\lambda = 2^{\lambda \cdot \lambda} = 2^\lambda = \kappa$, a contradiction. □

Example 12.68. We have seen in example 12.63 that $\mathrm{cof}(\aleph_\omega) = \aleph_0$, hence proposition 12.66, with $\kappa = \aleph_\omega$ and $\lambda = \aleph_0$, implies that $\aleph_\omega^{\aleph_0} > \aleph_\omega$ (cfr. exercise 12.48).

12.7 UNDECIDABLE STATEMENTS

There are some meaningful statements that we have not proved nor denied up to now. Let us analyse some of them. First of all, by Cantor's theorem 10.30, we have $2^\kappa > \kappa$ for every cardinal κ. This is equivalent to $2^\kappa \geq \kappa^+$ for every $\kappa \in \mathbf{Card}$. *For which infinite cardinals does the equality hold?* As a particular case, we consider $\kappa = \aleph_0$, getting $2^{\aleph_0} \geq \aleph_1$. Equivalently, $|\mathbb{R}| \geq \aleph_1$. *Does the equality hold in this case?* Another relevant question, that arose in section 12.6, is the following one: *do regular limit cardinals exist?* We formulate these questions in the form of conjectures:

- *Continuum hypothesis* (CH): $2^{\aleph_0} = \aleph_1$ (equivalently, $|\mathbb{R}| = \aleph_1$).
- *Generalised continuum hypothesis* (GCH): $2^\kappa = \kappa^+$ for every $\kappa \in \mathbf{Card}_\infty$.
- *Inaccessible cardinal conjecture* (ICC): there exists a regular limit cardinal.

These conjectures (among many others) are particularly interesting in set theory, but we cannot settle them with the tools developed up to now, since they are *undecidable* on the basis of the assumptions we have made (and we will make in the next chapter). This is a very relevant problem, that we will briefly discuss in the conclusions.

12.8 EXERCISES

1. Prove that the operations among cardinals *cannot* be defined by transfinite recursion as the ones among ordinals.

2. Prove that, assuming AC, if $|x_1| < |y_1|, \dots, |x_k| < |y_k|$, then $|x_1 \sqcup \dots \sqcup x_k| < |y_1 \sqcup \dots \sqcup y_k|$ and, if the sets x_i and y_j are not empty, $|x_1 \times \cdots \times x_k| < |y_1 \times \cdots \times y_k|$ (cfr. exercise 10.8, fourth item).

3. Given a set x and an ordinal α, we set $\mathrm{Func}^*(\alpha, x) := \{f : \beta \to x : \beta \in \alpha\}$. Given two cardinals κ and λ, we set $\kappa^{<\lambda} := |\mathrm{Func}^*(\lambda, \kappa)|$. Assuming AC, prove that, if κ is infinite, then

- $\kappa^{<n} = \kappa^{<\aleph_0} = \kappa$ for every $n \in \aleph_0 \setminus \{0, 1\}$; and
- $\kappa^{<\lambda} = \sup\{\kappa^\theta : \theta \in \lambda \cap \mathbf{Card}\}$.

4. Show that there exist infinitely many ordinals α such that $\alpha = \aleph_\alpha$, but $\aleph_\alpha$ is not a regular (limit) cardinal. *Hint:* See the resolution of exercise 5-i of section 11.6.

5. Prove that, if GCH holds,[4] then cardinal exponentiation can be easily computed as follows. If $\kappa, \lambda \supset 2$ and at least one of the two is infinite, then

(i) if $\kappa \subset \lambda$, then $\kappa^\lambda = \lambda^+$;

(ii) if $\mathrm{cof}(\kappa) \subset \lambda \in \kappa$, then $\kappa^\lambda = \kappa^+$; and

(iii) if $\lambda \in \mathrm{cof}(\kappa)$, then $\kappa^\lambda = \kappa$.

[4]One can prove that GCH implies AC.

CHAPTER 13

Well-Founded Sets

In this chapter we finally construct the reduced universe, whose necessity has become clear since chapter 2. In particular, we show how to select a minimal family of hereditary sets, in which all of the interesting mathematical operations can be performed. In this way, beyond excluding paradoxes like the Russell's one (that has already been ruled out through the distinction between sets and proper classes), we avoid certain pathologies, like sets that belong to themselves. Hopefully this universe is coherent, hence it seems to be the best environment in which the whole mathematics can be founded through set theory.

13.1 DEFINITION AND FIRST PROPERTIES

By transfinite recursion on ordinals we provide the following definition.

Definition 13.1. The set $R(\alpha)$, for every $\alpha \in \mathbf{Ord}$, is defined by:

- $R(0) := \emptyset$
- $R(s(\alpha)) := \mathcal{P}(R(\alpha))$
- if α is a limit ordinal, then $R(\alpha) := \bigcup\{R(\beta) : \beta \in \alpha\}$

A set x is called *well-founded* is there exists $\alpha \in \mathbf{Ord}$ such that $x \in R(\alpha)$. We denote by `wf` the class of well-founded sets—that is, $\mathtt{wf} := \bigcup\{R(\alpha) : \alpha \in \mathbf{Ord}\}$. ♢

Exercise 13.2. Formalise the definition of $R(\alpha)$ in 13.1 by applying the statement of proposition 11.48.

We remark that, restricting definition 13.1 to $\alpha \in \omega$, we get the hierarchy of finite sets constructed in section 2.4. Hence, we are extending that hierarchy to arbitrary ordinals.

Lemma 13.3. *(i) If x is a transitive set, then $\mathcal{P}(x)$ is transitive too. (ii) If x is a set and every $y \in x$ is transitive, then $\bigcup x$ is transitive too.*

DOI: 10.1201/9781003565291-13

Proof. (i) If $y \in a \in \mathcal{P}(x)$—that is, $y \in a \subset x$—then $y \in x$, hence by transitivity $y \subset x$—that is, $y \in \mathcal{P}(x)$. This shows that $\mathcal{P}(x)$ is transitive. Equivalently, by transitivity every element of x is a part of x too, hence a part of x is also a part of the parts of x—that is, $a \in \mathcal{P}(x)$ implies $a \subset \mathcal{P}(x)$.

(ii) If $w \in z \in \bigcup x$, then by definition of union there exists $y \in x$ such that $w \in z \in y$. Since y is transitive by hypothesis, we have $w \in y \in x$, hence $w \in \bigcup x$. □

Proposition 13.4. *For every $\alpha, \beta \in$ **Ord**:*

(i) the set $R(\alpha)$ is transitive;

(ii) if $\beta \subset \alpha$, then $R(\beta) \subset R(\alpha)$.

Proof. We apply transfinite induction on α. If $\alpha = 0$, then both (i) and (ii) are trivial. Let us fix $\alpha \in \mathbf{Ord}^*$ and suppose that (i) and (ii) hold for every $\beta \in \alpha$. We have two possibilities: I. α is a successor; II. α is a limit ordinal.

I. We set $\alpha = s(\alpha')$. By definition $R(\alpha) = \mathcal{P}(R(\alpha'))$ and by the inductive hypothesis $R(\alpha')$ is transitive. It follows from lemma 13.3-i that $R(\alpha)$ is transitive too. About item (ii), the statement is trivial if $\beta = \alpha$, hence we suppose $\beta \in \alpha$—that is, $\beta \subset \alpha'$. By the inductive hypothesis $R(\beta) \subset R(\alpha')$, that is, $R(\beta) \in \mathcal{P}(R(\alpha')) = R(\alpha)$. Since $R(\alpha)$ is transitive, this implies $R(\beta) \subset R(\alpha)$.

II. By definition $R(\alpha) := \bigcup\{R(\beta) : \beta \in \alpha\}$. By the inductive hypothesis, $R(\beta)$ is transitive for every $\beta \in \alpha$, hence it follows from lemma 13.3-ii that $R(\alpha)$ is transitive too. Item (ii) immediately follows from the definition of union. □

13.1.1 Rank

By definition, $x \in \mathtt{wf}$ if and only if the class $\mathsf{c} := \{\alpha \in \mathbf{Ord} : x \in R(\alpha)\}$ is not empty. Since **Ord** is well-ordered, such a class admits a minimum α'. Necessarily α' is a successor ordinal. In fact, it is impossible that $\alpha' = 0$, since otherwise $x \in R(0) = \emptyset$, a contradiction. Moreover, if α' is a limit ordinal, then by definition 13.1 there exists $\beta \in \alpha'$ such that $x \in R(\beta)$, hence α' is not the minimum of c. Therefore, there exist α such that $\alpha' = s(\alpha)$. We call α the *rank* of x, as the following definition states.

Definition 13.5. The *rank* of a well-founded sets x is the minimum ordinal α such that $x \in R(s(\alpha))$. We denote it by $\mathrm{rk}(x)$. ◇

Lemma 13.6. *If $\mathrm{rk}(x) = \alpha$, then*

- *$x \notin R(\beta)$ for every $\beta \subset \alpha$; and*
- *$x \in R(\gamma)$ for every $\gamma \supset s(\alpha)$.*

Proof. By definition or rank, we have $x \in R(s(\alpha))$, hence proposition 13.4-i implies that $x \subset R(s(\alpha))$. By proposition 13.4-ii, we have $R(s(\alpha)) \subset R(\gamma)$ for every $\gamma \supset s(\alpha)$, hence $x \in R(\gamma)$ for every such γ. Moreover, by definition of rank (in particular, by minimality of α), necessarily $x \notin R(\alpha)$. By proposition 13.4-ii, this implies that $x \notin R(\beta)$ for every $\beta \subset \alpha$. □

13.1.2 The Structure of the Class `wf`

Thanks to the notion of rank, we can investigate more deeply the structure of well-founded sets. In particular, we have $R(0) = \emptyset$, thus $R(1) = \{\emptyset\}$, the latter being the set of rank-0 wf sets; then, $R(2) = \{\emptyset, \{\emptyset\}\}$, the latter being the set of wf sets of rank at most 1, and so on until $R(\omega)$, that is the set of finite-rank wf sets. Afterwards, we have $R(\omega+1)$, that is formed by the wf sets of rank at most ω—that is, the parts of the set of finite-rank wf sets—and so on. In general, $R(\alpha)$ is the set of wf sets of rank lower than α, as the following proposition shows.

Proposition 13.7. $R(\alpha) = \{x \in \mathtt{wf} : \mathrm{rk}(x) \in \alpha\}$.

Proof. The following conditions are equivalent:

1. $\mathrm{rk}(x) \in \alpha$
2. $\min\{\beta \in \mathbf{Ord} : x \in R(s(\beta))\} \in \alpha$
3. there exists $\beta \in \alpha$ such that $x \in R(s(\beta))$
4. $x \in R(\alpha)$

Let us prove the equivalences. (1) ⇔ (2) By definition of rank. (2) ⇒ (3) The minimum of $\mathtt{c} := \{\beta \in \mathbf{Ord} : x \in R(s(\beta))\}$ satisfies 3. (2) ⇐ (3) If there exists $\beta \in \mathtt{c}$ such that $\beta \in \alpha$, then by minimality $\min(\mathtt{c}) \subset \beta \in \alpha$, hence $\min(\mathtt{c}) \in \alpha$. (3) ⇒ (4) Since $\beta \in \alpha$, we have $s(\beta) \subset \alpha$, hence $R(s(\beta)) \subset R(\alpha)$ by proposition 13.4-ii. Therefore, $x \in R(s(\beta)) \subset R(\alpha)$, thus $x \in R(\alpha)$. (3) ⇐ (4) If α is a successor, we set $\alpha = s(\beta)$ and 3 trivially holds. If α is a limit ordinal, then by definition 13.1 there exists $\beta \in \alpha$ such that $x \in R(\beta)$. Since $R(\beta) \subset R(s(\beta))$ by proposition 13.4-ii, we get $x \in R(s(\beta))$. □

The next proposition shows that the elements of a wf set are wf sets of lower rank.

Proposition 13.8. *If* $x \in \mathtt{wf}$, *then:*

(i) for every $y \in x$, *we have* $y \in \mathtt{wf}$ *and* $\mathrm{rk}(y) \in \mathrm{rk}(x)$;

(ii) if $x \neq \emptyset$, *then* $\mathrm{rk}(x) = \sup\{s(\mathrm{rk}(y)) : y \in x\}$.

Proof. (i) If $\mathrm{rk}(x) = \alpha$, then $x \in R(s(\alpha)) = \mathcal{P}(R(\alpha))$—that is, $x \subset R(\alpha)$. Hence, if $y \in x$, then $y \in R(\alpha)$, therefore y is well-founded and $\mathrm{rk}(y) \in \alpha$ by proposition 13.7. (ii) We set $\xi := \sup\{s(\mathrm{rk}(y)) : y \in x\}$, that exists by proposition 11.23-ii and assumption `(A6')`. If $y \in x$, then by item i we have

$s(\mathrm{rk}(y)) \subset \mathrm{rk}(x)$, therefore $\xi \subset \mathrm{rk}(x)$. Moreover, by definition of ξ, for every $y \in x$ we have $s(\mathrm{rk}(y)) \subset \xi$, hence $y \in R(s(\mathrm{rk}(y))) \subset R(\xi)$, thus $y \in R(\xi)$. It follows that $x \subset R(\xi)$—that is, $x \in \mathcal{P}(R(\xi)) = R(s(\xi))$. By definition of rank, we have $\mathrm{rk}(x) \subset \xi$, hence the equality holds. □

Corollary 13.9. *We have that*

(i) the class `wf` *is transitive;*

(ii) if $x \in$ `wf`*, then $x \notin x$;*

(iii) if $x, y \in$ `wf`*, then it is not possible that $y \in x$ and $x \in y$; and*

(iv) if $x \in$ `wf`*, then there is no infinite strictly decreasing sequence of the form $\cdots \in x_2 \in x_1 \in x_0 = x$.*

Proof. (i) It immediately follows from proposition 13.8-i. (ii) If $x \in$ `wf` and $x \in x$, then by proposition 13.8-i we have $\mathrm{rk}(x) \in \mathrm{rk}(x)$, a contradiction, since $\mathrm{rk}(x)$ is an ordinal (see proposition 11.20-ii). (iii) If $x, y \in$ `wf`, $y \in x$ and $x \in y$, then by proposition 13.8-i we have $\mathrm{rk}(y) \in \mathrm{rk}(x)$ and $\mathrm{rk}(x) \in \mathrm{rk}(y)$, a contradiction, since '$\in$' a strict order relation in **Ord** (see proposition 11.20-(ii, iii)). (iv) If such a sequence existed, then we would have $\cdots \in \mathrm{rk}(x_2) \in \mathrm{rk}(x_1) \in \mathrm{rk}(x_0)$, hence the set $\{\mathrm{rk}(x_n)\}_{n \in \mathbb{N}}$ would be a subset of **Ord** with no minimum, a contradiction (see proposition 11.20-iv). □

We remark that, in proposition 13.8-ii, we cannot replace $\sup\{s(\mathrm{rk}(y)) : y \in x\}$ by $s(\sup\{\mathrm{rk}(y) : y \in x\})$. In fact, suppose that $x = \omega$: then, $\sup\{s(\mathrm{rk}(y)) : y \in \omega\} = \omega$, while $s(\sup\{\mathrm{rk}(y) : y \in \omega\}) = s(\omega) \neq \omega$. The same argument holds for every x such that $\mathrm{rk}(x)$ is a limit ordinal.

Proposition 13.10. *For every set x, we have that $x \in$* `wf` *if and only if $x \subset$* `wf`*.*

Proof. ($\Rightarrow$) It is the statement of corollary 13.9-i. ($\Leftarrow$) By hypothesis, every element of x is well-founded, hence it admits a well-defined rank. We set $\xi := \sup\{s(\mathrm{rk}(y)) : y \in x\}$, that exists by proposition 11.23-ii and assumption `(A6')`. It follows that $s(\mathrm{rk}(y)) \subset \xi$ for every $y \in x$, hence $y \in R(s(\mathrm{rk}(y))) \subset R(\xi)$ for every $y \in x$, the first inclusion being due to the definition of rank and the second one to proposition 13.4-ii. This shows that $x \subset R(\xi)$, thus $x \in \mathcal{P}(R(\xi)) = R(s(\xi))$, hence $x \in$ `wf`. □

Corollary 13.11. *If $x \in$* `wf` *and $a \subset x$, then $a \in$* `wf`*.*

Proof. Since $x \in$ `wf`, we have $x \subset$ `wf` by transitivity (see proposition 13.9-i). Hence, $a \subset x \subset$ `wf`, that implies $a \subset$ `wf`. We deduce from proposition 13.10 that $a \in$ `wf`. □

Exercise 13.12. (i) Prove that $n \in$ `wf` and $\mathrm{rk}(n) = n$ for every natural number n. (ii) Prove that $\omega \in$ `wf` and $\mathrm{rk}(\omega) = \omega$. *Remark:* These statements are particular cases of proposition 13.34-(i, ii) below.

Exercise 13.13. Prove that, if $x \in \texttt{wf}$ and $\text{rk}(x)$ is finite, then x is finite. Show that the converse does not hold.

13.1.3 Equivalent Characterisation under AC

We will see in corollary 13.32 below that, assuming AC, the converse of corollary 13.9-iv holds too. This provides the following equivalent characterisation of well-founded sets under AC. Let us suppose to perform the following process, starting from a set x: we set $x_0 := x$; then, if x_0 is not empty, we choose $x_1 \in x_0$; then, if x_1 is not empty, we choose $x_2 \in x_1$, and so on. The set x is well-founded if and only if, after a finite number of steps, we necessarily reach the empty set.

In order to formalise this equivalence, we consider the set $\mathcal{S}$ of finite or infinite sequences $\{x_n\}$ such that $x_0 = x$ and, for every admissible i, we have $x_{i+1} \in x_i$ or $x_{i+1} = x_i = \emptyset$. Every finite sequence of choices is contained in such an infinite sequence. In fact, we partially order $\mathcal{S}$ by declaring that $\{x_n\} \prec \{y_n\}$ if and only if $\{x_n\}$ is finite and $\{y_n\}$ is an extension of the former on the right. The reader can verify that the hypothesis of ZL is satisfied, hence every sequence is contained in a maximal one, that is necessarily infinite. We anticipated that the set x is well-founded if and only if there exist no infinite strictly decreasing sequences; this is equivalent to stating that every maximal element of $\mathcal{S}$ is a finite sequence extended on the right through empty sets—that is, any sequence of choices terminates with the empty set in a finite number of steps.

We observe that there is no obstruction in forming an infinite strictly increasing sequence—for example, $0 \in 1 \in 2 \in \cdots$.

13.2 WELL-FOUNDED SETS AS THE UNIVERSE

Corollary 13.9 shows that, in the class of well-founded sets, certain pathologies are not possible. As it is clear from definition 13.1, the class $\texttt{wf}$ has been constructed by starting from the class **Ord** and applying elementary set-theoretical operations (in particular, power set and union). Hence, if $\mathbf{u}$ is any universe of hereditary sets, that contains the class of ordinals and in which all of the elementary operations can be applied, then necessarily $\texttt{wf} \subset \mathbf{u}$. Since $\texttt{wf}$ is small enough to avoid the paradoxes and the pathologies described in chapter 2, it is a good candidate to be the reduced universe we are looking for. In order to confirm this affirmation, we have to verify that mathematics can be thoroughly developed within $\texttt{wf}$. First of all, we check that all of the elementary set-theoretical operations can be performed within $\texttt{wf}$.

Proposition 13.14. *If $x_1, \ldots, x_n \in \texttt{wf}$, then $\{x_1, \ldots, x_n\} \in \texttt{wf}$ and $(x_1, \ldots, x_n) \in \texttt{wf}$.*

Proposition 13.15. *Starting from elements of $\texttt{wf}$, all of the operations introduced in definitions 2.8, 2.9, 4.1, and 4.7 produce elements of $\texttt{wf}$.*

Exercise 13.16. Prove propositions 13.14 and 13.15. *Hint:* Apply proposition 13.10 and corollary 13.11.

Exercise 13.17. Prove that, for most of the operations, the converse of proposition 13.15 holds too (for example, if $x \cup y \in \mathtt{wf}$, then $x, y \in \mathtt{wf}$). Show that there are some exceptions (for example, it may happen that $x \cap y \in \mathtt{wf}$ or $x \,\Delta\, y \in \mathtt{wf}$, even if $x, y \notin \mathtt{wf}$).

Exercise 13.18. (i) Prove that propositions 13.14 and 13.15 hold with respect to operations among *classes* contained in wf as well. (ii) Prove that, for most of the operations among classes, the converse of proposition 13.15 holds too.

Exercise 13.19. Prove that $\mathbb{N}, \mathbb{Z}, \mathbb{Q}, \mathbb{R}, \mathbb{C} \in \mathtt{wf}$.

Exercise 13.20. Prove that, if $f\colon \mathbf{c} \to \mathbf{d}$ is a surjective pre-function between classes and $f \in \mathtt{wf}$, then $\mathbf{c}, \mathbf{d} \subset \mathtt{wf}$ (hence, if $\mathbf{c}$ and $\mathbf{d}$ are sets, the surjective function $(\mathbf{c}, \mathbf{d}, f)$ belongs to wf).

We have seen that the elementary set-theoretical operations can be performed within the class of well-founded sets. Now it is easy to verify that all of the assumptions (A1)–(A8) are coherent within wf—that is, if we make these assumptions and we restrict our universe to wf, then they keep on holding without contradictions. In fact, (A1) can be assumed in any fixed universe; (A2) holds in wf by proposition 13.14; (A3) and (A5) hold by propositions 13.15. About (A4), if $a \in \mathtt{wf}$ and $x := \{y \in a : \varphi(y)\}$, then $x \subset a$, hence $x \in \mathtt{wf}$ by corollary 13.11. About (A6), if we apply it to a well-founded function f, then necessarily $\mathbf{c}, \mathbf{d} \subset \mathtt{wf}$ by exercise 13.20; the thesis states that $\mathbf{d}$ is a set, and it follows from proposition 13.10 that $\mathbf{d} \in \mathtt{wf}$. Assumption (A7) holds in wf, since $\mathbb{N} \in \mathtt{cs} \cap \mathtt{wf}$ (see exercise 13.19). About (A8), let us consider AC1. If $x \in \mathtt{wf}$, then $\bigcup x \in \mathtt{wf}$ by proposition 13.15; hence, $\mathrm{Func}(x, \bigcup x) \in \mathtt{wf}$, since it can be constructed from x and $\bigcup x$ through the previous assumptions. It follows that the choice function ξ, that we are assuming to exist, belongs to wf. The reader can verity that ZL, HMP, and WO are coherent in wf too.

The considerations above confirm that wf is a good candidate to be our reduced universe, hence it would be natural to add the assumption $\mathtt{u} = \mathtt{wf}$ to (A1)–(A8). Nevertheless, the statement $\mathtt{u} = \mathtt{wf}$ is quite elaborated, since we have to define wf by transfinite recursion on ordinals and impose that it coincides with the universe. It would be more elegant to formulate the equality $\mathtt{u} = \mathtt{wf}$ through a simple set-theoretical statement. We will achieve this aim in the next sections.

13.3 WELL-FOUNDED RELATIONS

The following definition generalises the notion of well-order.

Definition 13.21. A binary relation ρ in a set x is *well-founded* if every non-empty subset $a \subset x$ admits a ρ*-minimal* element—that is, an element $m \in a$ such that there not exits any $y \in a$ satisfying $y \,\rho\, m$. ◊

The expression "ρ-minimal" may induce to think of an order relation, but ρ can be any binary relation. Actually, it is necessarily anti-reflexive. In fact, if there exists $y \in x$ such that $y \, \rho \, y$, then the subset $a := \{y\}$ makes definition 13.21 fail. Nevertheless, ρ is not transitive in general.

Example 13.22. In the set $\mathbb{N}$, we define the relation $n \, \rho \, m$ if and only if $m = n+1$. This relation is clearly anti-reflexive, but it is not transitive, since $0 \, \rho \, 1$, $1 \, \rho \, 2$, but $0 \, \not\rho \, 2$. We observe that, if $n \, \rho \, m$, then $n < m$, hence, given a non-empty subset $a \subset \mathbb{N}$, the minimum of a with respect to the usual well-order '$<$' is a ρ-minimal element. It follows that ρ is well-founded, but it is not a strict order relation. We remark that a ρ-minimal element is not unique in general. For example, if $a = \{0, 1, 3, 4\}$, then 0 and 3 are ρ-minimal in a.

Example 13.23. In the set $\mathbb{N}^*$, we define the relation $n \, \rho \, m$ if and only if $n \mid m$ and $n \neq m$. This is a strict order relation. As in the previous example, if $n \, \rho \, m$, then $n < m$, hence, given a non-empty subset $a \subset \mathbb{N}$, the minimum of a with respect to the usual well-order '$<$' is a ρ-minimal element. It follows that ρ is well-founded.

It follows from definition 13.21 that, if ρ is a *total* strict order relation in a set x, then it is well-founded if and only if it is a well-order. In fact, in this case, an element $m \in a \subset x$ is ρ-minimal if and only if it is a minimum. Nevertheless, we have seen in the two previous examples that a well-founded relation is not necessarily an order relation and, if it is, it is not necessarily total. We conclude by observing that, if ρ is a well-founded relation in a set x and $a \subset x$, then the restriction $\rho|_{a \times a}$ is well-founded too, since a ρ-minimal element of $b \subset a$ is also $\rho|_{a \times a}$-minimal.

13.3.1 Well-Founded Relations and Strictly Decreasing Sequences

Assuming AC, we can characterise well-founded relations through infinite strictly decreasing sequences. Again the word "decreasing" induces to think of an order relation, but we can use it for generic relations too, as in the following statement.

Proposition 13.24 (AC$\Leftarrow$). *A relation ρ in a set x is well-founded if and only if there not exists an infinite strictly decreasing sequence in x—that is, a sequence $\{y_n\}_{n \in \mathbb{N}}$ such that $\cdots \rho \, y_2 \, \rho \, y_1 \, \rho \, y_0$.*

Proof. ($\Rightarrow$) If such a sequence exists, then the subset $\{y_n\}_{n \in \mathbb{N}}$ of x does not admit a ρ-minimal element. ($\Leftarrow$) If $x = \emptyset$, then the thesis is trivial, hence we assume $x \neq \emptyset$. Let us suppose that ρ is not well-founded. This means that there exists $a \subset x$ that is not empty and does not admit any ρ-minimal element. We fix a choice function $\xi \colon \mathcal{P}^*(a) \to a$ and any $y_0 \in a$. By recursion, let us suppose to have fixed $\{y_n, \ldots, y_0\} \subset a$ such that $y_n \, \rho \cdots \rho \, y_0$. Since there are no ρ-minimal elements in a, in particular y_n is not ρ-minimal, hence the set $b_n := \{y \in a : y \, \rho \, y_n\}$ is not empty. We set $y_{n+1} := \xi(b_n)$. In this way we get an infinite strictly decreasing sequence in a, hence in x too. □

13.3.2 Well-Founded Relations *vs* Well-Founded Sets

Let us see that the notions of well-founded relation and well-founded set are strictly related. The next proposition shows the "easy" part of this link.

Proposition 13.25. *If a set x is well-founded, then the relation '$\in$' is well-founded in x.*

Proof. If $x = \emptyset$, then the result is trivial, hence we assume $x \neq \emptyset$. Since $x \subset \texttt{wf}$ by transitivity, we get the function $\mathrm{rk}\colon x \to \mathbf{Ord}$ and we call r its image. Since **Ord** is well-ordered, the set r admits a minimum α. We choose any $y \in x$ such that $\mathrm{rk}(y) = \alpha$. The set y is $\in$-minimal in x, since, if there exists $z \in x$ such that $z \in y$, then we have $\mathrm{rk}(z) \in r$ and, by proposition 13.8-i, $\mathrm{rk}(z) \in \mathrm{rk}(y) = \alpha$, contradicting the minimality of α. □

Now it is natural to hope that the converse of proposition 13.25 holds too, but unfortunately we have no way to prove it. Actually, we cannot say that it is false, since we do not know if not well-founded sets exist, but we can convince ourselves that, if they exist in some sense, then the latter statement if false. In fact, let us suppose that in a certain universe, bigger than wf, there exist two sets x and y such that $x = \{y\}$, $y = \{x\}$, and $x \neq y$. These sets cannot be well-founded because of corollary 13.9-iii, but the relation '$\in$' is empty in x and in y, hence it is well-founded in both. Nevertheless, the following partial converse of proposition 13.25 holds.

Proposition 13.26. *If x is a* transitive *set, then x is well-founded if and only if '$\in$' is well-founded in x.*

Proof. ($\Rightarrow$) It is a particular case of proposition 13.25. ($\Leftarrow$) Let us suppose that $x \notin \texttt{wf}$. By proposition 13.10, this is equivalent to state that $x \not\subset \texttt{wf}$, hence $x \setminus \texttt{wf} \neq \emptyset$. Since '$\in$' is well-founded in x, there exists an $\in$-minimal element y of $x \setminus \texttt{wf}$. If $z \in y$, then:

- Since x is transitive and $z \in y \in x$, we have $z \in x$.
- The previous item implies that $z \notin x \setminus \texttt{wf}$ because of the minimality of y.

We proved that $z \in x$ and $z \notin x \setminus \texttt{wf}$, hence $z \in \texttt{wf}$. Since this is true for every $z \in y$, we have $y \subset \texttt{wf}$. By proposition 13.10 this implies $y \in \texttt{wf}$, a contradiction, because $y \in x \setminus \texttt{wf}$ by construction. □

13.3.3 An Alternative Proof of Proposition 13.26

Let us provide an alternative proof of the direction ($\Leftarrow$) of proposition 13.26, that hopefully helps to better understand why it holds; the uninterested reader can skip to the next section. We assume that $x \neq \emptyset$, otherwise the result is trivial, hence the whole set x admits an $\in$-minimal element y. Necessarily $y = \emptyset$, since an element of

y would contradict its $\in$-minimality by transitivity. This proves that $\emptyset \in x$—that is, x necessarily contains the unique well-founded set of rank 0.

If $x \neq \{\emptyset\}$, we set $x' := x \setminus \{\emptyset\}$. By hypothesis, there exists an $\in$-minimal element $y' \in x'$. Necessarily $y' = \{\emptyset\}$, since any other element of y' would contradict the minimality of y' in x' by transitivity.

If $x \neq \{\emptyset, \{\emptyset\}\}$, we set $x'' := x \setminus \{\emptyset, \{\emptyset\}\}$. By hypothesis, there exists an $\in$-minimal element $y'' \in x''$. Necessarily y'' is a well-founded set of rank 2. In fact, if there exists $z'' \in y''$, that is not a well-founded set of rank 0 or 1, then by transitivity $z'' \in x''$, contradicting the minimality of y''.

Inductively, let us suppose to have proven that x contains at least one set of rank β for every $\beta \in \alpha$, where α is any fixed ordinal. We call $a^{(\alpha)}$ the subset of x formed by the elements of rank less than α and, if $x \neq a^{(\alpha)}$, we set $x^{(\alpha)} := x \setminus a^{(\alpha)}$. By hypothesis, there exists an $\in$-minimal element $y^{(\alpha)} \in x^{(\alpha)}$. Necessarily $y^{(\alpha)}$ is a well-founded set of rank α. In fact, if there exists $z^{(\alpha)} \in y^{(\alpha)}$, that is not a well-founded set of rank less than α, then by transitivity $z^{(\alpha)} \in x^{(\alpha)}$, contradicting the minimality of $y^{(\alpha)}$.

Let us suppose that this process never stops—that is, $x \neq a^{(\alpha)}$ for every $\alpha \in \mathbf{Ord}$. Then, for every ordinal α, there exists $y^{(\alpha)} \in x$ that is well-founded of rank α. It follows that the function $\mathrm{rk}\colon x \to \mathbf{Ord}$ is surjective, hence $\mathbf{Ord}$ is a set by assumption `(A6)`, a contradiction. This proves that there exists α such that $x = a^{(\alpha)}$, thus x is well-founded of rank α. Actually, we also proved that, for every $\beta \in \mathrm{rk}(x)$, there exists an element of x of rank β. This statement can be also proved directly as a consequence of transitivity.

13.3.4 Transitive Closure

Given a set x, we can "close" it by transitivity—that is, there exists a set $\bar{x}$ such that $\bar{x}$ is transitive, it contains x and it is the minimum (with respect to inclusion) of all the transitive sets that contain x. Equivalently, $\bar{x}$ is the intersection of all the transitive sets that contain x. In principle, we could define $\bar{x}$ as such intersection, but we should prove that, for every x, there exists at least one transitive set y such that $x \subset y$, and this is not faster than constructing $\bar{x}$ directly as follows.

Definition 13.27. Given a set x, the *transitive closure* of x is the set

$$\bar{x} := \bigcup\{\bigcup^n x : n \in \mathbb{N}\}. \qquad \diamondsuit$$

This means that $\bar{x} = x \cup \left(\bigcup x\right) \cup \left(\bigcup^2 x\right) \cup \ldots$—that is, $\bar{x}$ is the set formed by the elements of x, the elements of the elements of x, and so on at any order.

Proposition 13.28. *For every set x:*

1. $x \subset \bar{x}$

2. *$\bar{x}$ is transitive.*

3. *If $x \subset y$ and y is transitive, then $\bar{x} \subset y$; equivalently, $\bar{x} = \bigcap\{y : (x \subset y) \wedge (y$ is transitive$)\}$.*

4. *If $y \in x$ or $y \subset x$, then $\bar{y} \subset \bar{x}$.*

5. $\bar{x} = x \cup \bigcup\{\bar{y} : y \in x\}$.

Exercise 13.29. Prove proposition 13.28.

The importance of the notion of transitive closure in the context of well-founded sets is due to the following proposition.

Proposition 13.30. *For every set x, the following conditions are equivalent:*

(i) x is well-founded

(ii) $\bar{x}$ is well-founded

(iii) '$\in$' is well-founded in $\bar{x}$

Proof. (i) $\Rightarrow$ (ii) Since $\bigcup^0 x = x$, by hypothesis $\bigcup^0 x$ is well-founded. Proposition 13.15 easily implies by induction on n that $\bigcup^n x$ is well-founded for every $n \in \mathbb{N}$. It follows that $\{\bigcup^n x : n \in \mathbb{N}\} \subset \mathtt{wf}$, hence $\{\bigcup^n x : n \in \mathbb{N}\} \in \mathtt{wf}$ by proposition 13.10. Again from proposition 13.15, we deduce that $\bar{x} \in \mathtt{wf}$. (i) $\Leftarrow$ (ii) By proposition 13.28-1 and by transitivity of $\mathtt{wf}$, we have $x \subset \bar{x} \subset \mathtt{wf}$, hence $x \subset \mathtt{wf}$. By proposition 13.10, we have $x \in \mathtt{wf}$. (ii) $\Leftrightarrow$ (iii) It immediately follows from propositions 13.28-2 and 13.26. □

The equivalence between items i and iii of proposition 13.30 is the full link between well-founded sets and well-founded relations. As an example, we have seen in the comments after proposition 13.25 that, if $x = \{y\}$, $y = \{x\}$ and $x \neq y$, then '$\in$' is well-founded in x and y, but they are not well-founded sets. Considering the transitive closures, we have $\bar{x} = \bar{y} = \{x, y\}$. Coherently with proposition 13.30, the relation '$\in$' is not well-founded in this set, since neither x nor y is $\in$-minimal.

Now we can prove the converse of corollary 13.9-iv assuming AC, as anticipated in section 13.1.3.

Lemma 13.31. *Let x be a set.*

(i) For any $n \in \mathbb{N}$, we have that $y \in \bigcup^n x$ if and only if there exists a finite sequence $y = y_n \in \cdots \in y_0 \in x$ such that $y_i \in \bigcup^i x$.

(ii) There exists an infinite strictly decreasing sequence of the form $\cdots \in x_2 \in x_1 \in x_0 = x$ if and only if there exists an infinite strictly decreasing sequence contained in $\bar{x}$.

Proof. (i) ($\Leftarrow$) Obvious, since $y = y_n \in \bigcup^n x$ by hypothesis. ($\Rightarrow$) The proof is straightforward by induction on n.

(ii) ($\Rightarrow$) Cutting x_0, we get the sequence $\cdots \in x_2 \in x_1$, that is contained in $\bar{x}$, since, by proposition 13.28-(i, ii), $x \subset \bar{x}$ (hence, $x_1 \in \bar{x}$) and $\bar{x}$ is transitive (hence, inductively, $x_i \in \bar{x}$ for every $i \in \mathbb{N}^*$). ($\Leftarrow$) An infinite strictly decreasing sequence in $\bar{x}$ is a sequence of the form $\cdots \in x_2 \in x_1 \in x_0 \in \bar{x}$. By definition of $\bar{x}$, there exists $n \in \mathbb{N}$ such that $x_0 \in \bigcup^n x$, hence, by applying item i to $y := x_0$, we get an infinite sequence of the form $\cdots \in x_2 \in x_1 \in x_0 = y_n \in \cdots \in y_0 \in x$. □

Corollary 13.32 (AC⇐). *A set x is well-founded if and only if there exists no infinite strictly decreasing sequence of the form $\cdots \in x_2 \in x_1 \in x_0 = x$.*

Proof. (⇒) It is the statement of corollary 13.9-iv. (⇐) By lemma 13.31-ii, the hypothesis is equivalent to state that there not exits an infinite strictly decreasing sequence in $\bar{x}$, thus the relation '$\in$' is well-founded in $\bar{x}$ by proposition 13.24. Hence, the thesis follows from the equivalence (i) ⇔ (iii) in proposition 13.30.[1] □

13.4 THE FOUNDATION ASSUMPTION

As we saw in the comments after proposition 13.25, if we require that '$\in$' is well-founded in a set x, this does not necessarily imply that x is well-founded, unless it is transitive. Nevertheless, if we require that '$\in$' is well-founded in *every* set, then, in particular, it is well-founded in the transitive closure of every set, hence every set is well-founded by proposition 13.30. Therefore, reducing the universe to $\mathtt{wf}$ is equivalent to assuming that '$\in$' is well-founded in every set. By definition 13.21, this means that every non-empty subset a of any set x admits a $\in$-minimal element. Since a is a set like x, it is enough to require that every non-empty set admits a $\in$-minimal element. We get the following proposition.

Proposition 13.33. *The following statements are equivalent:*

(i) Every non-empty set x admits a $\in$-minimal element.

(ii) The relation '$\in$' is well-founded in every set.

(iii) Every set is well-founded—that is, $\mathtt{u} = \mathtt{wf}$.

As we saw in section 13.2, we aim to impose statement iii of proposition 13.33 as our last assumption, but we prefer a formulation that can be expressed through a wf; statement i is suitable for this purpose, since it can be expressed as follows:

$$\forall x\big((x \neq \emptyset) \Rightarrow (\exists m \in x : \sim\exists y \in x : y \in m)\big).$$

[1] We can also prove the contrapositive of (⇐) as follows. For every $y \in \bar{x}$, we choose a choice function $\xi_y : \mathcal{P}^*(y) \to y$. Here we have a double application of AC: in fact, the set of choices function in y, for every $y \in \bar{x}$, is not empty by AC; hence, by applying AC again, we choose a choice function ξ_y for every y. If x is not well-founded, then we set $x_0 := x$. Let us suppose recursively to have constructed $x_n \in \cdots \in x_0 = x$ such that $x_i \notin \mathtt{wf}$ for every $i \in \{0, \ldots, n\}$. Since in particular $x_n \notin \mathtt{wf}$, by proposition 13.10 we have $x_n \not\subset \mathtt{wf}$, hence $x_n \setminus \mathtt{wf} \neq \emptyset$. We set $x_{n+1} := \xi_{x_n}(x_n \setminus \mathtt{wf})$ and we get the desired sequence.

Therefore, we state the following hypothesis, that we call *foundation assumption*:

(A9) Every non-empty set x admits a $\in$-minimal element.

Equivalently:

- For every non-empty set x, there exists an element $m \in x$ such that m does not contain any element of x.
- For every non-empty set x, there exists an element $m \in x$ such that x does not contain any element of m.
- For every non-empty set x, there exists an element $m \in x$ such that $x \cap m = \emptyset$.

13.5 WELL-FOUNDED SETS AND ORDINALS

The next proposition shows that every ordinal is well-founded and it coincides with its rank, as expected.

Proposition 13.34. *We have that:*

(i) $\mathbf{Ord} \subset \mathtt{wf}$

(ii) $\mathrm{rk}(\alpha) = \alpha$ *for every* $\alpha \in \mathbf{Ord}$

(iii) $R(\alpha) \cap \mathbf{Ord} = \alpha$ *for every* $\alpha \in \mathbf{Ord}$

Proof. (i) Transfinite induction on α. If $\alpha = 0$, then $0 \in R(1) \subset \mathtt{wf}$. If $\beta \in \mathtt{wf}$ for every $\beta \in \alpha$, then $\alpha \subset \mathtt{wf}$, hence $\alpha \in \mathtt{wf}$ by proposition 13.10.

(ii) Transfinite induction on α. Trivially $\mathrm{rk}(0) = 0$. If $\mathrm{rk}(\alpha) = \alpha$ and $\beta \in s(\alpha) = \alpha \cup \{\alpha\}$, we have two possibilities: I. $\beta \in \alpha$, hence $\mathrm{rk}(\beta) \in \mathrm{rk}(\alpha) = \alpha$ by proposition 13.8-i; II. $\beta = \alpha$, hence $\mathrm{rk}(\beta) = \mathrm{rk}(\alpha) = \alpha$. This shows that $\alpha = \max\{\mathrm{rk}(\beta) : \beta \in s(\alpha)\}$, hence $s(\alpha) = \max\{s(\mathrm{rk}(\beta)) : \beta \in s(\alpha)\}$. It follows from proposition 13.8-ii that $\mathrm{rk}(s(\alpha)) = s(\alpha)$, as required. If α is a limit ordinal and $\mathrm{rk}(\beta) = \beta$ for every $\beta \in \alpha$, then by proposition 13.8-ii we have $\mathrm{rk}(\alpha) = \sup\{s(\mathrm{rk}(\beta)) : \beta \in \alpha\} = \sup\{s(\beta) : \beta \in \alpha\} = \alpha$, the last equality being due to the fact that $\beta \in \alpha$ if and only if $s(\beta) \in \alpha$.

(iii) By item i and proposition 13.7 we have $R(\alpha) \cap \mathbf{Ord} = \{\beta \in \mathbf{Ord} : \mathrm{rk}(\beta) \in \alpha\}$. By item ii we get $R(\alpha) \cap \mathbf{Ord} = \{\beta \in \mathbf{Ord} : \beta \in \alpha\} = \alpha$. □

We anticipated in the comments before definition 11.14 that, in the class wf, a set is an ordinal if and only if it is properly hereditarily transitive. Let us prove this result.

Lemma 13.35. *If* $a \subset \mathbf{Ord}$ *is a transitive set, then* $a \in \mathbf{Ord}$.

Proof. The relation '$\in$' is a strict well-order in **Ord**, hence its restriction to a is a strict well-order too. It follows from definition 11.15 that a is an ordinal. □

Proposition 13.36. *A well-founded set is an ordinal if and only if it is hereditarily transitive.*

Proof. ($\Rightarrow$) It immediately follows from definition 11.14. ($\Leftarrow$) Let $x \in \mathtt{wf}$ (hence $x \subset \mathtt{wf}$) be hereditarily transitive. Properness, as required by definition 11.14, is automatic in $\mathtt{wf}$ by corollary 13.9-ii. Since x is transitive by hypothesis, by lemma 13.35 it is enough to prove that $x \subset \mathbf{Ord}$. Let us suppose by contradiction that $x \setminus \mathbf{Ord} \neq \emptyset$. We define $\mu := \min\{\alpha \in \mathbf{Ord} : \exists y \in x \setminus \mathbf{Ord} : \mathrm{rk}(y) = \alpha\}$. Equivalently, $\mu := \min(\mathrm{rk}(x \setminus \mathbf{Ord}))$. Let us fix $y \in x \setminus \mathbf{Ord}$ such that $\mathrm{rk}(y) = \mu$. We prove that $y \subset \mathbf{Ord}$. In fact, first of all, $y \in \mathtt{wf}$ (since $x \subset \mathtt{wf}$), hence $y \subset \mathtt{wf}$. If $z \in y$, then $\mathrm{rk}(z) \in \mathrm{rk}(y) = \mu$ (by proposition 13.8-i) and $z \in x$ by transitivity. By minimality of μ, we have $z \notin x \setminus \mathbf{Ord}$, thus z is an ordinal. This proves that $y \subset \mathbf{Ord}$. Since $y \in x$ and x is hereditarily transitive, y is transitive, hence it is an ordinal by lemma 13.35. This is a contradiction, since $y \in x \setminus \mathbf{Ord}$ by construction. □

Proposition 13.36 is reasonably false outside $\mathtt{wf}$. It is quite difficult to state such a result formally, since it is not clear what a non-well-founded set is. Anyway, we can imagine the following properly hereditarily transitive set: $x = \{y_n\}_{n \in \mathbb{Z}}$, where $y_n = \{y_m : m < n\}$. This means that $y_m \in y_n$ if and only if $m < n$, hence $(x, \in) \simeq (\mathbb{Z}, <)$, therefore $(x, \in)$ it is not a well-order. We can replace $(\mathbb{Z}, <)$ by any strictly ordered set $(I, \prec)$, by considering $x = \{y_i\}_{i \in I}$, with $y_i = \{y_j : j \prec i\}$. It follows that $(x, \in) \simeq (I, \prec)$.

13.6 WELL-FOUNDED SETS AND CARDINALS

Definition 13.37. We define the following function:

$$\begin{aligned} \beth \colon \mathbf{Ord} &\hookrightarrow \mathbf{Card}_\infty \\ \alpha &\mapsto \beth_\alpha, \end{aligned}$$

by transfinite recursion on ordinals as follows:

- $\beth_0 := \aleph_0$;
- $\beth_{s(\alpha)} := 2^{\beth_\alpha}$;
- if α is a limit ordinal, then $\beth_\alpha := \sup\{\beth_\beta : \beta \in \alpha\}$. ◇

In the third item, the supremum can be taken indifferently in **Ord** or in **Card**, since it is a cardinal anyway.

Proposition 13.38. *We have that:*

(i) $\beth$ *respects the order—that is,* $\alpha \in \beta$ *if and only if* $\beth_\alpha \in \beth_\beta$.

(ii) $\alpha \subset \aleph_\alpha \subset \beth_\alpha$ *for every* $\alpha \in \mathbf{Ord}$.

Item (i) immediately implies that $\beth$ *is injective.*

Proof. The same proof of proposition 12.25-(i, ii), replacing the function $\aleph$ by $\beth$, leads to i and $\alpha \subset \beth_\alpha$ for every $\alpha \in \mathbf{Ord}$. It remains to prove that $\aleph_\alpha \subset \beth_\alpha$. We apply transfinite induction on α. If $\alpha = 0$, then $\aleph_0 = \beth_0$ by definition. If $\aleph_\alpha \subset \beth_\alpha$, then $\aleph_{s(\alpha)} = \aleph_\alpha^+ \subset 2^{\aleph_\alpha} \subset 2^{\beth_\alpha} = \beth_{s(\alpha)}$, as required. If α is a limit ordinal and $\aleph_\beta \subset \beth_\beta$ for every $\beta \in \alpha$, then $\aleph_\alpha = \sup\{\aleph_\beta : \beta \in \alpha\} \subset \sup\{\beth_\beta : \beta \in \alpha\} = \beth_\alpha$. □

The next proposition shows that, assuming AC, the function $\beth$ describes the cardinality of the set of well-founded sets of a fixed infinite rank.

Proposition 13.39 (AC). *The cardinality of the set $R(\alpha)$, defined in 13.1, is the following one:*

- $|R(n)| = 2\hat{\ }n$ *for every $n \in \omega$, where $2\hat{\ }n := 0$ and $2\hat{\ }(n+1) := 2^{2\hat{\ }n}$*
- $|R(\omega + \alpha)| = \beth_\alpha$ *for every $\alpha \in$* **Ord**

Proof. The first item was proved in the resolution of exercise 12 in section 2.8. About the second one, we apply transfinite induction on α. If $\alpha = 0$, then by definition $R(\omega) := \bigcup\{R(n) : n \in \omega\}$. Since $R(n)$ is finite for every n and ω is denumerable, the set $R(\omega)$ is countable by proposition 10.22—that is, $|R(\omega)| \leq \aleph_0$. The reader can easily prove by induction that $2\hat{\ }n > n$ for every $n \in \omega$, hence by item i we get $|R(\omega)| > n$ for every $n \in \omega$—that is, $R(\omega)$ is infinite. It follows that $|R(\omega)| = \aleph_0 = \beth_0$, as required. If $|R(\omega + \alpha)| = \beth_\alpha$, then

$$|R(\omega + s(\alpha))| = |R(s(\omega + \alpha))| \overset{(\star)}{=} |\mathcal{P}(R(\omega + \alpha))| = 2^{|R(\omega+\alpha)|} \overset{(*)}{=} 2^{\beth_\alpha} = \beth_{s(\alpha)},$$

where $(\star)$ follows from definition 13.1 and $(*)$ follows from the inductive hypothesis. If α is a limit ordinal, then by definition $R(\alpha) = \bigcup\{R(\beta) : \beta \in \alpha\}$. Since $R(\beta) \subset R(\alpha)$ for every $\beta \in \alpha$, we have $\beth_\beta \leq |R(\alpha)|$ for every $\beta \in \alpha$, hence $\beth_\alpha = \sup\{\beth_\beta : \beta \in \alpha\} \leq |R(\alpha)|$. Conversely, by AC and the surjective function $\bar{\eta}$: $\bigsqcup\{R(\beta) : \beta \in \alpha\} \to \bigcup\{R(\beta) : \beta \in \alpha\}$, defined in (4.3), we have

$$\begin{aligned} |R(\alpha)| &\leq \left|\bigsqcup\{R(\beta) : \beta \in \alpha\}\right| \overset{(\star)}{=} \left|\bigsqcup\{\beth_\beta : \beta \in \alpha\}\right| \\ &\overset{(*)}{\leq} \left|\bigsqcup\{\beth_\alpha : \beta \in \alpha\}\right| = |\beth_\alpha \times \alpha| = \beth_\alpha \cdot \alpha \overset{(\#)}{=} \beth_\alpha, \end{aligned}$$

where $(\star)$ is due to the inductive hypothesis, $(*)$ to definition 13.37 and $(\#)$ to proposition 13.38-ii. □

Remark 13.40. The construction of `wf` has been performed without relying on the axiom of choice. In fact, the notion of ordinal number, used in definition 13.1, is well-defined independently of AC. Then, we assumed AC only in proposition 13.24 and in corollary 13.32, that provide an equivalent (but not necessary) characterisation of well-founded relations and sets. Hence, we can state that `wf` is our universe both assuming AC or not. Nevertheless, the structure of `wf` obviously depends on the acceptance of AC, since `wf` admits choice functions if AC holds, while it may not otherwise. That's because the structure of the class **Ord**, used to define `wf`, depends on AC; moreover, when we apply the power-set operation to a set x—in particular, when we define $R(s(\alpha)) := \mathcal{P}(R(\alpha))$—the structure of the set $\mathcal{P}(x)$ depends on AC too. For example, $\mathcal{P}(\mathcal{P}^*(x) \times x)$ contains a choice function for any x under AC, otherwise it may not. ♢

13.7 EXERCISES

1. Compute the rank of $\mathbb{N}$, $\mathbb{Z}$, $\mathbb{Q}$, $\mathbb{R}$, and $\mathbb{C}$ (see exercise 13.19).

2. Prove that, if x is a non-empty finite well-founded set, then $\mathrm{rk}(x)$ is a successor ordinal.

3. We say that a set is *hereditarily finite* if it is finite and, for every $n \in \mathbb{N}$, every element of $\bigcup^n x$ is finite. Assuming that x is well-founded, prove that the following conditions are equivalent:

(i) $\mathrm{rk}(x)$ is finite

(ii) x is hereditarily finite

(iii) the transitive closure $\bar{x}$ is finite

Remark: It follows that the finite sets in the hierarchy constructed in section 2.4 are all of the hereditarily finite sets in our universe.

4. Prove that the induction principle can be generalised to well-founded relations as follows. Let ρ be well-founded in x. *Hypothesis:* $a \subset x$ verifies the following condition for every $y \in x$: if, for every $z \in x$ such that $z \,\rho\, y$, we have $z \in a$, then $y \in a$. *Thesis:* $a = x$.

Remark: It is possible to generalise recursion too, but we omit the details here.

5. Prove that the following statements are equivalent:

(i) GCH (see section 12.7)

(ii) $\aleph_\alpha = \beth_\alpha$ for every $\alpha \in$ **Ord** (that is, $\aleph = \beth$ as functions)

(iii) $\beth$ is surjective (hence bijective)

Conclusions

The construction of well-founded sets completes the path presented in the preface, since the class `wf` is reasonably a good candidate to constitute the reduced universe we were looking for. Indeed, this class has been constructed on the basis of assumptions `(A1)`–`(A7)`, that are necessary in order to allow for the basic operations in ordinary mathematics. Hence, any universe that can host the whole mathematics has to include `wf`. Moreover, we have seen that every fundamental set-theoretical operation is well-defined in `wf`, thus we do not need anything more. This confirms the minimality of the universe of well-founded set. Since Russell's paradox and certain possible pathologies (like a set x such that $x = \{x\}$) are ruled out from `wf`, it is reasonable to hope that this reduced universe is coherent. As we anticipated in the preface, we cannot show any absolute proof of coherence, therefore we cannot exclude that certain pathologies or contradictions we are not aware of hold in `wf`. Nevertheless, we are not able to do anything better at the best of our knowledge.

The problem of coherence is not the only unsolved one. In fact, we have seen in section 12.7 that some meaningful statements are *undecidable* on the basis of the assumptions `(A1)`–`(A9)`. Let us analyse this issue more in detail.

UNDECIDABLE STATEMENTS

Let us consider the Continuum Hypothesis, that we called CH. What does it mean that it is "undecidable" under our assumptions? Basically, it means that there is no way to construct a proof of CH, nor of $\sim$CH, relying on `(A1)`–`(A9)`. Therefore, we cannot establish whether CH holds or not, unless new assumptions are made. How is it possible to prove concretely that CH is undecidable? In principle, we should show that, in the set of all possible proofs based on assumptions `(A1)`–`(A9)`, no one terminates with CH or with $\sim$CH. Nevertheless, this path seems quite difficult to follow, and it is ambiguous in the framework of naive set theory, since we have no formal notion of "proof". Hence, undecidability must be proven in another way.

Let us make a historical consideration. Plenty of mathematicians tried to prove that Euclid's fifth postulate (that we call P5) could be deduced from the first four ones (that we call P1–P4), until the discovery of non-Euclidean geometries in 19th century, which definitely showed the impossibility of such deduction. That's because a concrete realisation of hyperbolic geometry is a setting in which P1–P4 hold, but P5 fails. We say that such a setting is a *model* of P1–P4 and ~P5. On the contrary, a concrete realisation of Euclidean geometry is a model of P1–P5. Since both models are coherent, there is no general proof of P5 or ~P5 from P1–P4, otherwise one of the two models would be contradictory. Therefore, P5 is undecidable on the basis of P1–P4.

In the case of set theory, we use a similar argument—that is, we construct a model of `(A1)`–`(A9)` and CH on one side, and a model a model of `(A1)`–`(A9)` and ~CH on the other side, proving the undecidability of CH. Actually, the following example shows that this is not a new kind of argument for us, even if we had not focused on it explicitly before this chapter.

Example: Can we deduce assumption (A7) from assumptions (A1)–(A6) and (A8)–(A9)? In order to reply to this question, we call $\texttt{wf}_f$ the class of well-founded sets of finite rank, that coincides with the ones constructed in section 2.4. It is quite easy to verify that all of the assumptions `(A1)`–`(A6)` and `(A8)`–`(A9)` hold in $\texttt{wf}_f$, and ~`(A7)` holds too. On the contrary, in `wf` assumptions `(A1)`–`(A9)` hold. This proves that `(A7)` is undecidable on the basis of the other assumptions, hence, in order for it to hold, it must be assumed as a hypothesis.

The same kind of argument is used to prove that the axiom of choice is really an axiom. In section 9.1 we gave an intuitive justification of the independence of `(A8)` from the other assumptions, but, in order to make this justification more formal, we have to construct a model of `(A1)`–`(A7)`, `(A9)` and ~`(A8)`. Since `wf` is a model of `(A1)`–`(A9)`, the independence of `(A8)` follows. Unfortunately, in the case of `(A8)`, as well as for CH, GCH and ICC (see section 12.7), the construction of the desired models is not trivial at all, hence we cannot realise it in a basic course like the present one (see [6]). However, it is important to keep in mind that this is the kind of argument that concretely proves the independence of a certain statement with respect to certain assumptions.

MODELS OF SET THEORY

Some comments are in order about the notion of model in the framework of set theory. First of all, since `wf` is our universe, how can we construct different models of set theory? Are they distinct universes? The easiest answer is the following one: the distinct models we consider are *sub-universes* of `wf`. For example, in the case of assumption `(A7)`, we selected the two sub-universes $\mathtt{u}_1 = \texttt{wf}_f$ and $\mathtt{u}_2 = \texttt{wf}$. Similarly, in the case of CH, we select a sub-universe $\mathtt{u}_1 \subset \texttt{wf}$, in which `(A1)`–`(A9)` and CH hold, and a sub-universe $\mathtt{u}_2 \subset \texttt{wf}$, in which `(A1)`–`(A9)` and ~CH hold.

Is it all clear? Not exactly. In fact, what does it mean that $2^{\aleph_0} = \aleph_1$ in $\mathtt{u}_1$ and $2^{\aleph_0} > \aleph_1$ in $\mathtt{u}_2$? Shouldn't the statement $2^{\aleph_0} = \aleph_1$ be absolutely true or false, considering that its terms are well-defined? Moreover, even admitting that its truthfulness depends on the chosen sub-universe, is it true on false in the whole `wf`?

About the dependence on the model, first of all we observe that a bijection between $2^{\aleph_0}$ and $\aleph_1$ may belong to a sub-universe $\mathtt{u}_1$, but not to another sub-universe $\mathtt{u}_2$. Nevertheless, if this were the only possibility, then the bijection in $\mathtt{u}_1$ in particular would exist in wf, hence CH would be trivially true in wf. Unfortunately, the situation is more complicated. In fact, we know that, if the assumptions (A1)–(A9) hold in the sub-universes $\mathtt{u}_1, \mathtt{u}_2 \subset \mathtt{wf}$, then we can construct the classes of ordinals and cardinals in $\mathtt{u}_1$ and $\mathtt{u}_2$, but it is not necessary that a fixed cardinal is the same set in $\mathtt{u}_1$ and in $\mathtt{u}_2$. Hence, in both cases it can be different from the corresponding set in wf. Therefore, the existence of a bijection between $2^{\aleph_0}$ and $\aleph_1$ in a fixed sub-universe does not imply its existence in wf.

These considerations clarify that the undecidability of CH is not contradictory, but they do not establish whether CH holds in wf or not. The analogous problem can be formulated about GCH, ICC, and any other statement that is undecidable on the basis of (A1)–(A9). It is not easy to provide a clear solution. Let us briefly sketch why.

DOES THE UNIVERSE EXIST?

We ask the following ambiguous question: does the set-theoretical universe exist? There are at least two distinct viewpoints to be considered.

Platonic viewpoint. A so-called *platonist* believes that the universe of set-theory, that we identified with the universe of well-founded sets, actually exists in itself. Of course (s)he is not so naive to believe that it exists in the physical reality, but he thinks that it has a transcendental existence, that does not depend on our individual mind. Hence, while constructing such a universe, he thinks that we are exploring (not really constructing) a world that exists in itself and that we see with the eyes of our mind. In order to clarify this viewpoint, let us consider the axiom of choice. For a platonist, given a set x, a choice function $\xi\colon \mathcal{P}^*(x) \to x$ exists in the universe, hence the axiom of choice really holds in wf. It follows that, if we do not assume (A8), then we are only refraining from using choice functions, but we are not denying their existence. We can imagine that the universe is dark and it contains some switches to turn on the corresponding lights. Every time we make an assumption, we turn on a switch, so that we can see something we could not see before, but that it was there anyway. Hence, for example, assumption (A7) turns on the light on the set $\mathbb{N}$, assumption (A8) turns on the light on choice functions, and so on. The situation is different about assumption (A9), since it reduces the universe instead of extending it, but here we are thinking that only wf really exists, hence (A9) holds.

Following the platonic viewpoint, any statement is true or false in the existing universe. Hence, the fact that CH is undecidable from the assumptions we made simply means that (A1)–(A9) are not enough to turn on all of the necessary lights. Therefore, we have to look for other assumptions—that is, we have to turn on other lights—in order to discover whether a bijection between

$2^{\aleph_0}$ and $\aleph_1$ exists in **wf** or not. It follows that the undecidability of certain statements is only due to our ignorance, that hopefully will be overcome in future.

Anti-Platonic viewpoint. An alternative viewpoint consists in not thinking that the universe of set theory really exists, but that we are choosing how to construct it. Any other construction has the same dignity of the one we presented in this course, hence we are not enlightening any existing universe, but we are freely producing it through assumptions (A1)–(A9). Therefore, for example, we have no way to establish if choice functions exist or not: if we assume (A8), then we put them in our universe, otherwise we do not (at least not all of them), but there is no "real universe" with or without them.

Following this viewpoint, CH is true or false depending on what we choose: the fact that it is undecidable on the basis of (A1)–(A9) only means that in no one of these nine steps we put in the universe a bijection between $2^{\aleph_0}$ and $\aleph_1$, nor a cardinal between them. In principle, we can choose CH or $\sim$CH as our 10th assumption, so that we trivially overcome its undecidability. Why didn't we make such a choice? Simply because we had no reason to make it. Assumptions (A1)–(A6) were motivated by the practical necessity of allowing for the operations we need in ordinary mathematics; assumption (A7) was stated in order to make the set $\mathbb{N}$ available within the universe, and so on. On the contrary, we have no reason to choose between CH and $\sim$CH, hence we leave them undecidable, unless in future we find a reason to take a decision.

The problem of the platonic viewpoint is that, if we adopt it, we affirm the existence of an objective set-theoretical universe independently of our mind, but we make this assertion with our mind and we see this universe only with the eyes of our own mind. Hence, thinking of a universe outside our mind is similar to thinking of a reality before time, forgetting that before time there is no "before". On the contrary, the problem of the anti-platonic viewpoint is that, in order to construct a universe, we need a background in which this construction is performed. For example, if we want to construct a building, we need a plot of land, and we must rely on the physical laws to avoid that it falls down. Here the plot of land is provided by the intuitive notions we started from, and the physical laws are replaced by the intuitive laws of logic. This background cannot be the result of a construction, since it is the environment that makes any construction possible. Hence, it seems we only hide platonism under the carpet.

We are not going to carry on this philosophical discussion here, since it is not the focus of the present course. We just say for the sake of honesty that the author of this book usually thinks in a platonic way, but being aware of the problems mentioned above, that should be considered the starting point of a much deeper analysis.

APPENDIX A

Hereditary Sets

We saw in chapter 2 that in our universe every object is a set. It follows that the elements of a set in the universe are sets, whose elements are sets, and so on at any order. A set of this kind is called *hereditary*. In this appendix, we state the definition of hereditary set in a more formal way.

A.1 DEFINITION OF HEREDITARY SET

Let us consider sets whose elements can be objects of any kind. In this framework, we define the unary operation of union as follows.

Definition A.1. The *union* of a set x is the set

$$\bigcup x := \{z : \exists y \in x : (y \text{ is a set}) \wedge (z \in y)\}. \qquad \diamondsuit$$

This means that $\bigcup x$ contains the elements of the elements of x *that are sets*. Then, we can define iterated union as in section 4.4.1.

Definition A.2. A set x is called *hereditary* if, for every $n \in \mathrm{N}$, every element of $\bigcup^n x$ is a set. $\diamondsuit$

Proposition A.3. *Given a set x, the following conditions are equivalent:*

(i) x is hereditary;

(ii) every element of x is a hereditary set;

(iii) for every $n \in \mathrm{N}$, every element of $\bigcup^n x$ is a hereditary set.

Proof. (i) $\Rightarrow$ (iii) Let us suppose that there exist $n \in \mathrm{N}$ and $y \in \bigcup^n x$ such that y is not hereditary. This means that there exist $m \in \mathrm{N}$ and $z \in \bigcup^m y$ such that z is not a set. We deduce from lemma 4.12-iii that $z \in \bigcup^{n+m+1} x$, that is a contradiction, since every element of $\bigcup^{n+m+1} x$ must be a set by definition A.2. (iii) $\Rightarrow$ (ii) Obvious by choosing $n = 0$. (ii) $\Rightarrow$ (i) We have to prove that, for every $n \in \mathrm{N}$, every element of $\bigcup^n x$ is a set. If $n = 0$, then the thesis holds

DOI: 10.1201/9781003565291-A

by hypothesis, since a hereditary set is in particular a set. If $n > 0$—that is, $n = m + 1$—let us suppose that $z \in \bigcup^n x = \bigcup^{m+1} x$. By lemma 4.12-iii (with $n = 0$ in the statement) there exists a set $y \in x$ such that $z \in \bigcup^m y$. The set y is hereditary by hypothesis, hence z is a set by definition A.2. □

Corollary A.4. *If x is a hereditary set, then*

- *$\bigcup^n x$ is a hereditary set for every $n \in \mathrm{N}$; and*
- *if $y \subset x$, then y is hereditary too.*

Proof. By proposition A.3, every element of $\bigcup^n x$ and y is a hereditary set, hence $\bigcup^n x$ and y are hereditary. □

We impose that every object in our universe is a set. It is highly reasonable to state that, if x is a set in our universe, then the elements of x are objects of our universe too, hence they must be sets. By iterating this argument, we deduce that every element of $\bigcup^n x$, for every $n \in \mathrm{N}$, must be a set, thus x must be hereditary. Again, this implies that every element of $\bigcup^n x$ must be hereditary, but this is guaranteed by proposition A.3. Thus, *assuming that every object is a set is equivalent to assuming that every object is a hereditary set.*

A.2 REDUCING THE UNIVERSE

We saw in chapter 2 that, in order to eliminate pathologies and contradictions, we "restrict" the universe of hereditary sets to a suitable sub-universe. This process was completed in chapter 13 by constructing the universe of well-founded sets. Since we impose that every object of the universe is a well-founded set, every element of $\bigcup^n x$ must be a well-founded set too. This is guaranteed by the following proposition, analogous to A.3.

Proposition A.5. *Given a set x, the following conditions are equivalent:*

(i) x is well-founded;

(ii) every element of x is well-founded;

(iii) for every $n \in \mathrm{N}$, every element of $\bigcup^n x$ is well-founded.

Proof. (i) ⇔ (ii) It is the statement of proposition 13.10. (iii) ⇒ (ii) Obvious by choosing $n = 0$. (ii) ⇒ (iii) Induction on n. If $n = 0$, then iii coincides with ii. If iii holds for n and $y \in \bigcup^{n+1} x$, then by lemma 4.12-iii there exists $z \in \bigcup^n x$ such that $y \in z$. By hypothesis, z is well-founded, hence y is well-founded too by proposition 13.10. □

APPENDIX B

Fundamental Properties

As we anticipated at the end of chapter 5, the main properties of the operations among sets have been stated along the text and the exercises, so that a unitary exposition is missing up to now. Here we fill this gap, also adding some statements that did not fit in the previous chapters.

B.1 BASIC N-ARY OPERATIONS

We summarise the relevant properties of the basic n-ary operations among sets.

B.1.1 Unary and Binary Operations

We start from the operations introduced in definition 2.8. In the following statements, whenever we apply the operation of complement, we assume that it is applied within a bigger set X, that contains all of the sets involved in the operation.

Proposition B.1. *The unary operations of complement and power set satisfy the following properties:*

- $(x^{\complement})^{\complement} = x$
- $\emptyset^{\complement} = X$ *and* $X^{\complement} = \emptyset$
- $x \cap x^{\complement} = \emptyset$ *and* $x \cup x^{\complement} = X$
- $(x \subset y) \Leftrightarrow (y^{\complement} \subset x^{\complement})$
- $(y \subset x) \Rightarrow \big(\mathcal{P}(y) \subset \mathcal{P}(x)\big)$
- $\mathcal{P}^*(x^{\complement}) \subset \mathcal{P}(x)^{\complement}$, *considering* $\mathcal{P}(x) \subset \mathcal{P}(X)$

The proof can be found in the resolutions of the exercises in section 2.8. In particular, the first four items can be found in exercise 5-(1–4), the fifth one

DOI: 10.1201/9781003565291-B

in exercise 6-5, and the last one is a particular case of exercise 7-3, replacing x and y in the latter respectively by X and x.

Proposition B.2. *Binary union and intersection satisfy the following properties:*

- *they are associative and commutative, that is:*
  - $(x \cup y) \cup z = x \cup (y \cup z)$ *and* $(x \cap y) \cap z = x \cap (y \cap z)$
  - $x \cup y = y \cup x$ *and* $x \cap y = y \cap x$
- *each one is distributive over the other, that is:*
  - $(x \cup y) \cap z = (x \cap z) \cup (y \cap z)$
  - $(x \cap y) \cup z = (x \cup z) \cap (y \cup z)$
- $x \cup \emptyset = x \cup x = x \cap x = x$ *and* $x \cap \emptyset = \emptyset$
- $x \cap y \subset x \subset x \cup y$ *(it follows that* $x \cap y \subset x \cup y$*)*
- $x \cap (x \cup y) = x \cup (x \cap y) = x$
- $(y \subset x) \Leftrightarrow (x \cup y = x) \Leftrightarrow (x \cap y = y)$

In section 2.8, the first item can be found in exercise 1, the second in exercise 3, and the last four in exercise 4-(1–4). Now we show the main properties that link binary union and intersection on one side and unary operations on the other side.

Proposition B.3. *Binary union and intersection are related to unary operations by the following properties:*

- $(x \cup y)^{\complement} = x^{\complement} \cap y^{\complement}$
- $(x \cap y)^{\complement} = x^{\complement} \cup y^{\complement}$
- $\mathcal{P}(x \cup y) \supset \mathcal{P}(x) \cup \mathcal{P}(y)$
- $\mathcal{P}(x \cap y) = \mathcal{P}(x) \cap \mathcal{P}(y)$

The first two items are called De Morgan's laws.

In section 2.8, the first two items can be found in exercise 5-5, and the last two in exercise 7-(1, 2), with $n = 2$ in the latter. Now we show the main properties of binary Cartesian product and disjoint union.

Proposition B.4. *Binary Cartesian product and disjoint union satisfy the following properties:*

- *$((x, \mathrm{id}_x) \times (y, \mathrm{id}_y)) \times (z, \mathrm{id}_z)$ and $(x, \mathrm{id}_x) \times ((y, \mathrm{id}_y) \times (z, \mathrm{id}_z))$ are both Cartesian products of x, y, and z, hence they are related by a unique natural bijection; similarly, $((x, \mathrm{id}_x) \sqcup (y, \mathrm{id}_y)) \sqcup (z, \mathrm{id}_z)$ and $(x, \mathrm{id}_x) \sqcup ((y, \mathrm{id}_y) \sqcup (z, \mathrm{id}_z))$ are both disjoint unions of x, y, and z, hence they are related by a unique natural bijection.*

- *$(x, \mathrm{id}_x) \times (y, \mathrm{id}_y)$ is a Cartesian product of x and y and $(y, \mathrm{id}_y) \times (x, \mathrm{id}_x)$ is a Cartesian product of y and x, hence they are related by a unique natural bijection that respects the order; similarly, $(x, \mathrm{id}_x) \sqcup (y, \mathrm{id}_y)$ is a disjoint union of x and y and $(y, \mathrm{id}_y) \sqcup (x, \mathrm{id}_x)$ is a disjoint union of y and x, hence they are related by a unique natural bijection that respects the order.*

- *Cartesian product is distributive over disjoint union—that is, the function $(x \times z) \sqcup (y \times z) \to (x \sqcup y) \times z$, naturally induced by the two universal properties, is a bijection, its inverse being denoted by*

$$\Theta_{x,y,z} \colon (x \sqcup y) \times z \xrightarrow{\approx} (x \times z) \sqcup (y \times z).$$

A sketch of the proof can be found in section 5.4. The reader can find other properties of these operations in section 2.8 and in chapter 5, some of them involving set difference and symmetric difference too, but here we just summarised the main ones.

B.1.2 n-Ary Operations

We now consider the operations introduced in definition 2.9.

Proposition B.5. *The following properties hold:*

- *Binary union and intersection, extended to n-ary operations by associativity, coincide with the corresponding operations directly defined as n-aries.*

- *By commutativity, the order of the n entries is immaterial.*

- *Each one is distributive over the other, that is:*
  - $(x_1 \cup \ldots \cup x_n) \cap z = (x_1 \cap z) \cup \ldots \cup (x_n \cap z)$
  - $(x_1 \cap \ldots \cap x_n) \cup z = (x_1 \cup z) \cap \ldots \cap (x_n \cup z)$

The proof can be found in the resolution of exercises 2 and 3 of section 2.8.

Proposition B.6. *n-Ary union and intersection are related to unary operations by the following properties:*

- $(x_1 \cup \ldots \cup x_n)^{\complement} = x_1^{\complement} \cap \ldots \cap x_n^{\complement}$

- $(x_1 \cap \ldots \cap x_n)^{\complement} = x_1^{\complement} \cup \ldots \cup x_n^{\complement}$
- $\mathcal{P}(x_1 \cup \ldots \cup x_n) \supset \mathcal{P}(x_1) \cup \ldots \cup \mathcal{P}(x_n)$
- $\mathcal{P}(x_1 \cap \ldots \cap x_n) = \mathcal{P}(x_1) \cap \ldots \cap \mathcal{P}(x_n)$

The first two items are called De Morgan's laws.

The proof can be found in the resolution of exercises 5-7 and 7-(1, 2) of section 2.8.

Proposition B.7. *The following properties hold:*

- *Binary Cartesian product and disjoint union, extended to n-ary operations by associativity, essentially coincide with the corresponding operations directly defined as n-aries, up to a unique bijection provided by the corresponding universal property.*
- *By commutativity, the order of the n entries is immaterial, up to a unique bijection.*
- *Cartesian product is distributive over disjoint union—that is, the function $(x_1 \times z) \sqcup \ldots \sqcup (x_n \times z) \to (x_1 \sqcup \ldots \sqcup x_n) \times z$, naturally induced by the two universal properties, is a bijection, its inverse being denoted by*

$$\Theta_{x_1,\ldots,x_n,z} \colon (x_1 \sqcup \ldots \sqcup x_n) \times z \xrightarrow{\approx} (x_1 \times z) \sqcup \ldots \sqcup (x_n \times z).$$

The proof follows from the content of section 5.4, but we omit the unnecessary technical details here.

B.2 OPERATIONS ON SETS AND INDEXED SETS

Various properties of the fundamental operations on family of sets and indexed sets have been stated in section 4.5. Here we recall them and add some relevant ones. We mainly consider indexed sets, since the statements are often more natural in this framework. The analogous formulations for families of sets can be thought of as a particular case (see remark 4.5), but we will state them explicitly anyway in the exercises at the end of the chapter or we will provide suitable references to the exercises above.

Proposition B.8. *The operations of union and intersection of an indexed set $x = \{x_i\}_{i \in I}$ satisfy the following properties:*

- *generalised associativity—that is, if $p \subset \mathcal{P}(I)$ is a partition of I, then:*
  - $\bigcup_{i \in I} x_i = \bigcup_{a \in p} \bigcup_{i \in a} x_i$
  - $\bigcap_{i \in I} x_i = \bigcap_{a \in p} \bigcap_{i \in a} x_i$

- *generalised distributivity over each other—that is, for every set y:*
  - $y \cap \bigcup_i x_i = \bigcup_i (y \cap x_i)$
  - $y \cup \bigcap_i x_i = \bigcap_i (y \cup x_i)$
- $\bigcap \bigcup_i x_i \subset \bigcup \bigcap_i x_i$

In section 4.5, the first item can be found in exercise 18, the second one in exercise 14-(3, 4), and the last one in exercise 13.

Exercise B.9. Show that associativity and distributivity of binary union and intersection can be described as particular cases of the corresponding generalised versions, shown in proposition B.8.

In this case, there is no reason to deal with commutativity, since the set of indices I is not supposed to be ordered.

Exercise B.10. Show that commutativity of binary union and intersection can be deduced from the properties of union and intersection of indexed sets, like associativity and distributivity.

Proposition B.11. *Union and intersection are related to unary operations by the following properties:*

- $(\bigcup_i x_i)^{\complement} = \bigcap_i x_i^{\complement}$
- $(\bigcap_i x_i)^{\complement} = \bigcup_i x_i^{\complement}$
- $\mathcal{P}(\bigcup_i x_i) \supset \bigcup_i \mathcal{P}(x_i)$
- $\mathcal{P}(\bigcap_i x_i) = \bigcap_i \mathcal{P}(x_i)$
- $x = \bigcup \mathcal{P}(x)$
- $x \subset \mathcal{P}(\bigcup x)$

The first two items are called De Morgan's laws.

In section 4.5, the first two items can be found in exercise 16, the third and the fourth in exercise 15-(2, 3), and the last two in exercise 7-(1, 2).

About Cartesian product and disjoint union, we can state the analogous properties of generalised associativity and distributivity, based on the universal property. Nevertheless, as in the case of n-ary operations, the rigorous formulation of these properties rely on the definitions of Cartesian product and disjoint union acting respectively on Cartesian products and disjoint unions. Again, we omit the technical details.

Proposition B.12. *The operations of Cartesian product and disjoint union of an indexed set $x = \{x_i\}_{i \in I}$ satisfy the following properties:*

- $\prod_{i\in I}(x_i, \{\mathrm{id}_{x_i}\})$ *and* $\prod_{a\in p}\prod_{i\in a}(x_i, \{\mathrm{id}_{x_i}\})$ *are both Cartesian products of* x*, hence they are related by a unique natural bijection.*

- $\bigsqcup_{i\in I}(x_i, \{\mathrm{id}_{x_i}\})$ *and* $\bigsqcup_{a\in p}\bigsqcup_{i\in a}(x_i, \{\mathrm{id}_{x_i}\})$ *are both disjoint unions of* x*, hence they are related by a unique natural bijection.*

- *Cartesian product satisfies generalised distributivity over disjoint union—that is, for every set* y*, the function* $\bigsqcup_i(y \times x_i) \to y \times \bigsqcup_i x_i$*, naturally induced by the two universal properties, is a bijection, its inverse being denoted by:*

$$\Theta_{x,y} \colon y \times \bigsqcup_i x_i \xrightarrow{\approx} \bigsqcup_i (y \times x_i).$$

Again, there is no reason to deal with commutativity, since the set of indices I is not supposed to be ordered.

In the case of (non-indexed) sets, the first item of proposition B.8 can be found in the first two items of exercise 4 in section 4.5. In fact, given $p \in \mathcal{P}(x)$, from the latter exercise we get $\bigcup\bigcup p = \bigcup\bar{\bigcup} p$—that is, $\bigcup x = \bigcup\bar{\bigcup} p$, where $\bar{\bigcup} p = \{\bigcup a : a \in p\}$. The same holds for intersection. The second item of proposition B.8 can be found in the last two items of exercise 6 and the third item in exercise 5. About proposition B.11, the first two items can be found in exercise 8; the third and fourth ones can be found in the last two items of exercise 7; and the last two items are already formulated for a (non-indexed) set x.

Exercise B.13. State the proposition analogous to B.12 in the framework of (non-indexed) sets.

B.3 RELATIONS AND FUNCTIONS

We recall that a function $f \colon x \to y$ induces the following two functions: $\dot{f} \colon \mathcal{P}(x) \to \mathcal{P}(y)$, $a \mapsto \{f(z) : z \in a\}$, and $\ddot{f} \colon \mathcal{P}(y) \to \mathcal{P}(x)$, $b \mapsto \{z \in x : f(z) \in b\}$. We summarise the basic properties of $\dot{f}$ and $\ddot{f}$, that have been stated partially in section 2.8 and partially in section 4.5. Moreover, we show some properties that we could not state in chapter 3, since they involve the operations introduced in chapter 4.

Proposition B.14. *The functions* $\dot{f}$ *and* $\ddot{f}$ *satisfy the following fundamental properties:*

- $a \subset \ddot{f}(\dot{f}(a))$ *for every* $a \in \mathcal{P}(x)$
- $\dot{f}(\ddot{f}(b)) \subset b$ *for every* $b \in \mathcal{P}(y)$
- $a = \ddot{f}(\dot{f}(a))$ *for every* $a \in \mathcal{P}(x)$ *if and only if* f *is injective*
- $\dot{f}(\ddot{f}(b)) = b$ *for every* $b \in \mathcal{P}(y)$ *if and only if* f *is surjective*

- $\dot{f}(a_1 \cup a_2) = \dot{f}(a_1) \cup \dot{f}(a_2)$ *and* $\dot{f}(a_1 \cap a_2) \subset \dot{f}(a_1) \cap \dot{f}(a_2)$ *for every* $a_1, a_2 \subset x$
- $\dot{f}(a_1 \cap a_2) = \dot{f}(a_1) \cap \dot{f}(a_2)$ *for every* $a_1, a_2 \subset x$ *if and only if* f *is injective*
- $\ddot{f}(b_1 \cup b_2) = \ddot{f}(b_1) \cup \ddot{f}(b_2)$ *and* $\ddot{f}(b_1 \cap b_2) = \ddot{f}(b_1) \cap \ddot{f}(b_2)$ *for every* $b_1, b_2 \subset y$
- $\ddot{f}(y \setminus b) = x \setminus \ddot{f}(b)$ *for every* $b \subset y$
- $\dot{f}(x \setminus a) \subset y \setminus \dot{f}(a)$ *for every* $a \subset x$ *if and only if* f *is injective*
- $y \setminus \dot{f}(a) \subset \dot{f}(x \setminus a)$ *for every* $a \subset x$ *if and only if* f *is surjective*

Moreover:

- $\dot{f}$ *is injective if and only if* f *is injective*
- $\dot{f}$ *is surjective if and only if* f *is surjective*
- $\ddot{f}$ *is injective if and only if* f *is surjective*
- $\ddot{f}$ *is surjective if and only if* f *is injective*
- *if* f *is bijective, then* $\ddot{f} = (\dot{f})^{-1} = (f^{-1})^{\cdot}$ *and* $\dot{f} = (\ddot{f})^{-1} = (f^{-1})^{\cdot\cdot}$

The proof can be found in the resolution of exercises 7–10 of section 3.5. We now consider two indexed sets $a = \{a_i\}_{i \in I}$ e $b = \{b_j\}_{j \in J}$ such that $a \subset \mathcal{P}(x)$ and $b \subset \mathcal{P}(y)$. We get the indexed sets $\dot{f}(a) = \{\dot{f}(a_i)\}_{i \in I}$ and $\ddot{f}(b) = \{\ddot{f}(b_j)\}_{j \in J}$.

Proposition B.15. *We have that:*

- $\dot{f}(\bigcup_i a_i) = \bigcup_i \dot{f}(a_i)$ *and* $\dot{f}(\bigcap_i a_i) \subset \bigcap_i \dot{f}(a_i)$ *for every* $a = \{a_i\}_{i \in I}$ *such that* $a \subset \mathcal{P}(x)$ *as a set.*
- $\dot{f}(\bigcap_i a_i) = \bigcap_i \dot{f}(a_i)$ *for every* $a = \{a_i\}_{i \in I}$ *such that* $a \subset \mathcal{P}(x)$ *as a set if and only if* f *is injective.*
- $\ddot{f}(\bigcup_j b_j) = \bigcup_j \ddot{f}(b_j)$ *and* $\ddot{f}(\bigcap_j b_j) = \bigcap_j \ddot{f}(b_j)$ *for every* $b = \{b_j\}_{j \in J}$ *such that* $b \subset \mathcal{P}(y)$ *as a set.*

The proof can be found in the resolution of exercise 17 in section 4.5.

Formulations of Recursion

In this appendix we state and prove the various formulations of the recursion principle that are usually applied in mathematics.

C.1 FIRST FORM

The basic formulation of the first form of recursion is the statement of proposition 7.8, but it may happen that the recursive definition of $f(\boldsymbol{s}(\boldsymbol{n}))$ depends on $\boldsymbol{n}$ too, not only on $f(\boldsymbol{n})$ (see exercise C.5 below). This seems a stronger formulation, but it can be deduced from the previous one as follows.

Corollary C.1. *Given:*

- *a class* c*;*
- *a fixed element* $a \in \mathsf{c}$*;*
- *a function* $h\colon \mathbb{N} \times \mathsf{c} \to \mathsf{c}$*;*

there exists a unique function $f\colon \mathbb{N} \to \mathsf{c}$ *such that:*

I. $f(\mathbf{0}) = a$*;*

II. $f(\boldsymbol{s}(\boldsymbol{n})) = h(\boldsymbol{n}, f(\boldsymbol{n}))$ *for every* $\boldsymbol{n} \in \mathbb{N}$.

Proof. We define the function

$$\begin{aligned} h'\colon \mathbb{N} \times \mathsf{c} &\to \mathbb{N} \times \mathsf{c} \\ (\boldsymbol{n}, w) &\mapsto (\boldsymbol{s}(\boldsymbol{n}), h(\boldsymbol{n}, w)). \end{aligned} \tag{A}$$

By applying proposition 7.8, replacing c by $\mathbb{N} \times \mathsf{c}$ in the statement, we construct $f'\colon \mathbb{N} \to \mathbb{N} \times \mathsf{c}$ such that:

I'. $f'(\mathbf{0}) = (\mathbf{0}, a)$;

II'. $f'(\boldsymbol{s}(\boldsymbol{n})) = h'(f'(\boldsymbol{n}))$ for every $\boldsymbol{n} \in \mathbb{N}$.

DOI: 10.1201/9781003565291-C

Now we set $f := \pi \circ f'$, where $\pi\colon \mathbb{N} \times \mathsf{c} \to \mathsf{c}$ is the natural projection. We prove by induction that

$$f'(\boldsymbol{n}) = (\boldsymbol{n}, f(\boldsymbol{n})). \tag{B}$$

In fact, by definition of f, the second component of $f'(\boldsymbol{n})$ is $f(\boldsymbol{n})$. About the first component, for $\boldsymbol{n} = 0$, the result immediately follows from I'. Supposing that (B) holds for $\boldsymbol{n}$, we get

$$f'(\boldsymbol{s}(\boldsymbol{n})) \overset{\text{II'}}{=} h'(f'(\boldsymbol{n})) \overset{\text{(B)}}{=} h'(\boldsymbol{n}, f(\boldsymbol{n})) \overset{\text{(A)}}{=} (\boldsymbol{s}(\boldsymbol{n}), h(\boldsymbol{n}, f(\boldsymbol{n}))), \tag{C}$$

proving (B) for $\boldsymbol{s}(\boldsymbol{n})$ about the first component too. Finally, let us prove I and II. The former immediately follows from I' and (B). The latter follows by comparing (B), replacing $\boldsymbol{n}$ by $\boldsymbol{s}(\boldsymbol{n})$ in the statement, and (C). □

It is also possible to define recursively a function $f(z, \boldsymbol{n})$, where z ranges within a fixed class d (as in the definition of sum, product, and power; see section 7.2.3). In this case, the function h may depend on z too. We get the following statement.

Corollary C.2. *Given:*

- *two classes* c *and* d*;*
- *a function* $g\colon \mathsf{d} \to \mathsf{c}$*;*
- *a function* $h\colon \mathsf{d} \times \mathbb{N} \times \mathsf{c} \to \mathsf{c}$*;*

there exists a unique function $f\colon \mathsf{d} \times \mathbb{N} \to \mathsf{c}$ *such that:*

I. $f(z, \mathbf{0}) = g(z)$ *for every* $z \in \mathsf{d}$*;*

II. $f(z, \boldsymbol{s}(\boldsymbol{n})) = h(z, \boldsymbol{n}, f(z, \boldsymbol{n}))$ *for every* $z \in \mathsf{d}$ *and* $\boldsymbol{n} \in \mathbb{N}$.

Proof. For every $z \in \mathsf{d}$ fixed, we define $h_z\colon \mathbb{N} \times \mathsf{c} \to \mathsf{c}$, $(\boldsymbol{n}, w) \mapsto h(z, \boldsymbol{n}, w)$, and $f_z\colon \mathbb{N} \to \mathsf{c}$, $\boldsymbol{n} \mapsto f(z, \boldsymbol{n})$. In this way, conditions I and II can be expressed as $f_z(\mathbf{0}) = g(z)$ and $f_z(\boldsymbol{s}(\boldsymbol{n})) = h_z(\boldsymbol{n}, f_z(\boldsymbol{n}))$. Hence, by applying corollary C.1 (replacing a, h and f respectively by $g(z)$, h_z and f_z in the statement), we get f_z. Since this construction holds for every $z \in \mathsf{d}$ fixed, we get f defined by $f(z, \boldsymbol{n}) := f_z(\boldsymbol{n})$. □

The instance of the function h in condition II may depend not only on z and $f(z, \boldsymbol{n})$, but, more generally, on z and the whole function $w \mapsto f(w, \boldsymbol{n})$ (see exercise C.6 below). In this case, the domain of h is $\mathsf{d} \times \mathbb{N} \times \mathrm{Func}(\mathsf{d}, \mathsf{c})$. Nevertheless, exercise 7 in section 6.5 shows that, in order for this domain to be a well-defined class, we must suppose that d is a set. Summarizing, we can state recursion as follows.

- We fix a set x and a class c, and we establish what is $f(z, \mathbf{0}) \in \mathsf{c}$ for every $z \in x$. This means that $f(z, \mathbf{0}) := g(z)$, where $g\colon x \to \mathsf{c}$ is a given function.

- Assuming to know what is $f(z, \boldsymbol{n})$ for every $z \in x$, we define $f(z, \boldsymbol{s}(\boldsymbol{n}))$ for every $z \in x$. This can be formalised as follows. We consider the function $f_{\boldsymbol{n}}: x \to \mathtt{c}$, $z \mapsto f(z, \boldsymbol{n})$, that is known by hypothesis. Then we define $f(z, \boldsymbol{s}(\boldsymbol{n})) := h(z, \boldsymbol{n}, f_{\boldsymbol{n}})$, where h is a given function depending on z, $\boldsymbol{n}$ and $f_{\boldsymbol{n}}$. Since $f_{\boldsymbol{n}} \in \mathrm{Func}(x, \mathtt{c})$, it follows that $h: x \times \mathbb{N} \times \mathrm{Func}(x, \mathtt{c}) \to \mathtt{c}$.

We get the following statement.

Corollary C.3. *Given:*

- *a class* $\mathtt{c}$ *and a set* x*;*
- *a function* $g: x \to \mathtt{c}$*;*
- *a function* $h: x \times \mathbb{N} \times \mathrm{Func}(x, \mathtt{c}) \to \mathtt{c}$*;*

there exists a unique function $f: x \times \mathbb{N} \to \mathtt{c}$ *such that, setting* $f_{\boldsymbol{n}}: x \to \mathtt{c}$, $z \mapsto f(z, \boldsymbol{n})$, *we have:*

I. $f_{\mathbf{0}} = g$*;*

II. $f_{\boldsymbol{s}(\boldsymbol{n})}(z) = h(z, \boldsymbol{n}, f_{\boldsymbol{n}})$ *for every* $z \in x$ *and* $\boldsymbol{n} \in \mathbb{N}$.

Proof. In the statement, we have $h \in \mathrm{Func}(x \times \mathbb{N} \times \mathrm{Func}(x, \mathtt{c}), \mathtt{c})$ and $f \in \mathrm{Func}(x \times \mathbb{N}, \mathtt{c})$. Since cartesian product is commutative up to canonical bijection, we think of $h \in \mathrm{Func}(\mathbb{N} \times \mathrm{Func}(x, \mathtt{c}) \times x, \mathtt{c})$ and $f \in \mathrm{Func}(\mathbb{N} \times x, \mathtt{c})$. By applying exercise 9 in section 6.5, we get $h \in \mathrm{Func}(\mathbb{N} \times \mathrm{Func}(x, \mathtt{c}), \mathrm{Func}(x, \mathtt{c}))$ and $f \in \mathrm{Func}(\mathbb{N}, \mathrm{Func}(x, \mathtt{c}))$, i.e. $h: \mathbb{N} \times \mathrm{Func}(x, \mathtt{c}) \to \mathrm{Func}(x, \mathtt{c})$ and $f: \mathbb{N} \to \mathrm{Func}(x, \mathtt{c})$. Moreover, $g \in \mathrm{Func}(x, \mathtt{c})$. Hence, we can apply corollary C.1, replacing $\mathtt{c}$ and a respectively by $\mathrm{Func}(x, \mathtt{c})$ and g in the statement. □

In proposition 7.8 and in corollaries C.1, C.2, and C.3, we could replace the fixed class $\mathtt{c}$ by the whole universe $\mathtt{u}$ without loss of generality. In fact, considering for example $h: \mathtt{c} \to \mathtt{c}$ in 7.8, we can we trivially extend it to $h': \mathtt{u} \to \mathtt{u}$ by declaring $h'(x) := \emptyset \ \forall x \notin \mathtt{c}$. By 7.8 we get $f': \mathbb{N} \to \mathtt{u}$. Since $h'(x) = h(x) \in \mathtt{c}$ for every $x \in \mathtt{c}$, it follows that $\mathrm{Im}(f') \subset \mathtt{c}$, hence we set $f := f' \|_{\mathtt{c}}$. An analogous consideration holds about C.1, C.2, and C.3. Nevertheless, we consider more natural to fix the class $\mathtt{c}$, since in most of the applications the function f has a fixed codomain, smaller than $\mathtt{u}$.

Exercise C.4. Show in detail the recursive definition of sum, product, and power in $\mathbb{N}$ using corollary C.2.

Exercise C.5. Apply corollary C.1 to define $\boldsymbol{n}!$ recursively.

Exercise C.6. Apply corollary C.3 to define $\binom{\boldsymbol{n}}{\boldsymbol{k}}$ recursively as $\binom{\boldsymbol{n}}{\mathbf{0}} := \mathbf{1}$ and $\binom{\boldsymbol{s}(\boldsymbol{n})}{\boldsymbol{s}(\boldsymbol{k})} := \frac{\boldsymbol{s}(\boldsymbol{n}) \cdot \binom{\boldsymbol{n}}{\boldsymbol{k}}}{\boldsymbol{s}(\boldsymbol{k})}$.

C.2 SECOND FORM

The formulations of the recursion principle considered above hold about the second form too. We get the following statements.

Corollary C.7. *Given:*

- *two classes* c *and* d*;*
- *a function* $g\colon \mathsf{d} \to \mathsf{c}$*;*
- *a function* $h\colon \mathsf{d} \times \mathrm{FSeq}(\mathsf{c}) \to \mathsf{c}$*;*

there exists a unique function $f\colon \mathsf{d} \times \mathbb{N} \to \mathsf{c}$ *such that:*

I. $f(z, \mathbf{0}) = g(z)$ *for every* $z \in \mathsf{d}$*;*

II. $f(z, \boldsymbol{s}(\boldsymbol{n})) = h(z, (f(z, \mathbf{0}), \ldots, f(z, \boldsymbol{n})))$ *for every* $z \in \mathsf{d}$ *and* $\boldsymbol{n} \in \mathbb{N}$.

Corollary C.8. *Given:*

- *a class* c *and a set* x*;*
- *a function* $g\colon x \to \mathsf{c}$*;*
- *a function* $h\colon x \times \mathrm{FSeq}(\mathrm{Func}(x, \mathsf{c})) \to \mathsf{c}$*;*

there exists a unique function $f\colon x \times \mathbb{N} \to \mathsf{c}$ *such that, setting* $f_{\boldsymbol{n}}\colon x \to \mathsf{c}$, $z \mapsto f(z, \boldsymbol{n})$, *we have:*

I. $f_{\mathbf{0}} = g$*;*

II. $f_{\boldsymbol{s}(\boldsymbol{n})}(z) = h(z, (f_{\mathbf{0}}, \ldots, f_{\boldsymbol{n}}))$ *for every* $\boldsymbol{n} \in \mathbb{N}$.

Exercise C.9. Deduce corollaries C.7 and C.8 from proposition 7.17.

Exercise C.10. State transfinite recursion in the form that generalises to ordinal numbers corollaries C.1, C.2, C.3, C.7, and C.8.

Exercise C.11. Formalise definitions 11.62, 11.63, and 11.64, by applying exercise C.10.

Bibliography

[1] B. Bukh. Notes on the axiom of choice: `http://www.borisbukh.org/MathStudiesAlgebra1718/notes_ac.pdf`.

[2] A. Hajnal and P. Hamburger. *Set Theory*. Cambridge University Press, 1999.

[3] P. Halmos. *Naive Set Theory*. UTM, Springer, 1974.

[4] K. Hrbacek and T. Jech. *Introduction to Set Theory*. Marcel Dekker, Inc., 1999.

[5] T. Jech. *Set Theory*. Springer, 2003.

[6] K. Kunen. *Set Theory—An Introduction to Independence Proofs*. North-Holland Publishing Company, 1980.

[7] K. Kunen. *Set Theory*. College Publications, 2013.

[8] A. Levy. *Basic Set Theory*. Springer-Verlag, 1979.

[9] Y. Moschovakis. *Notes on Set Theory*. North-Holland, 2006.

[10] M. Potter. *Set Theory and its Philosophy*. Oxford University Press, 2004.

[11] E. Schimmerling. *A Course on Set Theory*. Cambridge University Press, 2011.

[12] D. Velleman. *How to Prove It*. Cambridge University Press, 2006.

Index

axiom of choice, 103

bijection, 28
bounded
  above, 35, 97
  below, 35

cardinal number, 153
  limit (weak), 157
  operations, 163
  successor, 156
  weakly inaccessible, 174
cardinality
  equal, 115
  lower, 115
  lower or equal, 115
Cartesian product
  n-ary, 22
  binary, 20
  of a set, 40
  of an indexed set, 44
  of classes, 73
chain
  in an ordered set, 105
class, 72
  proper, 72
class difference, 73
codomain, 28
cofinality, 170
complement, 20
composition, 29
comprehension hypothesis, 24
continuum hypothesis (generalised), 175

decimal notation, 10
diagram, 29
  commutative, 29
disjoint union
  n-ary, 22
  binary, 20
  of a set, 40
  of an indexed set, 44
  of classes, 73
domain
  of a binary relation, 27

element, 2
embedding, 28
empty set, 2
equality, 2
equivalence class, 32

foundation assumption, 188
function, 28
  $\aleph$ or ω, 159
  bijective, 28
  cofinal, 170
  identity, 30
  inclusion, 31
  injective, 28
  inverse, 30
  left inverse, 30
  projection, 33
  right inverse, 30
  surjective, 28

graph
  of a binary pre-relation, 27
  of a function, 28
  of an n-ary pre-relation, 28

Hausdorff Maximal Principle, 107
hereditary set, 14, 196

image
  of a binary relation, 27
  of a function, 28

inaccessible cardinal conjecture, 175
indexed class, 74
indexed set, 43
induction principle, 3
    second form, 6
infimum, 35
infinity hypothesis, 79
initial segment (proper), 97
intersection
    n-ary, 22
    binary, 20
    of a set, 40
    of an indexed set, 44
    of classes, 73
isomorphism
    of well-ordered sets, 95

lemma
    König's, 174
logical connectives, 1
lower bound, 35
    strict, 35

maximal element, 35, 106
maximum, 35
minimal element, 35
minimum, 35

natural number, 3
number-theoretic k-ary operation, 7
number-theoretic function, 7

ordered
    n-uple, 20
    pair, 19
ordinal number, 132
    limit, 139
    operations, 147
    regular, 173
    successor, 139

partition, 33
partition principle, 104
power, 4
power set, 20
pre-function, 28
    between classes, 73
pre-relation
    n-ary, 28
    between classes, 73
    binary, 27
product, 4
property
    as an unary relation, 28
proposition, 1

quantifiers, 2
quotient set, 33

rank
    of a well-founded set, 178
recursion principle, 3
    second form, 6
relation
    n-ary, 28
    antireflexive, 34
    antisymetric, 34
    binary, 27
    equivalence, 32
        generated, 52
    internal, 27
    order, 34
    reflexive, 32
    strict order, 34
    symmetric, 32
    transitive, 32
    well-founded, 182
replacement hypothesis, 75
restriction
    of a binary relation, 27
    of an n-ary relation, 28
Russell's paradox, 18

separation
    of a set, 40
    of an indexed set, 44
sequence, 7
    in set theory, 83
set, 2
    closed under successor, 79
    countable, 85

denumerable, 85
finite, 85
hereditarily transitive (properly), 131
infinite, 85
transitive, 129
well-founded, 177
with n elements, 85
set difference, 20
set-theoretical property, 15
subset, 3
proper, 3
successor, 3
sum, 4
supremum, 35
symmetric difference, 20
of classes, 73

theorem
Cantor, 126
Cantor-Bernstein, 124
transfinite induction
first form, 142
second form, 142
transfinite recursion
first form, 143
second form, 144
transitive closure, 185

union
n-ary, 22
binary, 20
of a set, 40
of an indexed set, 44
of classes, 73
universal property
of n-ary Cartesian product, 60
of n-ary disjoint union, 62
of Cartesian product of a set, 68
of Cartesian product of an indexed set, 69
of disjoint union of a set, 69
of disjoint union of an indexed set, 70
upper bound, 35, 105
strict, 35
upper bound (strict), 97

variable, 1

well-founded formula, 14
well-order, 93
well-ordering principle, 109

Zorn's lemma, 106

FSC
www.fsc.org
MIX
Papier | Fördert
gute Waldnutzung
FSC® C083411

Kolibri 360 GmbH
Ferdinand-Jühlke-Straße 7
99095 Erfurt, Deutschland
produktsicherheit@kolibri360.de